Synthesis Lectures on Mathematics & Statistics

Series Editor

Steven G. Krantz, Department of Mathematics, Washington University, Saint Louis, MO, USA

This series includes titles in applied mathematics and statistics for cross-disciplinary STEM professionals, educators, researchers, and students. The series focuses on new and traditional techniques to develop mathematical knowledge and skills, an understanding of core mathematical reasoning, and the ability to utilize data in specific applications.

Chao Wang · Ravi P. Agarwal

Dynamic Equations and Almost Periodic Fuzzy Functions on Time Scales

Springer

Chao Wang
Department of Mathematics
Yunnan University
Yunnan, China

Ravi P. Agarwal
Department of Mathematics
Texas A&M University-Kingsville
Kingsville, TX, USA

ISSN 1938-1743 ISSN 1938-1751 (electronic)
Synthesis Lectures on Mathematics & Statistics
ISBN 978-3-031-11238-6 ISBN 978-3-031-11236-2 (eBook)
https://doi.org/10.1007/978-3-031-11236-2

This Springer imprint is published by the registered company Springer Nature Switzerland AG
The registered company address is: Gewerbestrasse 11, 6330 Cham, Switzerland

We dedicate this book to our wives Wei Du and Sadhna Agarwal

Preface

Stefan Hilger introduced the concept of time scales in 1988 which can be applied to unify the study of differential equations for continuous case and difference equations for discrete case (see [1]). Since then, time scale theory has developed rapidly (see [2]) and it was used to develop function calculus in various mathematical fields such as the calculus for real-valued functions (see [2, 4]), quaternion dynamic equations (see [5, 6]), measure theory (see [7]), set-valued functions (see [8, 9]) and fuzzy-valued functions (see [10, 11, 12]), etc. When studying the dynamical behavior of an object on time scales it is significant to unify a larger range of time scales to obtain more comprehensive results (see [13, 14, 15]).

For a more accurate description of the real world phenomena, it is necessary to consider a number of uncertain factors and this leads naturally to fuzzy dynamical models (see [16, 17, 18, 19, 20, 21, 22, 23, 24]) and set-valued functions, fuzzy-valued functions and their related applications to dynamic equations on time scales (see [10, 11, 12]).

Almost periodic theory was initiated by Bohr during the period 1923–1925 (see [25, 26]) which is an important theory to precisely describe almost periodic approximation phenomena in engineering, life sciences, information sciences and control theory (see [27, 28, 29, 30, 31, 32, 33]) and in particular it arises in celestial mechanics, bioengineering and electronic circuits (see [28, 30]). The study of almost periodicity of solutions on time scales was considered in [34, 35, 36]. In 2001, Park, Jung and Lee considered the existence and stability of almost periodic solutions for fuzzy functional differential equations (see [37]), then in 2004, Bede and Gal developed a theory of almost periodic fuzzy number-valued functions and studied a class of almost periodic fuzzy dynamical systems (see [38]). In 2016, Hong and Peng extended the concept of almost periodic functions to set-valued functions on periodic time scales and established the existence of almost periodic solutions for set-valued dynamic equations (see [39]). Unfortunately, the validity of all results involved above must be restricted to traditional periodic time scales (i.e., periodic time scale under a translation) which have a nice closedness under translations. It is worth noting that these results cannot be applied to some important irregular time scales like $\overline{(-q)^{\mathbb{Z}}} = \{(-q)^n : q > 1, n \in \mathbb{Z}\} \cup \{0\}$ (which has important applications in quantum theory), $\pm\mathbb{N}^{\frac{1}{2}}$ and $\mathbb{T} = \mathbb{T}_n$, the space of harmonic numbers, because the concept of almost periodic functions in these works become unsuitable and inapplicable for these irregular time scales which are without closedness under translations. To overcome this difficulty, in 2017, inspired by Advar's work [40], Wang and Agarwal introduced the concept of relatively dense set under shift operators and proposed almost periodic stochastic processes. This development makes it possible to study almost periodic problems on many irregular time scales involving q-difference equations and more general types of dynamic equations (see [15, 41]).

On the other hand, fuzzy arithmetic is also a complex and intractable problem including the multiplication and division of fuzzy numbers and fuzzy vectors. Fortunately, Stefanini et al., opened a feasible avenue to fuzzy arithmetic and introduced a generalized Hukuhara difference and division for interval and fuzzy arithmetic (see [42, 24]). Moreover, they also initiated the unidimensional and multidimensional boxes which motivated us to introduce fuzzy (box) vectors and establish the calculus of fuzzy vector-valued functions on time scales (see [43]). Further, in [43], we introduced a new multiplication determined by a determinant algorithm, based on which the classical fundamental formulas of calculus can be derived under fuzzy background, and this provides a powerful calculus tool to solve classical problems of multidimensional fuzzy dynamic systems in various research fields.

We organize this book into six chapters.

In Chap. 1, some necessary knowledge of interval and fuzzy arithmetic is presented. A generalization of the Hukuhara difference is introduced. First, the case of compact convex sets is investigated which are applied to generalize the Hukuhara difference of fuzzy numbers by using their compact and convex level-cuts. Moreover, a similar approach is presented to propose a generalization of division for real intervals and fuzzy numbers. Some applications are provided to solve interval and fuzzy linear equations and fuzzy differential equations.

In Chap. 2, an embedding theorem for fuzzy multidimensional space is established and two new types of multiplication of fuzzy vectors are introduced and studied.

In Chap. 3, we introduce the basic notions of gH-Δ-derivatives of fuzzy vector-valued functions on time scales and obtain their fundamental properties. Moreover, the Δ-integral of fuzzy vector-valued functions is introduced and studied. Some basic results related to calculus of fuzzy vector-valued functions are established on time scales.

In Chap. 4, some necessary knowledge of shift operators and a generalized periodic time scales is presented. A notion of shift almost periodic fuzzy vector-valued functions is addressed and studied on complete-closed time scales under non-translational shifts, some fundamental results of shift almost periodic fuzzy vector-valued functions are established.

In Chap. 5, some basic results of fuzzy multidimensional spaces are demonstrated and a new division of multidimensional intervals and fuzzy vectors induced by a determinant algorithm is introduced and studied.

In Chap. 6, we develop a theory of almost periodic fuzzy multidimensional dynamic systems on time scales and several applications are provided. In particular, a new type of fuzzy dynamic systems called fuzzy q-dynamic systems (i.e., fuzzy quantum dynamic systems) is proposed and studied.

This book will establish an almost periodic theory of multidimensional fuzzy dynamic equations and fuzzy vector-valued functions on complete-closed time scales under non-translational shifts including some commonly irregular time scales, and it involves an almost periodic theory of fuzzy functions on quantum-like time scales. Our results are not only effective on periodic time scales (i.e., $\mathbb{T} = \mathbb{Z}, \mathbb{R}$ or $h\mathbb{Z}$, etc.) but also are valid for irregular time scales $\overline{q^{\mathbb{Z}}}$, $-\overline{q^{\mathbb{Z}}}$, $\pm\mathbb{N}^{\frac{1}{2}}$, $\mathbb{N}^{\frac{1}{3}}$, $\overline{(-q)^{\mathbb{Z}}}$, etc.

The book is written at a graduate level and is intended for university libraries. Graduate students and researchers working in the field of fuzzy dynamic equations on time scales will be able to stimulate further research. The book is also a good reference material for those undergraduates who are interested in fuzzy dynamic equations and functions on time scales and familiar with fuzzy sets and systems and ordinary differential equations.

We acknowledge with gratitude the support of National Natural Science Foundation of China (11961077), CAS "Light of West China" Program of Chinese Academy of Sciences and Educational Reform Research Project of Yunnan University (No. 2021Y10).

Kunming, China — Chao Wang
Kingsville, USA — Ravi P. Agarwal

Contents

Generalized Hukuhara Difference and Division for Interval and Fuzzy Arithmetic

1

1.1 Generalized Hukuhara Difference and Properties

In this Section, we will present some basic knowledge of generalized Hukuhara difference and properties which was established in the literature [75] and it will be used in our later chapters.

Consider a metric vector space $\mathbb{X}$ with the induced topology and in particular the space $\mathbb{X} = \mathbb{R}^n$, $n \geq 1$, of real vectors equipped with standard addition and scalar multiplication operations. Following Diamond and Kloeden (see [28]), denote by $\mathscr{K}(\mathbb{X})$ and $\mathscr{K}_C(\mathbb{X})$ the spaces of nonempty compact and compact convex sets of $\mathbb{X}$. Given two subsets $A, B \subseteq \mathbb{X}$ and $k \in \mathbb{R}$, Minkowski addition and scalar multiplication are defined by $A + B = \{a + b | a \in A, b \in B\}$ and $kA = \{ka | a \in A\}$ and it is well known that addition is associative and commutative and with neutral element $\{0\}$. If $k = -1$, scalar multiplication gives the opposite $-A = (-1)A = \{-a | a \in A\}$ but, in general, $A + (-A) \neq \{0\}$, i.e., the opposite of A is not the inverse of A in Minkowski addition (unless $A = \{a\}$ is a singleton). Minkowski difference is $A - B = A + (-1)B = \{a - b | a \in A, b \in B\}$. A first implication of this fact is that, in general, even if it is true that $(A + C = B + C) \Longleftrightarrow A = B$, addition/subtraction simplification is not valid, i.e., $(A + B) - B \neq A$.

To partially overcome this situation, Hukuhara [44] introduced the following H-difference:

$$A \ominus B = C \Longleftrightarrow A = B + C \tag{1.1}$$

and an important property of $\ominus$ is that $A \ominus A = \{0\}$, $\forall A \in \mathscr{K}(\mathbb{X})$ and $(A + B) \ominus B = A$, $\forall A, B \in \mathscr{K}(\mathbb{X})$; H-difference is unique, but a necessary condition for $A \ominus B$ to exist is that A contains a translate $\{c\} + B$ of B. In general, $A - B \neq A \ominus B$.

C. Wang and R. P. Agarwal, *Dynamic Equations and Almost Periodic Fuzzy Functions on Time Scales*, Synthesis Lectures on Mathematics & Statistics,
https://doi.org/10.1007/978-3-031-11236-2_1

From an algebraic point of view, the difference of two sets A and B may be interpreted both in terms of addition as in (1.1) or in terms of negative addition, i.e.,

$$A \boxminus B = C \Longleftrightarrow B = A + (-1)C, \tag{1.2}$$

where $(-1)C$ is the opposite set of C. Conditions (1.1) and (1.2) are compatible to each other and this suggests a generalization of Hukuhara difference:

Definition 1.1 *(see* [75]) Let $A, B \in \mathscr{K}(\mathbb{X})$; we define the generalized Hukuhara difference of A and B as the set $C \in \mathscr{K}(\mathbb{X})$ such that

$$A \ominus_{gH} B = C \Longleftrightarrow \begin{cases} \text{(i) } A = B + C, \\ \text{or (ii) } B = A + (-1)C. \end{cases} \tag{1.3}$$

Proposition 1.1 (see [75]) (Unicity of $A \ominus_{gH} B$) *If $C = A \ominus_{gH} B$ exists, it is unique and if also $A \ominus B$ exists then $A \ominus_{gH} B = A \ominus B$.*

Proof If $C = A \ominus_{gH} B$ exists in case (i), we obtain $C = A \ominus B$ which is unique. Suppose that case (ii) is satisfied for C and D, i.e., $B = A + (-1)C$ and $B = A + (-1)D$; then $A + (-1)C = A + (-1)D \Longrightarrow (-1)C = (-1)D \Longrightarrow C = D$. If case (i) is satisfied for C and case (ii) is satisfied for D, i.e., $A = B + C$ and $B = A + (-1)D$, then $B = B + C + (-1)D \Longrightarrow \{0\} = C - D$ and this is possible only if $C = D = \{c\}$ is a singleton. This completes the proof. □

The generalized Hukuhara difference $A \ominus_{gH} B$ will be called the gH-difference of A and B.

Remark 1.1 A necessary condition for $A \ominus_{gH} B$ to exist is that either A contains a translate of B (as for $A \ominus B$) or B contains a translate of A. In fact, for any given $c \in C$, we get $B + \{c\} \subseteq A$ from (i) or $A + \{-c\} \subseteq B$ from (ii).

Remark 1.2 It is possible that $A = B + C$ and $B = A + (-)C$ hold simultaneously; in this case, A and B translate into each other and C is a singleton. In fact, $A = B + C$ implies $B + \{c\} \subseteq A \; \forall c \in C$ and $B = A + (-1)C$ implies $A - \{c\} \subseteq B \; \forall c \in C$ i.e., $A \subseteq B + \{c\}$; it follows that $A = B + \{c\}$ and $B = A + \{-c\}$. On the other hand, if $c', c'' \in C$ then $A = B + \{c'\} = B + \{c''\}$ and this requires $c' = c''$.

Remark 1.3 If $A \ominus_{gH} B$ exists, then $B \ominus_{gH} A$ exists and $B \ominus_{gH} A = -(A \ominus_{gH} B)$.

Proposition 1.2 (see [75]) *The gH-difference $\ominus_{gH}$ has the following properties:*

(1) $A \ominus_{gH} A = \{0\}$.
(2) (a) $(A + B) \ominus_{gH} B = A$; (b) $A \ominus_{gH} (A - B) = B$; (c) $A \ominus_{gH} (A + B) = -B$.
(3) $A \ominus_{gH} B$ *exists if and only if* $B \ominus_{gH} A$ *and* $(-B) \ominus_{gH} (-A)$ *exist and* $A \ominus_{gH} B = (-B) \ominus_{gH} (-A) = -(B \ominus_{gH} A)$.
(4) *In general,* $B \ominus_{gH} A = A \ominus_{gH} B$ *does not imply* $A = B$; *but* $A \ominus_{gH} B = B \ominus_{gH} A = C$ *if and only if* $C = -C$ *and, in particular,* $C = \{0\}$ *if and only if* $A = B$.
(5) *If* $B \ominus_{gH} A$ *exists then either* $A + (B \ominus_{gH} A) = B$ *or* $B - (B \ominus_{gH} A) = A$ *and both equalities hold if and only if* $B \ominus_{gH} A$ *is a singleton set.*
(6) *If* $B \ominus_{gH} A = C$ *exists, then for all* $D \in \mathcal{K}(\mathbb{X})$ *either* $(B + D) \ominus_{gH} A = C + D$ *or* $B \ominus_{gH} (A + D) = C - D$.

Proof Property 1 is immediate. To prove 2(a) if $C = (A + B) \ominus_{gH} B$ then either $A + B = C + B$ or $B = (A + B) + (-1)C = B + (A + (-1)C)$; in the first case it follows that $C = A$, in the second case $A + (-1)C = \{0\}$ and A and C are singleton sets so $A = C$. With a similar argument, 2(b) and 2(c) can be proved. To prove the first part of (3) let $C = A \ominus_{gH} B$ according to case (i), i.e., $A = B + C$, then $A = B - (-C)$ and $B \ominus_{gH} A = -C$ according to case (ii); if instead $C = A \ominus_{gH} B$ according to case (ii), i.e., $B = A - C$, then $B = A + (-C)$ and $B \ominus_{gH} A = -C$ according to case (i); on the other hand, if $A = B + C$ or $B = A - C$, then $-A = -B + (-C)$ or $-B = -A + C$ and this means $(-B) \ominus_{gH} (-A) = C$. To see the first part of (4) consider for example the unidimensional case $A = [a^-, a^+]$, $B = [b^-, b^+]$; equality $A - B = B - A$ is valid if $a^- + a^+ = b^- + b^+$ and this does not require $A = B$ (unless A and B are singletons). For the second part of (4), from $(A \ominus_{gH} B) = (B \ominus_{gH} A) = C$, considering the four combinations derived from (3), one of the following four cases is valid: ($A = B + C$ and $B = A + C$) or ($A = B + C$ and $A = B - C$) or ($B = A + (-1)C$ and $B = A + C$) or ($B = A + (-1)C$ and $A = B + (-1)C$); in all of them we deduce $C = -C$ and $C = \{0\}$ if and only if $A = B$. To see (5), consider that if $(B \ominus_{gH} A)$ exists in the sense of (i) the first equality is valid and if it exists in the sense of (ii) the second one is valid. To prove (6), if $B = A + C$ then $B + D = A + C + D$ and $(B + D) \ominus_{gH} A = C + D$ according to case (i); if $A = B - C$ then $A + D = B - C + D = B - (C - D)$ and $B \ominus_{gH} (A + D) = C - D$ according to case (ii). □

Remark 1.4 The equivalence $(A \ominus B) = C \iff (A \ominus C) = B$ is valid only for $\ominus$, or for $\ominus_{gH}$ in case (i). In fact, if $A \ominus_{gH} B = C$ in the sense (ii), then $B = A - C$ and this does not imply $A = B + C$ nor $C = A - B$ unless $B = B + C - C$ (i.e., $C = \{\widehat{c}\}$) or $B = B + A - A$ (i.e., $A = \{\widehat{a}\}$). Note also that, in general, $A + (B \ominus_{gH} A) \neq A$ and $A - (A \ominus_{gH} B) \neq B$.

For sets $A, B \in \mathcal{K}(\mathbb{X})$ over a normed space $(\mathbb{X}, \|\cdot\|)$ the Hausdorff distance is defined as usual by

$$H(A, B) = \max\{d_H(A, B), d_H(B, A)\},$$

where

$$d_H(A, B) = \sup_{a \in A} \inf_{b \in B} \|a - b\| \quad \text{and} \quad d_H(B, A) = \sup_{b \in B} \inf_{a \in A} \|a - b\|,$$

we denote $\|A\|_H = H(A, \{0\}) = \sup_{a \in A} \|a\|$.

If $\mathbb{X} = \mathbb{R}^n$, $n \geq 1$ is the real n-dimensional vector space with internal product $\langle x, y\rangle$ and corresponding norm $\|x\| = \sqrt{\langle x, x\rangle}$, we denote by $\mathcal{K}^n$ and $\mathcal{K}_C^n$ the spaces of (nonempty) compact and compact convex sets of $\mathbb{R}^n$, respectively.

If $A \subseteq \mathbb{R}^n$ and $S^{n-1} = \{p | p \in \mathbb{R}^n, \|p\| = 1\}$ is the unit sphere, the support function associated to A is

$$s_A : \mathbb{R}^n \to \mathbb{R} \quad \text{defined by}$$
$$s_A(p) = \sup\{\langle p, a\rangle | a \in A\}, \quad p \in \mathbb{R}^n.$$

If $A \neq \emptyset$ is compact, then $s_A(p) \in \mathbb{R}$, $\forall p \in S^{n-1}$. The following properties are well known (see e.g. [28] or [55]):

- Any function $s : \mathbb{R}^n \to \mathbb{R}$ which is continuous (or, more generally, upper semicontinuous), positively homogeneous $s(tp) = ts(p)$, $\forall t \geq 0$, $\forall p \in \mathbb{R}^n$ and subadditive $s(p' + p'') \leq s(p') + s(p'')$, $\forall p', p'' \in \mathbb{R}^n$ is a support function of a compact convex set; the restriction $\widehat{s}$ of s to S^{n-1} is such that $\widehat{s}(p/\|p\|) = (1/\|p\|)s(p)$, $\forall p \in \mathbb{R}^n$, $p \neq 0$ and we can consider s restricted to S^{n-1}. It also follows that $s : S^{n-1} \to \mathbb{R}$ is a convex function.
- If $A \in \mathcal{K}_C^n$ is a compact convex set, then it is characterized by its support function and

$$A = \{x \in \mathbb{R}^n | \langle p, x\rangle \leq s_A(p), \forall p \in \mathbb{R}^n\} = \{x \in \mathbb{R}^n | \langle p, x\rangle \leq s_A(p), \forall p \in S^{n-1}\}.$$

- For $A, B \in \mathcal{K}_C^n$ and $\forall p \in S^{n-1}$ we have $s_{\{0\}}(p) = 0$ and

$$A \subseteq B \Longrightarrow s_A(p) \leq s_B(p); \quad A = B \Longleftrightarrow s_A = s_B,$$

$$s_{kA}(p) = ks_A(p), \forall k \geq 0; \quad s_{kA+hB}(p) = s_{kA}(p) + s_{hB}(p), \quad \forall k, h \geq 0$$

and in particular

$$s_{A+B}(p) = s_A(p) + s_B(p).$$

- If s_A is the support function of $A \in \mathcal{K}_C^n$ and s_{-A} is the support function of $-A \in \mathcal{K}_C^n$, then $\forall p \in S^{n-1}$, $s_{-A}(p) = s_A(-p)$;
- If v is a measure on $\mathbb{R}^n$ such that $v(S^{n-1}) = \int_{S^{n-1}} v(\mathrm{d}p) = 1$, a distance is defined by

$$\rho_2(A, B) = \|s_A - s_B\| = \left(n \int_{S^{n-1}} [s_A(p) - s_B(p)]^2 v(\mathrm{d}p)\right)^{\frac{1}{2}},$$

the distance $\rho_2(\cdot, \cdot)$ induces the norm on $\mathscr{K}_C^n$ defined by $\|A\| = \rho_2(A, \{0\})$.

- The Steiner point of $A \in \mathscr{K}_C^n$ is defined by $\sigma_A = n \int_{S^{n-1}} p s_A(p) v(\mathrm{d}p)$ and $\sigma_A \in A$.

We can express the generalized Hukuhara difference of compact convex sets $A, B \in \mathscr{K}_C^n$ by the use of the support functions. Consider $A, B, C \in \mathscr{K}_C^n$ with $C = A \ominus_{gH} B$ as defined in (1.3); let s_A, s_B, s_C and $s_{(-1)C}$ be the support functions of A, B, C, and $(-1)C$, respectively. In case (i) we have $s_A = s_B + s_C$ and in case (ii) we have $s_B = s_A + s_{(-1)C}$. So, $\forall p \in S^{n-1}$

$$s_C(p) = \begin{cases} s_A(P) - s_B(P) & \text{in cases (i),} \\ s_B(-P) - s_A(-P) & \text{in cases (ii),} \end{cases}$$

i.e.,

$$s_C(p) = \begin{cases} s_A(P) - s_B(P) & \text{in cases (i),} \\ s_{(-1)B}(P) - s_{(-1)A}(P) & \text{in cases (ii).} \end{cases} \tag{1.4}$$

Now, s_C in (1.4) is a correct support function if it is continuous (upper semicontinuous), positively homogeneous and subadditive and this requires that, in the corresponding cases (i) and (ii), $s_A - s_B$ and/or $s_{-B} - s_{-A}$ be support functions, assuming that s_A and s_B are support functions.

Consider $s_1 = s_A - s_B$ and $s_2 = s_B - s_A$. Continuity of s_1 and s_2 is obvious. To see their positive homogeneity let $t \geq 0$; we have $s_1(tp) = s_A(tp) - s_B(tp) = ts_A(p) - ts_B(p) = ts_1(p)$ and similarly for s_2. But s_1 and/or s_2 may fail to be subadditive and the following four cases, related to the definition of gH-difference, are possible.

Proposition 1.3 (see [75]) *Let s_A and s_B be the support functions of $A, B \in \mathscr{K}_C^n$ and consider $s_1 = s_A - s_B$ and $s_2 = s_B - s_A$; the following four cases apply:*

1 *If s_1 and s_2 are both subadditive, then $A \ominus_{gH} B$ exists;* (i) *and* (ii) *are satisfied simultaneously and $A \ominus_{gH} B = \{c\}$.*
2 *If s_1 is subadditive and s_2 is not, then $C = A \ominus_{gH} B$ exists,* (i) *is satisfied and $s_C = s_A - s_B$.*
3 *If s_1 is not subadditive and s_2 is, then $C = A \ominus_{gH} B$ exists,* (ii) *is satisfied and $s_C = s_{-B} - s_{-A}$.*
4 *If s_1 and s_2 are both not subadditive, then $A \ominus_{gH} B$ does not exist.*

Proof In case 1 subadditivity of s_1 and s_2 means that, $\forall p', p'' \in S^{n-1}$

$$s_1 : s_A(p' + p'') - s_B(p' + p'') \leq s_A(p') + s_A(p'') - s_B(p') - s_B(p'')$$
$$s_2 : s_B(p' + p'') - s_A(p' + p'') \leq s_B(p') + s_B(p'') - s_A(p') - s_A(p''),$$

it follows that

$$s_A(p'+p'')-s_A(p')-s_A(p'')\le s_B(p'+p'')-s_B(p')-s_B(p'')$$
$$s_B(p'+p'')-s_B(p')-s_B(p'')\le s_A(p'+p'')-s_A(p')-s_A(p'')$$

so that equality holds:

$$s_B(p'+p'')-s_A(p'+p'')=s_B(p')+s_B(p'')-s_A(p')-s_A(p'')$$

Taking $p'=-p''=p$ produces, $\forall p\in S^{n-1}$, $s_B(p)+s_B(-p)=s_A(p)+s_A(-p)$ i.e., $s_B(p)+s_{-B}(p)=s_A(p)+s_{-A}(p)$ i.e., $s_{B-B}(p)=s_{A-A}(p)$ and $B-B=A-A$ (A and B translate into each other); it follows that $\exists c\in\mathbb{R}^n$ such that $A=B+\{c\}$ and $B=A+\{-c\}$ so that $A\ominus_{gH}B=\{c\}$.

In case 2 we have that being s_1 a support function it characterizes a nonempty set $C\in\mathcal{K}_C^n$ and $s_C(p)=s_1(p)=s_A(p)-s_B(p)$, $\forall p\in S^{n-1}$; then $s_A=s_B+s_C=s_{B+C}$ and $A=B+C$ from which (i) is satisfied.

In case 3 we have that s_2 the support function of a nonempty set $D\in\mathcal{K}_C^n$ and $s_D(p)=s_B(p)-s_A(p),\forall p\in S^{n-1}$ so that $s_B=s_A+s_D=s_{A+D}$ and $B=A+D$. Defining $C=(-1)D$ (or $D=(-1)C$) we obtain $C\in\mathcal{K}_C^n$ with $s_C(p)=s_{-D}(p)=s_D(-p)=s_B(-p)-s_A(-p)=s_{-B}(p)-s_{-A}(p)$ and (ii) is satisfied.

In case 4 there is no $C\in\mathcal{K}_C^n$ such that $A=B+C$ (otherwise $s_1=s_A-s_B$ is a support function) and there is no $D\in\mathcal{K}_C^n$ such that $B=A+D$ (otherwise $s_2=s_B-s_A$ is a support function); it follows that (i) and (ii) cannot be satisfied and $A\ominus_{gH}B$ does not exist. This completes the proof. □

Proposition 1.4 (see [75]) *If $C=A\ominus_{gH}B$ exists, then $\|A\ominus_{gH}B\|=\rho_2(A,B)$; it follows that $\|A\ominus_{gH}B\|=0\Longleftrightarrow A=B$.*

Proof In fact $\rho_2(A,B)=\|s_A-s_B\|$ and, if $A\ominus_{gH}B$ exists, then either $s_C=s_A-s_B$ or $s_C=s_{-B}-s_{-A}$; but $\|s_A-s_B\|=\|s_{-A}-s_{-B}\|$ as, changing variable p into $-q$ and recalling that $s_{-A}(p)=s_A(-p)$, we have

$$\|s_{-A}-s_{-B}\|=\int_{S^{n-1}}[s_{-A}(p)-s_{-B}(p)]^2v(dp)=\int_{S^{n-1}}[s_A(-p)-s_B(-p)]^2v(dp)$$
$$\int_{S^{n-1}}[s_A(q)-s_B(q)]^2v(-dq)=\|s_A-s_B\|, \tag{1.5}$$

the last property follows from the fact that $\|A\ominus_{gH}B\|=0$ implies $\rho_2(A,B)=0$ so that $A=B$; on the other hand, for $A=B$, $A\ominus_{gH}A=\{0\}$. This completes the proof. □

An interesting property relates the Steiner point of $A \ominus_{gH} B$ to the Steiner points of A and B.

Proposition 1.5 (see [75]) *If $C = A \ominus_{gH} B$ exists, let σ_A, σ_B and σ_C be the Steiner points of A, B and C, respectively; then $\sigma_C = \sigma_A - \sigma_B$.*

Proof For the Steiner points, we have

$$\sigma_A = n \int_{S^{n-1}} p s_A(p) v(\mathrm{d}p), \quad \sigma_B = n \int_{S^{n-1}} p s_B(p) v(\mathrm{d}p) = -n \int_{S^{n-1}} q s_B(q) v(-\mathrm{d}q) \tag{1.6}$$

and

$$\sigma_C = \begin{cases} n \int_{S^{n-1}} p[s_A(p) - s_B(p)] v(\mathrm{d}p) & \text{or} \\ n \int_{S^{n-1}} q[s_A(q) - s_B(q)] v(-\mathrm{d}q), \end{cases} \tag{1.7}$$

the result follows from the additivity of the integral. This completes the proof. □

1.2 The Case of Compact Intervals in $\mathbb{R}^n$

In this section, the gH-difference of compact intervals in $\mathbb{R}^n$ will be considered. For more details, one may consult [75]. If $n = 1$, i.e., for unidimensional compact intervals , the gH-difference always exists. In fact, let $A = [a^-, a^+]$ and $B = [b^-, b^+]$ be two intervals; the gH-difference is

$$[a^-, a^+] \ominus_{gH} [b^-, b^+] = [c^-, c^+] \Longleftrightarrow \begin{cases} (1) \begin{cases} a^- = b^- + c^-, \\ a^+ = b^+ + c^+, \end{cases} \\ \text{or } (2) \begin{cases} b^- = a^- - c^+, \\ b^+ = a^+ - c^-, \end{cases} \end{cases}$$

so that $[a^-, a^+] \ominus_{gH} [b^-, b^+] = [c^-, c^+]$ is always defined by

$$c^- = \min\{a^- - b^-, a^+ - b^+\}, \; c^+ = \max\{a^- - b^-, a^+ - b^+\},$$

i.e.,

$$[a, b] \ominus_{gH} [c, d] = [\min\{a - c, b - d\}, \max\{a - c, b - d\}].$$

Conditions (i) and (ii) are satisfied simultaneously if and only if the two intervals have the same length and $c^- = c^+$. Also, the result is $\{0\}$ if and only if $a^- = b^-$ and $a^+ = b^+$.

Two simple examples on real compact intervals illustrate the generalization (from [28, p. 8]); $[-1, 1] \ominus [-1, 0] = [0, 1]$ as in fact (i) is $[-1, 0] + [0, 1] = [-1, 1]$ but $[0, 0] \ominus_{gH} [0, 1] = [-1, 0]$ and $[0, 1] \ominus_{gH} [-\frac{1}{2}, 1] = [0, \frac{1}{2}]$ satisfy (ii).

Of interest are the symmetric intervals $A = [-a, a]$ and $B = [-b, b]$ with $a, b \geq 0$; it is well known that Minkowski operations with symmetric intervals are such that $A - B = B - A = A + B$ and, in particular, $A - A = A + A = 2A$. We have $[-a, a] \ominus_{gH} [-b, b] = [-|a - b|, |a + b|]$.

As $S^0 = \{-1, 1\}$ and the support functions satisfy $s_A(-1) = -a^-$, $s_A(1) = a^+$, $s_B(-1) = -b^-$, $s_B(1) = b^+$, the same results as before can be deduced from definition (1.4).

Remark 1.5 An alternative representation of an interval $A = [a^-, a^+]$ is by the use of the midpoint $\widehat{a} = a^- + a^+/2$ and the (semi)width $\bar{a} = a^+ - a^-/2$ and we can write $A = (\widehat{a}, \bar{a})$, $\bar{a} \geq 0$, so that $a^- = \widehat{a} - \bar{a}$ and $a^+ = \widehat{a} + \bar{a}$. If $B = (\widehat{b}, \bar{b})$, $\bar{b} \geq 0$ is a second interval, the Minkowski addition is $A + B = (\widehat{a} + \widehat{b}, \bar{a} + \bar{b})$ and the gH-difference is obtained by $A \ominus_{gH} B = (\widehat{a} - \widehat{b}, |\bar{a} - \bar{b}|)$. We see immediately that $A \ominus_{gH} A = \{0\}$, $A = B \iff A \ominus_{gH} B = \{0\}$, $(A + B) \ominus_{gH} B = A$, but $A + (B \ominus_{gH} A) = B$ only if $\bar{a} \leq \bar{b}$.

Let now $A = \times_{i=1}^{n} A_i$ and $B = \times_{i=1}^{n} B_i$ where $A_i = [a_i^-, a_i^+]$, $B_i = [b_i^-, b_i^+]$ are real compact intervals ($\times_{i=1}^{n}$ denotes the cartesian product).

If $A \ominus_{gH} B$ exists, then the following equality holds:

$$A \ominus_{gH} B = \times_{i=1}^{n} (A_i \ominus_{gH} B_i).$$

In fact, consider the support function of A (and similarly for B), defined by

$$s_A(p) = \max_x \{\langle p, x \rangle | a_i^- \leq x_i \leq a_i^+\}, \quad p \in S^{n-1}, \tag{1.8}$$

it can be obtained simply by $s_A(p) = \sum_{p_i > 0} p_i a_i^+ + \sum_{p_i < 0} p_i a_i^-$ as the box-constrained maxima of the linear objective functions $\langle p, x \rangle$ above are attained at vertices $\widehat{x}(p) = (\widehat{x}_1(p), \ldots, \widehat{x}_i(p), \ldots, \widehat{x}_n(p))$ of A, i.e., $\widehat{x}_i(p) \in \{a_i^-, a_i^+\}$, $i = 1, 2, \ldots, n$. Then

$$s_A(p) - s_B(p) = \sum_{p_i > 0} p_i (a_i^+ - b_i^+) + \sum_{p_i < 0} p_i (a_i^- - b_i^-) \tag{1.9}$$

and, being $s_{-A}(p) = s_{-B}(p) = -\sum_{p_i < 0} p_i a_i^+ - \sum_{p_i > 0} p_i a_i^-$,

$$s_{-B}(p) - s_{-A}(p) = \sum_{p_i > 0} p_i (a_i^- - b_i^-) + \sum_{p_i < 0} p_i (a_i^+ - b_i^+). \tag{1.10}$$

From the relations above, we deduce that

$$A \ominus_{gH} B = C \Longleftrightarrow \begin{cases} \text{(i)} \begin{cases} C = \times_{i=1}^n [a_i^- - b_i^-, a_i^+ - b_i^+] \\ \text{provided that } a_i^- - b_i^- \le a_i^+ - b_i^+, \quad \forall i, \end{cases} \\ \text{or} \quad \text{(ii)} \begin{cases} C = \times_{i=1}^n [a_i^+ - b_i^+, a_i^- - b_i^-] \\ \text{provided that } a_i^- - b_i^- \ge a_i^+ - b_i^+, \quad \forall i, \end{cases} \end{cases}$$

and:

Proposition 1.6 (see [75]) *The gH-difference $A \ominus_{gH} B$ exists if and only if one of the two conditions is satisfied:*

$$\begin{aligned} &\text{(i)} \ \ a_i^- - b_i^- \ge a_i^+ - b_i^+, \ i = 1, 2, \ldots, n \\ &\quad or \\ &\text{(ii)} \ \ a_i^- - b_i^- \le a_i^+ - b_i^+, \ i = 1, 2, \ldots, n. \end{aligned} \tag{1.11}$$

Example 1.2.1 1. case (i): $A_1 = [5, 10]$, $A_2 = [1, 3]$, $B_1 = [3, 6]$, $B_2 = [2, 3]$ for which $(A_1 \ominus_{gH} B_1) = [2, 4]$, $(A_2 \ominus_{gH} B_2) = [-1, 0]$ and $A \ominus_{gH} B = C = [2, 4] \times [-1, 0]$ exists with $B + C = A$, $A + (-1)C \neq B$.

2. case (ii): $A_1 = [3, 6]$, $A_2 = [2, 3]$, $B_1 = [5, 10]$, $B_2 = [1, 3]$ for which $(A_1 \ominus_{gH} B_1) = [-4, -2]$, $(A_2 \ominus_{gH} B_2) = [0, 1]$ and $A \ominus_{gH} B = C = [-4, -2] \times [0, 1]$ exists with $B + C \neq A$, $A + (-1)C = B$.

3. case (iii): $A_1 = [3, 6]$, $A_2 = [2, 3]$, $B_1 = [5, 8]$, $B_2 = [3, 4]$ for which $(A_1 \ominus_{gH} B_1) = [-2, -2] = \{-2\}$, $(A_2 \ominus_{gH} B_2) = [-1, -1] = \{-1\}$ and $A \ominus_{gH} B = C = \{(-2, -1)\}$ exists with $B + C = A$, $A + (-1)C = B$. □

We have seen that, for general $A, B \in \mathcal{K}_C^n$, a necessary condition for $A \ominus_{gH} B$ to exist is that either A contains a translate of B or B contains a translate of A. In the case of multidimensional intervals , the same condition is also sufficient.

Proposition 1.7 (see [75]) *Let $A = \times_{i=1}^n A_i$ and $B = \times_{i=1}^n B_i$ where $A_i = [a_i^-, a_i^+]$, $B_i = [b_i^-, b_i^+]$. If A contains a translate of B or if B contains a translate of A (in particular if $A \subseteq B$ or if $B \subseteq A$), then $A \ominus_{gH} B = \times_{i=1}^n (A_i \ominus_{gH} B_i)$.*

Proof Consider first the case $B \subseteq A$, i.e., $[b_i^-, b_i^+] \subseteq [a_i^-, a_i^+]$ and $b_i^- \ge a_i^-$, $b_i^+ \le a_i^+ \ \forall i$, then Proposition 1.6 applies and $A \ominus_{gH} B$ exists according to case (i). Analogously, if $A \subseteq B$, i.e., $[a_i^-, a_i^+] \subseteq [b_i^-, b_i^+]$ and $b_i^- \le a_i^-$, $b_i^+ \ge a_i^+ \ \forall i$, then Proposition 1.6 applies and $A \ominus_{gH} B$ exists according to case (ii). If A contains a translate of B, i.e., $\exists \widehat{b} = (\widehat{b}_1, \widehat{b}_2, \ldots, \widehat{b}_n) \in \mathbb{R}^n$ such that $\{\widehat{b}\} + B \subseteq A$, then $A \ominus_{gH} (\{\widehat{b}\} + B) = C'$ exists according to case (ii) with $a_i^- - b_i^- - \widehat{b}_i \le a_i^+ - b_i^+ - \widehat{b}_i, \forall i$; if follows that $a_i^- - b_i^- \le a_i^+ - b_i^+$

and by Proposition 1.6 $A \ominus_{gH} B$ exists according to case (i). Finally, by a similar reasoning, if B contains a translate $\{\widehat{a}\} + A$ of A, then $(\{\widehat{a}\} + A) \ominus_{gH} B$ exists according to case (ii) so that $a_i^- - b_i^- \geq a_i^+ - b_i^+$, $\forall i$ and by Proposition 1.6 $A \ominus_{gH} B$ exists according to case (ii). This completes the proof. □

1.3 gH-Difference of Fuzzy Numbers

A general fuzzy set over a given set (or space) $\mathbb{X}$ of elements (the universe) is usually defined by its membership function $\mu : \mathbb{X} \to \mathbb{T} \subseteq [0, 1]$ and a fuzzy (sub)set u of $\mathbb{X}$ is uniquely characterized by the pairs $(x, \mu_u(x))$ for each $x \in \mathbb{X}$; the value $\mu_u(x) \in [0, 1]$ is the membership grade of x to the fuzzy set u and μ_u is the membership function of a fuzzy set u over $\mathbb{X}$ (see [121–123] for the origins of Fuzzy Set Theory). The support of u is the (crisp) subset of points of $\mathbb{X}$ at which the membership grade $\mu_u(x)$ is positive: $supp(u) = \{x | x \in \mathbb{X}, \mu_u(x) > 0\}$. For $\alpha \in [0, 1]$, the α-level cut of u (or simply the α-cut) is defined by $[u]_\alpha = \{x | x \in \mathbb{X}, \mu_u(x) \geq \alpha\}$ and for $\alpha = 0$ (or $\alpha \to +0$) by the closure of the support $[u]_0 = cl\{x | x \in \mathbb{X}, \mu_u(x) > 0\}$.

We will consider the case $\mathbb{X} = \mathbb{R}^n$ with $n \geq 1$. A particular class of fuzzy sets u is when the support is a convex set and the membership function is quasi-concave (i.e., $\mu_u((1-t)x' + tx'') \geq \min\{\mu_u(x'), \mu_u(x'')\}$ for every $x', x'' \in supp(u)$ and $t \in [0, 1]$). Equivalently, μ_u is quasi-concave if the level sets $[u]_x$ are convex sets for all $\alpha \in [0, 1]$. We will also require that the level-cuts $[u]_x$ are closed sets for all $\alpha \in [0, 1]$ and that the membership function is normal, i.e., the core $[u]_1 = \{x | \mu_u(x) = 1\}$ is compact and nonempty.

The following properties characterize the normal, convex and upper semicontinuous fuzzy sets (in terms of the level-cuts):

(F_1) $[u]_\alpha \in \mathscr{K}_C(\mathbb{R}^n)$ for all $\alpha \in [0, 1]$;
(F_2) $[u]_\alpha \subseteq [u]_\beta$ for $\alpha \geq \beta$ (i.e., they are nested);
(F_3) $[u]_\alpha = \bigcap_{k=1}^{\infty} [u]_{\alpha_k}$ for all increasing sequences $\alpha_k \uparrow \alpha$ converging to α.

Furthermore, any family $\{U_\alpha | \alpha \in [0, 1]\}$ satisfying conditions (F1) – (F3) represents the level-cuts of a fuzzy set u having $[u]_\alpha = U_\alpha$.

We will denote by $\mathscr{F}^n$ the set of the fuzzy sets with the properties above (also called fuzzy quantities). The space $\mathscr{F}^n$ of real fuzzy quantities is structured by an addition and a scalar multiplication, defined either by the level sets or, equivalently, by the Zadeh extension principle .

Let $u, v \in \mathscr{F}^n$ have membership functions μ_u, μ_v and α-cuts $[u]_\alpha$, $[v]_\alpha$, $\alpha \in [0, 1]$, respectively. In the unidimensional case $u \in \mathscr{F}$, we will denote by $[u]_\alpha = [u_\alpha^-, u_\alpha^+]$ the compact intervals forming the α-cuts and the fuzzy quantities will be called fuzzy numbers.

The addition $u + v \in \mathscr{F}^n$ and the scalar multiplication $ku \in \mathscr{F}^n$ have level cuts

$$[u + v]_\alpha = [u]_\alpha + [v]_\alpha = \{x + y | x \in [u]_\alpha, y \in [v]_\alpha\}, \tag{1.12}$$

$$[ku]_\alpha = k[u]_\alpha = \{kx | x \in [u]_\alpha\}, \tag{1.13}$$

In the fuzzy or in the interval arithmetic contexts, equation $u = v + w$ is not equivalent to $w = u - v = u + (-1)v$ or $v = u - w = u + (-1)w$ and this has motivated the introduction of the following Hukuhara difference [28, 44, 55].

Definition 1.2 (*see* [75]) Given $u, v \in \mathscr{F}^n$, the H-difference is defined by

$$u \ominus v = w \Longleftrightarrow u = v + w.$$

Clearly, $u \ominus v = \{0\}$; if $u \ominus v$ exists, it is unique.

In the unidimensional case ($n = 1$), the α-cuts of H-difference are $[u \ominus v]_\alpha = [u_\alpha^- - v_\alpha^-, u_\alpha^+ - v_\alpha^+]$ where $[u]_\alpha = [u_\alpha^-, u_\alpha^+]$ and $[v]_\alpha = [v_\alpha^-, v_\alpha^+]$.

The Hukuhara difference is also motivated by the problem of inverting the addition: if x, y are crisp numbers then $(x + y) - y = x$ but this is not true if x, y are fuzzy. It is possible to see that (see [18]), if u and v are fuzzy numbers (and not in general fuzzy sets), then $(u + v) \ominus v = u$ i.e., the H-difference inverts the addition of fuzzy numbers.

The gH-difference for fuzzy numbers can be defined as follows:

Definition 1.3 Given $u, v \in \mathscr{F}^n$, the gH-difference is the fuzzy quantity $w \in \mathscr{F}^n$, if it exists, such that

$$u \ominus_g v \Longleftrightarrow \begin{cases} \text{(i) } u = v + w, \\ \text{or} \quad \text{(ii) } v = u + (-1)w. \end{cases} \tag{1.14}$$

If $u \ominus_{gH} v$ and $u \ominus v$ exist, $u \ominus v = u \ominus_{gH} v$; if (i) and (ii) are satisfied simultaneously, then w is a crisp quantity. Also, $u \ominus_{gH} u = u \ominus u = \{0\}$.

1.3.1 Support Functions and Fuzzy gH-Difference

An equivalent definition of $w = u \ominus_{gH} v$ for multidimensional fuzzy numbers can be obtained in terms of support functions in a way similar to Eq. (1.4)

$$s_w(p, \alpha) = \begin{cases} s_u(p, \alpha) - s_v(p, \alpha) & \text{in case(i),} \\ s_{(-1)v}(p, \alpha) - s_{(-1)u}(p, \alpha) & \text{in case(ii),} \end{cases} \quad \alpha \in [0, 1], \tag{1.15}$$

where, for a fuzzy quantity u, the support functions are considered for each α-cut and defined to characterize the (compact) α-cuts $[u]_\alpha$:

$$s_u : \mathbb{S}^{n-1} \times [0, 1] \to \mathbb{R} \quad \text{defined by}$$

$$s_u(p; \alpha) = \sup\{\langle p, x\rangle | x \in [u]_\alpha\} \quad \text{for each} \quad p \in \mathbb{S}^{n-1}, \alpha \in [0, 1].$$

As a function of α, $s_u(p, \cdot)$ is nonincreasing for all $p \in \mathbb{S}^{n-1}$, due to the nesting property of the α-cuts.

Proposition 1.8 (see [75]) *Let $s_u(p, \alpha)$ and $s_v(p, \alpha)$ be the support functions of two fuzzy quantities $u, v \in \mathcal{F}^n$. Consider $s_1 = s_u - s_v$, $s_2 = s_v - s_u$; the following four cases apply:*

1. *If s_1 and s_2 are both subadditive in p for all $\alpha \in [0, 1]$ and are nonincreasing for all p, then $u \ominus_{gH} v$ exists;* (i) *and* (ii) *in* (1.15) *are satisfied simultaneously and $u \ominus_{gH} v$ is crisp.*
2. *If s_1 is subadditive in p for all $\alpha \in [0, 1]$ and nonincreasing for all p and s_2 is not, then $w = u \ominus_{gH} v$ exists,* (i) *is satisfied and $s_w = s_u - s_v$.*
3. *If s_2 is subadditive in p for all $\alpha \in [0, 1]$ and nonincreasing for all p and s_1 is not, then $w = u \ominus_{gH} v$ exists,* (ii) *is satisfied and $s_w = s_{-v} - s_{-u}$.*
4. *If s_1 and s_2 are both not subadditive and nonincreasing for all p, then $u \ominus_{gH} v$ does not exist.*

Proof The proof is similar to the proof of Proposition 1.3. For $\alpha \in]0, 1]$ consider the sets $W_\alpha = \{x \in \mathbb{R}^n | \langle p, x\rangle \leq s_u(p; \alpha) \text{ for all } p \in \mathbb{R}^n \text{ with } \|p\| = \alpha\}$, and $W_0 = cl(\bigcup_{\alpha \in]0,1]} W_\alpha)$. Property (F1) for $\{W_\alpha\}$ is ensured by the validity of the subadditive property for all $\alpha \in [0, 1]$. The monotonicity condition $\forall p$ ensures the nesting property (F2) of the α-cuts . It remains to show property (F3), i.e., $W_\alpha = \bigcap_{k=1}^\infty W_{\alpha_k}$ for all increasing sequences $\alpha_k \uparrow \alpha$ converging to $\alpha \in]0, 1]$. As $W_\alpha \subseteq W_{\alpha_k}$ we have $W_\alpha \subseteq \bigcap_{k=1}^\infty W_{\alpha_k}$; let now $x \in \bigcap_{k=1}^\infty W_{\alpha_k}$, for all p having $\|p\| = \alpha$ and all $k = 1, 2, \ldots$ we have $\langle(\alpha_k/\alpha)p, x\rangle \leq s_u((\alpha_k/\alpha)p; \alpha)$ and taking the limit for $k \to \infty$ we obtain, as s_w is continuous (upper semicontinuous), $\langle p, x\rangle = \lim\langle(\alpha_k/\alpha)p, x\rangle \leq \limsup s_u((\alpha_k/\alpha)p; \alpha) = s_u(p; \alpha)$ and $x \in W_\alpha$. The proof is completed. □

It immediately follows a necessary and sufficient condition for $u \ominus_{gH} v$ to exist:

Proposition 1.9 (see [75]) *Let $u, v \in \mathcal{F}^n$ be given with support functions $s_u(p, \alpha)$ and $s_v(p, \alpha)$; then $u \ominus_{gH} v \in \mathcal{F}^n$ exists if and only if at least one of the two functions $s_u - s_v$, $s_{-v} - s_{-u}$ is a support function and is nonincreasing with α for all p.*

In the unidimensional case, the conditions of the definition of $w = u \ominus_{gH} v$ are

$$[w]_\alpha = [w_\alpha^-, w_\alpha^+] = [u]_\alpha \ominus_{gH} [v]_\alpha$$

and

$$\begin{cases} w_\alpha^- = \min\{u_\alpha^- - v_\alpha^-, u_\alpha^+ - v_\alpha^+\}, \\ w_\alpha^+ = \max\{u_\alpha^- - v_\alpha^-, u_\alpha^+ - v_\alpha^+\} \end{cases} \tag{1.16}$$

provided that w_α^- is nondecreasing, w_α^+ is nonincreasing and $w_1^- \le w_1^+$; in particular, for $\alpha \in [0, 1]$,

$$\begin{aligned} &(1) \begin{cases} w_\alpha^- = u_\alpha^- - v_\alpha^- \\ w_\alpha^+ = u_\alpha^+ - v_\alpha^+ \end{cases} \quad \text{if len([u])} \le \text{len([v])}, \\ &(2) \begin{cases} w_\alpha^- = u_\alpha^+ - v_\alpha^+ \\ w_\alpha^+ = u_\alpha^- - v_\alpha^- \end{cases} \quad \text{if len([u])} \ge \text{len([v])}, \end{aligned} \tag{1.17}$$

where $len([u]_\alpha) = u_\alpha^+ - u_\alpha^-$ is the length of the α-cuts of u (similarly $len([v]_\alpha)$ for v).

Proposition 1.10 (see [75]) *Let $u, v \in \mathcal{F}$ be two fuzzy numbers with α-cuts given by $[u]_\alpha$ and $[v]_\alpha$ respectively; the gH-difference $u \ominus_{gH} v \in \mathcal{F}$ exists if and only if one of the two conditions is satisfied:*

$$(a) \begin{cases} len([u]_\alpha) \ge len([v]_\alpha) & \text{for all } \alpha \in [0, 1], \\ u_\alpha^- - v_\alpha^- & \text{is increasing with respect to } \alpha, \\ u_\alpha^+ - v_\alpha^+ & \text{is decreasing with respect to } \alpha \end{cases}$$

or

$$(b) \begin{cases} len([u]_\alpha) \le len([v]_\alpha) & \text{for all } \alpha \in [0, 1], \\ u_\alpha^+ - v_\alpha^+ & \text{is increasing with respect to } \alpha, \\ u_\alpha^- - v_\alpha^- & \text{is decreasing with respect to } \alpha. \end{cases}$$

Proof In fact, consider the support function of u(and similarly for v), obtained by

$$\begin{aligned} s_u(p; \alpha) &= \max_x \{px | u_\alpha^- \le x \le u_\alpha^+\}, \\ p &\in S^0 = \{-1, 1\}, \\ \alpha &\in [0, 1], \end{aligned} \tag{1.18}$$

i.e., simply by $s_u(-1; \alpha) = -u_\alpha^-$ and $s_u(1; \alpha) = u_\alpha^+$. Then

$$s_u(p; \alpha) - s_v(p; \alpha) = \begin{cases} -(u_\alpha^- - v_\alpha^-) & \text{if } p = -1, \\ u_\alpha^+ - v_\alpha^+ & \text{if } p = 1 \end{cases} \tag{1.19}$$

and

$$s_{-v}(p;\alpha) - s_{-u}(p;\alpha) = \begin{cases} -(u_\alpha^+ - v_\alpha^+) & \text{if } p = -1, \\ u_\alpha^- - v_\alpha^- & \text{if } p = 1. \end{cases} \tag{1.20}$$

From the relations above and from Proposition 1.9. we deduce that (symbols $\nearrow$ and $\searrow$ mean that the function is increasing or decreasing, respectively):

$$u \ominus_{gH} v = w \Longleftrightarrow \begin{cases} \text{(a)} \begin{cases} [w]_\alpha = [u_\alpha^- - v_\alpha^-, u_\alpha^+ - v_\alpha^+] \\ \text{provided that } u_\alpha^- - v_\alpha^- \le u_\alpha^+ - v_\alpha^+,\ \forall\alpha \\ \text{and } u_\alpha^- - v_\alpha^- \nearrow, u_\alpha^+ - v_\alpha^+ \searrow \end{cases} \\ or \\ \text{(b)} \begin{cases} [w]_\alpha = [u_\alpha^+ - v_\alpha^+, u_\alpha^- - v_\alpha^-] \\ \text{provided that } u_\alpha^- - v_\alpha^- \ge u_\alpha^+ - v_\alpha^+,\ \forall\alpha \\ \text{and } u_\alpha^+ - v_\alpha^+ \nearrow, u_\alpha^- - v_\alpha^- \searrow \end{cases} \end{cases} \tag{1.21}$$

and the proof is completed. □

The monotonicity of $u_\alpha^- - v_\alpha^-$ and $u_\alpha^+ - v_\alpha^+$ according to (a) or (b) in (1.21) is an important condition for the existence of $u \ominus_{gH} v$ and is to be verified explicitly as in fact it may not be satisfied. Consider $[u]_\alpha = [5 + 4\alpha, 11 - 2\alpha]$ and $[v]_\alpha = [12 + 3\alpha, 19 - 4\alpha]$; then $u_\alpha^- - v_\alpha^- = -6 + \alpha$, $u_\alpha^+ - v_\alpha^+ = -8 + 2\alpha$ and $u_\alpha^+ - v_\alpha^+ < u_\alpha^- - v_\alpha^-$ but $u_\alpha^- - v_\alpha^-$ is not decreasing as required by (1.21)(b).

Remark 1.6 Conditions (a) and (b) of the above proposition are both valid if $len([u]_\alpha) = len([v]_\alpha)$ for all $\alpha \in [0, 1]$; in this case, $u \ominus_{gH} v$ is a crisp quantity.

Example 1.3.1 (*Case of linear membership*) If u and v are trapezoidal linear shaped fuzzy numbers, denoted by $u = \langle u_0^-, u_1^-, u_1^+, u_0^+\rangle$ with $u_0^- \le u_1^- \le u_1^+ \le u_0^+$ (similarly for v) and α-cuts $[u]_\alpha = [u_0^- + \alpha(u_1^- - u_0^-), u_0^+ + \alpha(u_1^+ - u_0^+)]$, then $u \ominus_{gH} v$ exists if and only if

$$(a)\quad u_0^- - v_0^- \le u_1^- - v_1^- \le u_1^+ - v_1^+ \le u_0^+ - v_0^+$$

or

$$(b)\quad u_0^- - v_0^- \ge u_1^- - v_1^- \ge u_1^+ - v_1^+ \ge u_0^+ - v_0^+.$$

In particular, if u and v are (linear) triangular fuzzy numbers(i.e., $u_1^- = u_1^+$ and $v_1^- = v_1^+$) denoted by $u = \langle u_0^-, \widehat{u}, u_0^+\rangle$ and $v = \langle v_0^-, \widehat{v}, v_0^+\rangle$ then $w = u \ominus_g v$ exists if and only if

$$(a)\quad u_0^- - v_0^- \le \widehat{u} - \widehat{v} \le u_0^+ - v_0^+$$

or

$$(b)\quad u_0^- - v_0^- \ge \widehat{u} - \widehat{v} \ge u_0^+ - v_0^+.$$

To illustrate: $\langle 12, 15, 19\rangle \ominus_{gH} \langle 5, 9, 11\rangle$ does not exist as $u_\alpha^- - v_\alpha^- = -6 + \alpha$ and $u_\alpha^+ - v_\alpha^+ = -8 + 2\alpha$ are both increasing with respect to $\alpha \in [0, 1]$ and conditions (1.21) are not satisfied; $\langle 12, 15, 19\rangle \ominus_{gH} \langle 5, 7, 10\rangle = \langle 7, 8, 9\rangle$, according to (a); $\langle 12, 15, 19\rangle \ominus_{gH} \langle 9, 13, 18\rangle = \langle 1, 2, 3\rangle$, according to (b). □

If $u \ominus_{gH} v$ is a fuzzy number, it has the same properties illustrated in Sect. 1.1 for intervals. In particular, the same properties as in Proposition 1.2 are also immediate.

Proposition 1.11 (see [75]) *Let $u, v \in \mathcal{F}$. If $u \ominus_{gH} v$ exists, it is unique and has the following properties* (0 *denotes the crisp set* $\{0\}$)*:*

(1) $u \ominus_{gH} v = 0$.
(2) *(a)* $(u + v) \ominus_{gH} v = u$; *(b)* $u \ominus_{gH} (u - v) = v$.
(3) If $u \ominus_{gH} v$ exists then also $(-v) \ominus_{gH} (-u)$ does and $0 \ominus_{gH} (u \ominus_{gH} v) = (-v) \ominus_{gH} (-u)$.
(4) $u \ominus_{gH} v = v \ominus_{gH} u = w$ *if and only if* $w = -w$*(in particular* $w = 0$ *if and only if* $u = v$*).*
(5) If $v \ominus_{gH} u$ exists then either $u + (v \ominus_{gH} u) = u$ or $v - (v \ominus_{gH} u) = u$ and if both equalities hold then $v \ominus_{gH} u$ is a crisp set.

1.3.2 A Decomposition of Fuzzy Numbers and gH-Difference

In the literature [72] the authors proposed a decomposition of fuzzy numbers (or intervals) which is useful in the study of fuzzy arithmetic operations. The same decomposition can help in performing the gH-difference. Its basic elements are summarized as follows.

Given $u \in \mathcal{F}$ with α-cuts $[u]_\alpha = \left[u_\alpha^-, u_\alpha^+\right]$, define the following quantities:

$$\widehat{u} = \left[\widehat{u}^-, \widehat{u}^+\right] \tag{1.22}$$

be the core $\left[u_1^-, u_1^+\right]$, corresponding to the ($\alpha = 1$)-cut;

$$\widetilde{u}_\alpha = \frac{u_\alpha^- + u_x^+}{2} - \frac{\widehat{u}^- + \widehat{u}^+}{2} \tag{1.23}$$

be the symmetry profile (briefly the profile) of u;

$$\bar{u}_\alpha = \frac{u_\alpha^+ - u_\alpha^-}{2} - \frac{\widehat{u}^+ - \widehat{u}^-}{2} \tag{1.24}$$

be the symmetric fuzzy component of u.

Clearly, we have

$$u_\alpha^- = \widehat{u}^- + \widetilde{u}_\alpha - \bar{u}_\alpha, \quad u_\alpha^+ = \widehat{u}^+ + \widetilde{u}_\alpha + \bar{u}_\alpha$$

and, in interval notation,

$$[u]_\alpha = \widehat{u} + \{\widetilde{u}_\alpha\} + [-\bar{u}_\alpha, \bar{u}_\alpha], \quad \alpha \in [0, 1]. \tag{1.25}$$

We then obtain a decomposition of $u \in \mathcal{F}$ in terms of three components:

1. $\widehat{u} = \left[\widehat{u}^-, \widehat{u}^+\right] \in \mathbb{I}$ is a standard compact real interval;
2. $\widetilde{u} : [0, 1] \to \mathbb{R}$ is a given function such that, for $\alpha = 1$, $\widetilde{u}_1 = 0$; denote with $\mathbb{P}$ the set of all such (profile) functions;
3. $\bar{u}$ is a symmetric fuzzy number with core given by the singleton $\{0\}$; denote by $\mathbb{S}_0$ the family of all such fuzzy numbers.

By the use of the three previous elements, any fuzzy number or interval can be represented by a triplet (with a small abuse of notation we denote by $\bar{u}$ both the function in (1.24) and the 0-symmetric fuzzy number obtained with it):

$$\begin{aligned} & u = (\widehat{u}, \widetilde{u}, \bar{u}) \in \mathbb{I} \times \mathbb{P} \times \mathbb{S}_0, \\ & \text{i.e.,} \\ & \widehat{u} \in \mathbb{I} \quad \text{(crisp number or interval)}, \\ & \widetilde{u} \in \mathbb{P} \quad \text{(crisp symmetry profile)}, \\ & \bar{u} \in \mathbb{S}_0 \quad \text{(0-symmetric fuzzy number)}. \end{aligned}$$

The profile function $\widetilde{u} \in \mathbb{P}$ and the symmetric fuzzy number $\bar{u} \in \mathbb{S}_0$ form what we call a valid pair:

Definition 1.4 (*see* [75]) A pair of elements $(\widetilde{u}, \bar{u}) \in \mathbb{P} \times \mathbb{S}_0$ is said to form a valid pair if it represents a fuzzy number having α-cuts $[\widetilde{u}_\alpha - \bar{u}_\alpha, \widetilde{u}_\alpha + \bar{u}_\alpha]$, i.e., if the following conditions are satisfied:

$$\alpha' < \alpha'' \Longrightarrow \begin{cases} \widetilde{u}_{\alpha'} - \bar{u}_{\alpha'} \le \widetilde{u}_{\alpha''} - \bar{u}_{\alpha''}, \\ \widetilde{u}_{\alpha''} + \bar{u}_{\alpha'} \ge \widetilde{u}_{\alpha''} + \bar{u}_{\alpha''}, \end{cases}.$$

i.e., if $\widetilde{u}_\alpha - \bar{u}_\alpha$ is a nondecreasing function and $\widetilde{u}_\alpha + \bar{u}_\alpha$ is a nonincreasing function (note that for $\alpha = 1$ both $\widetilde{u}$ and $\bar{u}$ are zero).

By the definition above, we can define the following decomposition of the fuzzy numbers (intervals) $u \in \mathcal{F}$:

Proposition 1.12 (see [75]) *Any fuzzy number (interval) $u \in \mathscr{F}$ with α-cuts $[u]_\alpha = [u_\alpha^-, u_\alpha^+]$ can be represented in the form $u = (\widehat{u}, \widetilde{u}, \bar{u}) \in \mathbb{I} \times \mathbb{P} \times \mathbb{S}_0$ defined in* (1.22)-(1.24) *and $(\widetilde{u}, \bar{u})$ is a valid pair. Vice versa, valid triplet $(\widehat{u}, \widetilde{u}, \bar{u}) \in \mathbb{I} \times \mathbb{P} \times \mathbb{S}_0$ (i.e., $(\widetilde{u}, \bar{u})$ is a valid pair) represents a fuzzy number (interval) $u \in \mathscr{F}$ with α-cuts given by* (1.25).

Proof See [72]. □

Definition 1.5 (*see* [75]) We call

$$u = (\widehat{u}, \widetilde{u}, \bar{u}) \in \mathbb{I} \times \mathbb{P} \times \mathbb{S}_0$$

(where we assume $(\widetilde{u}, \bar{u})$ be a valid pair) as the CPS-decomposition of $u \in \mathscr{F}$ (C = Crisp, P = Profile, S = 0 − Symmetricfuzzy). We write this as

$$\mathscr{F} = \mathbb{I} \times \mathbb{V}(\mathbb{P}, \mathbb{S}_0),$$

where we assume $\mathbb{V}(\mathbb{P}, \mathbb{S}_0) \subset \mathbb{P} \times \mathbb{S}_0$ is the set of all valid pairs.

Now, we can find the gH-difference $u \ominus_{gH} v$ of $u, v \in \mathscr{F}$ in terms of the CPS decomposition.

Proposition 1.13 (see [75]) *Given $u = (\widehat{u}, \widetilde{u}, \bar{u}) \in \mathscr{F}$ and $v = (\widehat{v}, \widetilde{v}, \bar{v}) \in \mathscr{F}$, the gH-difference $w = u \ominus_{gH} v$ is given by $w = (\widehat{w}, \widetilde{w}, \bar{w}) \in \mathscr{F}$ with*

$$\begin{aligned} \widehat{w} &= \widehat{u} \ominus_{gH} \widehat{v}, \\ \widetilde{w}_\alpha &= \widetilde{u}_\alpha - \widetilde{v}_\alpha, \\ \bar{w}_\alpha &= \begin{cases} \bar{u}_\alpha - \bar{v}_\alpha & \text{if } \bar{u} - \bar{v} \in \mathbb{S}_0, \\ \bar{v}_\alpha - \bar{u}_\alpha & \text{if } \bar{v} - \bar{u} \in \mathbb{S}_0 \end{cases} \end{aligned}$$

and $u \ominus_{gH} v$ exists if and only if one of the two conditions are satisfied:

(1) $\bar{u} - \bar{v} \in \mathbb{S}_0, \widehat{u} \ominus_{gH} \widehat{v}$ *exists in case* (i) *and* $(\widetilde{u} - \widetilde{v}, \bar{u} - \bar{v}) \in \mathbb{V}(\mathbb{P}, \mathbb{S}_0)$ *or*
(2) $\bar{v} - \bar{u} \in \mathbb{S}_0, \widehat{u} \ominus_{gH} \widehat{v}$ *exists in case* (ii) *and* $(\widetilde{u} - \widetilde{v}, \bar{v} - \bar{u}) \in \mathbb{V}(\mathbb{P}, \mathbb{S}_0)$.

In particular, if $u = (\widehat{u}, \widetilde{u}, \bar{u}) \in \mathscr{F}$ and $v = (\widehat{v}, \widetilde{v}, \bar{v}) \in \mathscr{F}$ are fuzzy numbers, i.e., $\widehat{u}^- = \widehat{u}^+ = \widehat{u}$ and $\widehat{v}^- = \widehat{v}^+ = \widehat{v}$, then $u \ominus_{gH} v$ exists (in this case $\widehat{w} = \widehat{u} - \widehat{v}$ is crisp) if and only if one of the two conditions are satisfied:

(1′) $\bar{u} - \bar{v} \in \mathbb{S}_0$ *and* $(\widetilde{u} - \widetilde{v}, \bar{u} - \bar{v}) \in \mathbb{V}(\mathbb{P}, \mathbb{S}_0)$ *or*
(2′) $\bar{v} - \bar{u} \in \mathbb{S}_0$ *and* $(\widetilde{u} - \widetilde{v}, \bar{v} - \bar{u}) \in \mathbb{V}(\mathbb{P}, \mathbb{S}_0)$.

Proof First, as $\widehat{u}, \widehat{v} \in \mathbb{I}$, then $\widehat{w} = \widehat{u} \ominus_{gH} \widehat{v} \in \mathbb{I}$ always exists and if $\widetilde{u}, \widetilde{v} \in \mathbb{P}$ then $\widetilde{w} = \widetilde{u} - \widetilde{v} \in \mathbb{P}$. If $u \ominus_{gH} v$ exists then clearly $\bar{u} - \bar{v} \in \mathbb{S}_0$ or $\bar{v} - \bar{u} \in \mathbb{S}_0$ (as in fact $\bar{w} = \bar{u} - \bar{v}$ or $\bar{w} = \bar{v} - \bar{u}$ from Eqs. (1.17)) and $(\widetilde{u} - \widetilde{v}, \bar{u} - \bar{v})$ or $(\widetilde{u} - \widetilde{v}, \bar{v} - \bar{u})$ is a valid pair. Vice versa, if $\bar{u} - \bar{v} \in \mathbb{S}_0$ and $(\widetilde{u} - \widetilde{v}, \bar{u} - \bar{v})$ is a valid pair(or if $\bar{v} - \bar{u} \in \mathbb{S}_0$ and $(\widetilde{u} - \widetilde{v}, \bar{v} - \bar{u})$ is a valid pair), then the triplet $(\widehat{u} \ominus_{gH} \widehat{v}, \widetilde{u} - \widetilde{v}, \bar{u} - \bar{v})$ (or the triplet $(\widehat{u} \ominus_{gH} \widehat{v}, \widetilde{u} - \widetilde{v}, \bar{v} - \bar{u})$) represents a fuzzy number (interval) and by (1.16) it is $\widehat{u} \ominus_{gH} \widehat{v}$. The second part is immediate as the gH-difference $\widehat{u} \ominus_{gH} \widehat{v}$ is the standard crisp difference and $\widehat{w}$ is crisp. This completes the proof. □

Remark 1.7 If the functions u_α^- and u_α^+ are differentiable with respect to α, then $\widetilde{u}_\alpha$ and $\bar{u}_\alpha$ are differentiable and the condition for $(\widetilde{u}, \bar{u})$ to be a valid pair is $|\widetilde{u}'_\alpha| \le -\bar{u}'_\alpha \ \forall \alpha \in]0, 1[$ (recall that $\bar{u}_\alpha$ is a decreasing function).

Example 1.3.2 (*see* [75]) If we have symmetric fuzzy numbers $u, v \in \mathscr{F}$ (i.e., $\widehat{u}^- = \widehat{u}^+ = \widehat{u}, \widehat{v}^- = \widehat{v}^+ = \widehat{v}$, and $\widetilde{u}_\alpha = \widetilde{v}_\alpha = 0 \ \forall \alpha \in [0, 1]$) then (only) condition $|\bar{u} - \bar{v}| \in \mathbb{S}_0$ i.e., ($\bar{u} - \bar{v} \in \mathbb{S}_0$ or $\bar{v} - \bar{u} \in \mathbb{S}_0$) is necessary and sufficient for $u \ominus_{gH} v$ to exist and $(\widehat{u}, 0, \bar{u}) \ominus_{gH} (\widehat{v}, 0, \bar{v}) = (\widehat{u} - \widehat{v}, 0, |\bar{u} - \bar{v}|)$ i.e., with α-cuts

$$[u \ominus_{gH} v]_\alpha = [(\widehat{u} - \widehat{v}) - |\bar{u}_\alpha - \bar{v}_\alpha|, (\widehat{u} - \widehat{v}) + |\bar{u}_\alpha - \bar{v}_\alpha|].$$

For example, $(1, 0, 1 - \alpha^2) \ominus_{gH} (0, 0, 2 - 2\alpha) = (1, 0, (1 - \alpha)^2)$, according to case (ii) of gH-difference, has α-cuts $[1 - (1 - \alpha)^2, 1 + (1 - \alpha)^2]$. □

Example 1.3.3 (*see* [75]) If we have symmetric triangular (with linear membership) fuzzy numbers $u = \langle \widehat{u} - \Delta u, \widehat{u}, \widehat{u} + \Delta u \rangle$ and $v = \langle \widehat{v} - \Delta v, \widehat{v}, \widehat{v} + \Delta v \rangle$ then $w = u \ominus_{gH} v$ always exists and is the triangular symmetric fuzzy number

$$w = \langle \widehat{w} - \Delta w, \widehat{w}, \widehat{w} + \Delta w \rangle \quad \text{where}$$
$$\widehat{w} = \widehat{u} - \widehat{v} \quad \text{and} \quad \Delta w = |\Delta u - \Delta v|.$$

In fact, in this case, $|\bar{u}_\alpha - \bar{v}_\alpha| = |\Delta u - \Delta v|(1 - \alpha)$, i.e., $\bar{u} - \bar{v} \in \mathbb{S}_0$ if $|\Delta u - \Delta v| \ge 0$ or $\bar{v} - \bar{u} \in \mathbb{S}_0$ if $|\Delta v - \Delta u| \le 0$ so that

$$[u \ominus_{gH} v]_\alpha = [(\widehat{u} - \widehat{v}) - |\Delta u - \Delta v|(1 - \alpha), (\widehat{u} - \widehat{v}) + |\Delta u - \Delta v|(1 - \alpha)].$$

In particular, if $\Delta u = \Delta v$ then $u \ominus_{gH} v$ is a crisp number. To illustrate, $\langle 2, 4, 6 \rangle \ominus_{gH} \langle -2, 1, 4 \rangle = \langle 2, 3, 4 \rangle$, according to case (ii) of gH-difference, and in fact $\langle 2, 4, 6 \rangle + (-1)\langle 2, 3, 4 \rangle = \langle 2, 4, 6 \rangle + \langle -4, -3, -2 \rangle = \langle -2, -1, 4 \rangle$; $\langle -2, -1, 4 \rangle \ominus_{gH} \langle 2, 4, 6 \rangle = \langle -4, -3, -2 \rangle$, according to case (i) of gH-difference, and in fact

$$\langle 2, 4, 6 \rangle + \langle -4, -3, -2 \rangle = \langle -2, -1, 4 \rangle.$$

1.3.3 Approximated Fuzzy gH-Difference

The approximated fuzzy gH-difference was proposed and studied [75].

If the gH-differences $[u]_\alpha \ominus_{gH} [v]_\alpha$ do not define a proper fuzzy number, we can use the nested property of the α-cuts and obtain a proper fuzzy number by

$$[u\widetilde{\ominus}v]_\alpha := cl\Bigg(\bigcup_{\beta\geq\alpha}([u]_\beta \ominus_{gH} [v]_\beta)\Bigg) \quad \text{for } \alpha \in [0, 1]. \tag{1.26}$$

As each gH-difference $[u]_\beta \ominus_{gH} [v]_\beta$ exists for $\beta \in [0, 1]$ and (1.26) defines a proper fuzzy number, it follows that $z = u\widetilde{\ominus}v$ can be considered as a generalization of Hukuhara difference for fuzzy numbers, existing for any u, v.

Example 1.3.4 (*see* [75]) $\langle 12, 15, 19\rangle \ominus_{gH} \langle 5, 9, 11\rangle$ does not exists; we obtain

$$[u]_\beta \ominus_{gH} [v]_\beta = [8 - 2\beta, 7 - \beta]$$

and $[u\widetilde{\ominus}v]_\alpha = [6, 7 - \alpha]$. □

A discretized version of $z = u\widetilde{\ominus}v$ can be obtained by choosing a partition $0 = \alpha_0 < \alpha_1 < \cdots < \alpha_N = 1$ of $[0, 1]$ and from $[w_i^-, w_i^+] = [u]_{\alpha_i} \ominus_{gH} [v]_{\alpha_i}$ by the following backward iteration:

$$z_N^- = w_N^-, \quad z_N^+ = w_N^+.$$

$$\text{For } k = N-1, \ldots, 0: \begin{cases} z_k^- = \min\{z_{k+1}^-, w_k^-\}, \\ z_k^+ = \max\{z_{k+1}^+, w_k^+\}. \end{cases}$$

A third possibility for a gH-difference of fuzzy numbers may be obtained by defining $z = \widetilde{u\ominus_{gH}v}$ to be the fuzzy number whose α-cuts are as near as possible to the gH-differences $[u]_\alpha \ominus_{gH} [v]_\alpha$, for example by minimizing the functional ($w_\alpha \geq 0$ and $\gamma_\alpha \geq 0$ are weighting functions)

$$G(z|u, v) = \int_0^1 (w_\alpha[z_\alpha^- - (u \ominus_{gH} v)_\alpha^-]^2 + \gamma_\alpha[z_\alpha^+ - (u \ominus_{gH} v)_\alpha^+]^2)\mathrm{d}\alpha \tag{1.27}$$

such that z_α^- is increasing with α, z_α^+ is decreasing with α and $z_\alpha^- \leq z_\alpha^+\ \forall\alpha \in [0, 1]$.

A discretized version of $G(z|u, v)$ can be obtained by choosing a partition $0 = \alpha_0 < \alpha_1 < \cdots < \alpha_N = 1$ of $[0, 1]$ and defining the discretized $G(z|u, v)$ as

$$G_N(z|u, v) = \sum_{i=0}^{N} w_i[z_i^- - (u \ominus_{gH} v)_i^-]^2 + \gamma_i[z_i^+ - (u \ominus_{gH} v)_i^+]^2,$$

we minimize $G_N(z|u,v)$ with the given data $(u \ominus_{gH} v)_i^- = \min\{u_{\alpha_i}^- - v_{\alpha_i}^-, u_{\alpha_i}^+ - v_{\alpha_i}^+\}$ and $(u \ominus_{gH} v)_i^+ = \max\{u_{\alpha_i}^- - v_{\alpha_i}^-, u_{\alpha_i}^+ - v_{\alpha_i}^+\}$, subject to the constraints $z_0^- \le z_1^- \le \cdots \le z_N^- \le z_N^+ \le z_{N-1}^+ \le \cdots \le z_0^+$. We obtain a linearly constrained least squares minimization of the form

$$\min_{z \in \mathbb{R}^{2N+2}} (z-w)^T D^2 (z-w) \quad s.t. Ez \ge 0, \tag{1.28}$$

where $z = (z_0^-, z_1^-, \ldots, z_N^-, z_N^+, z_{N-1}^+, \ldots, z_0^+)$, $w_i^- = (u \ominus_{gH} v)_i^-$, $w_i^+ = (u \ominus_{gH} v)_i^+$,

$$w = (w_0^-, w_1^-, \ldots, w_N^-, w_N^+, w_{N-1}^+, \ldots, w_0^+),$$

$D = \text{diag}\{\sqrt{w_0}, \ldots, \sqrt{w_N}, \sqrt{\gamma_N}, \ldots, \sqrt{\gamma_0}\}$ and E is the $(N, N+1)$ matrix

$$E = \begin{bmatrix} -1 & 1 & 0 & \cdots & \cdots & 0 \\ 0 & -1 & 1 & 0 & \cdots & 0 \\ \vdots & \vdots & \vdots & \vdots & \cdots & \vdots \\ 0 & 0 & \cdots & \cdots & -1 & 1 \end{bmatrix},$$

which can be solved by standard efficient procedures (see the classical book [56, Chap. 23]). If, at solution z^*, we have $z^* = w$, then we obtain the gH-difference as defined in (1.14).

1.4 Generalized Division

An ideal similar to the gH-difference can be used to introduce a division of real intervals and fuzzy numbers. One may consult [75] for more details.

Consider first the case of real compact intervals $A = [a^-, a^+]$ and $B = [b^-, b^+]$ with $b^- > 0$ or $b^+ < 0$ (i.e., $0 \notin B$).

The interval $C = [c^-, c^+]$ defining the multiplication $C = AB$ is given by

$$c^- = \min\{a^-b^-, a^-b^+, a^+b^-, a^+b^+\}, \quad c^+ = \max\{a^-b^-, a^-b^+, a^+b^-, a^+b^+\}$$

and the multiplicative "inverse" (it is not the inverse in the algebraic sense) of an interval B is defined by $B^{-1} = [1/b^+, 1/b^-]$.

Definition 1.6 (*see* [75]) For $A = [a^-, a^+]$ and $B = [b^-, b^+]$ we define the generalized division (g-division) $\div_g$ as follows:

$$A \div_g B = C \Longleftrightarrow \begin{cases} & \text{(i) } A = BC \\ \text{or} & \text{(ii) } B = AC^{-1}. \end{cases} \tag{1.29}$$

If both cases (i) and (ii) are valid, we have $CC^{-1} = C^{-1}C = \{1\}$, i.e., $C = \{\widehat{c}\}$, $C^{-1} = \{1/\widehat{c}\}$ with $\widehat{c} \neq 0$. It is immediate to see that $A \div_g B$ always exists and is unique for given $A = [a^-, a^+]$ and $B = [b^-, b^+]$ with $0 \notin B$.

It is easy to see that the following six cases are possible (see [75]), with the indicated rules:

Case 1: If $0 < a^- \leq a^+$ and $b^- \leq b^+ < 0$ then

$$\begin{aligned} &\text{if } a^-b^- \geq a^+b^+ \text{ then } c^- = \frac{a^+}{b^-},\ c^+ = \frac{a^-}{b^+} \text{ and (i) is satisfied,} \\ &\text{if } a^-b^- \leq a^+b^+ \text{ then } c^- = \frac{a^-}{b^+},\ c^+ = \frac{a^+}{b^-} \text{ and (ii) is satisfied.} \end{aligned} \tag{1.30}$$

Case 2: If $0 < a^- \leq a^+$ and $0 < b^- \leq b^+$ then

$$\begin{aligned} &\text{if } a^-b^+ \leq a^+b^- \text{ then } c^- = \frac{a^-}{b^-},\ c^+ = \frac{a^+}{b^+} \text{ and (i) is satisfied,} \\ &\text{if } a^-b^+ \geq a^+b^+ \text{ then } c^- = \frac{a^+}{b^+},\ c^+ = \frac{a^-}{b^-} \text{ and (ii) is satisfied.} \end{aligned} \tag{1.31}$$

Case 3: If $a^- \leq a^+ < 0$ and $b^- \leq b^+ < 0$ then

$$\begin{aligned} &\text{if } a^+b^- \leq a^-b^+ \text{ then } c^- = \frac{a^+}{b^+},\ c^+ = \frac{a^-}{b^-} \text{ and (i) is satisfied,} \\ &\text{if } a^+b^- \geq a^-b^+ \text{ then } c^- = \frac{a^-}{b^-},\ c^+ = \frac{a^+}{b^+} \text{ and (ii) is satisfied.} \end{aligned} \tag{1.32}$$

Case 4: If $a^- \leq a^+ < 0$ and $0 < b^- \leq b^+$ then

$$\begin{aligned} &\text{if } a^-b^- \leq a^+b^+ \text{ then } c^- = \frac{a^-}{b^+},\ c^+ = \frac{a^+}{b^-} \text{ and (i) is satisfied,} \\ &\text{if } a^-b^- \geq a^+b^+ \text{ then } c^- = \frac{a^+}{b^-},\ c^+ = \frac{a^-}{b^+} \text{ and (ii) is satisfied.} \end{aligned} \tag{1.33}$$

Case 5: If $a^- \leq 0$, $a^+ \geq 0$ and $b^- \leq b^+ < 0$ then the solution does not depend on b^+,

$$c^- = \frac{a^+}{b^-},\quad c^+ = \frac{a^-}{b^-} \text{ and (i) is satisfied.} \tag{1.34}$$

Case 6: If $a^- \leq 0$, $a^+ \geq 0$ and $0 < b^- \leq b^+$ then the solution does not depend on b^-,

$$c^- = \frac{a^-}{b^+},\quad c^+ = \frac{a^+}{b^+} \text{ and (i) is satisfied.} \tag{1.35}$$

Remark 1.8 If $0 \in]b^-, b^+[$ the g-division is undefined: for intervals $B = [0, b^+]$ or $B = [b^-, 0]$ the division is possible but obtaining unbounded results C of the form $C =]-\infty, c^+]$ or $C =]c^-, +\infty[$: we work with $B = [\varepsilon, b^+]$ or $B = [b^-, -\varepsilon]$ and we obtain the result by the limit for $\varepsilon \to 0^+$. Example: for $[-2, -1] \div_g [0, 3]$ we consider $[-2, -1] \div_g [\varepsilon, 3] = [c_\varepsilon^-, c_\varepsilon^+]$ with (case 2) $c_\varepsilon^- = \min\{\frac{-2}{3}, -1/\varepsilon\}$ and $c_\varepsilon^+ = \max\{-2/\varepsilon, \frac{-1}{3}\}$ and obtain the result $C = [-\infty, -\frac{1}{3}]$ at the limit $\varepsilon \to 0^+$.

The following properties are immediate.

Proposition 1.14 (see [75]) *For any $A = [a^-, a^+]$ and $B = [b^-, b^+]$ with $0 \notin B$, we have (here 1 is the same as $\{1\}$):*

1. $B \div_g B = 1$, $B \div_g B^{-1} = \{b^- b^+\} (= \{\widehat{b}^2\}$ *if* $b^- = b^+ = \widehat{b})$.
2. $(AB) \div_g B = A$.
3. $1 \div_g B = B^{-1}$ *and* $1 \div_g B^{-1} = B$.
4. *At least one of the equality $B(A \div_g B) = A$ or $A(A \div_g B)^{-1} = B$ is valid and both hold if and only if $A \div_g B$ is a singleton.*

1.4.1 The Fuzzy Case

The search for alternative definitions of the division operator between fuzzy numbers has received some attention in recent literature, with the objective of inverting multiplication; some recent studies are in [14, 24, 48, 62, 63, 72]. In particular, Boukezzoula et al. [24] have analyzed the fuzzy division in a way similar to our setting (considering case (i) of the division operator). Using our notation, the generalized division introduced in [24] is $z = u \oslash v$ where the α-cuts $[z_\alpha^-, z_\alpha^+]$ of z are defined by

$$z_\alpha^- = A_\alpha^- / B_\alpha^-, \quad z_\alpha^+ = A_\alpha^+ / B_\alpha^+,$$

with

$$A_\alpha^- = \begin{cases} u_\alpha^- \text{ if } v_\alpha^- > 0, \\ u_\alpha^+ \text{ if } v_\alpha^+ < 0, \end{cases} \qquad B_\alpha^- = \begin{cases} v_\alpha^- \text{ if } A_\alpha^- \geq 0, \\ v_\alpha^+ \text{ if } A_\alpha^- < 0 \end{cases}$$

and

$$A_\alpha^+ = \begin{cases} u_\alpha^+ \text{ if } v_\alpha^- > 0, \\ u_\alpha^- \text{ if } v_\alpha^+ < 0, \end{cases} \qquad B_\alpha^+ = \begin{cases} v_\alpha^+ \text{ if } A_\alpha^+ \geq 0, \\ v_\alpha^- \text{ if } A_\alpha^+ < 0. \end{cases}$$

Here, we suggest an approach to division as inverse operator of fuzzy multiplication, similar to gH-difference (as inverse operator of fuzzy addition).

Definition 1.7 (*see* [75]) Let $u, v \in \mathcal{F}$ have α-cuts $[u]_\alpha = [u_\alpha^-, u_\alpha^+]$, $[v]_\alpha = [v_\alpha^-, v_\alpha^+]$ with $0 \notin [v]_\alpha$ $\forall \alpha \in [0, 1]$. The g-division $\div_g$ is the operator that calculates the fuzzy number $w = u \div_g v \in \mathcal{F}$ having level cuts $[w]_\alpha = [w_\alpha^-, w_\alpha^+]$ (here $[w]_\alpha^{-1} = [1/w_\alpha^+, 1/w_\alpha^-]$) defined by

$$[u]_\alpha \div_g [v]_\alpha = [w]_\alpha \Longleftrightarrow \begin{cases} \quad \text{(i) } [u]_\alpha = [v]_\alpha [w]_\alpha, \\ \text{or} \quad \text{(ii) } [v]_\alpha = [u]_\alpha [w]_\alpha^{-1} \end{cases} \tag{1.36}$$

provided that w is a proper fuzzy number, where the multiplications between intervals are performed in the standard interval arithmetic setting.

The fuzzy g-division $\div_g$ is well defined if the α-cuts $[w]_\alpha$ are such that $w \in \mathcal{F}$ (w_α^- is nondecreasing, w_α^+ is non-increasing, $w_1^- \le w_1^+$).

Clearly, if $u \div_g v \in \mathcal{F}$ exists, it has the properties already illustrated for the interval case.

Proposition 1.15 (see [75]) *Let $u, v \in \mathcal{F}$ (here 1 is same as $\{1\}$). We have:*

1. *if $0 \notin [u]_\alpha$ $\forall\alpha$, then $u \div_g u = 1$;*
2. *if $0 \notin [v]_\alpha$ $\forall\alpha$, then $(uv) \div_g v = u$;*
3. *if $0 \notin [u]_\alpha$ $\forall\alpha$, then $1 \div_g v = v^{-1}$ and $1 \div_g v^{-1} = v$;*
4. *if $v \div_g u$ exists then either $u(v \div_g u) = v$ or $v(v \div_g u)^{-1} = u$ and both equalities hold if and only if $v \div_g u$ is crisp.*

It is easy to see that if $w = u \div_g v$ exists according to case (i) then also $z = u \oslash v$ of [24] exists and $w = z$; but the existence of $u \div_g v$ according to case (ii) is not allowed for $u \oslash v$.

We illustrate the g-division $\div_g$ with some examples (the fuzzy numbers are defined in terms of their α-cuts). The following examples can be found in [75].

1. $[1 + 2\alpha, 7 - 4\alpha] \div_g [-3 + \alpha, -1 - \alpha] = [(7 - 4\alpha)/(-3 + \alpha), (1 + 2\alpha)/(-1 - \alpha)]$ according to case (i); the same result is obtained by $u \oslash v$ of [24].
2. $[-3 + 3\alpha, 2 - 2\alpha] \div_g [3 + 2\alpha, 8 - 3\alpha] = [(-3 + 3\alpha)/(8 - 3\alpha), (2 - 2\alpha)/(8 - 3\alpha)]$ according to case (i) (the same result is obtained by $u \oslash v$); note that in this case, the result does not depend on v_α^-.
3. $[1 + 0.5\alpha, 5 - 3.5\alpha] \div_g [-4 + 2\alpha, -1 - \alpha]$ does not exist (the same as in [24]).
4. $[-7 + 2\alpha, -4 - \alpha] \div_g [12 + 5\alpha, -4 - 3\alpha] = [(-7 + 2\alpha)/(-12 + 5\alpha), (-4 - \alpha)/(-4 - 3\alpha)]$ according to case (ii).
5. $[-5 + \alpha, -3 - \alpha] \div_g [4 + 2\alpha, 11 - 5\alpha] = [(-3 - \alpha)/(4 + 2\alpha), (-5 + \alpha)/(11 - 5\alpha)]$ according to case (ii).

Note that for examples 4 and 5 the division $u \oslash v$ does not exist.

1.4.2 Approximated Fuzzy g-Division

If the g-divisions $[u]_\alpha \div_g [v]_\alpha$ do not define a proper fuzzy number, we can proceed similarly to what is done in Sect. 1.3.3 (see [75]) and obtain an approximated fuzzy division with α-cuts

$$[u\widetilde{\div}v]_\alpha := cl\Big(\bigcup_{\beta\geq\alpha}([u]_\beta \div_g [v]_\beta)\Big). \tag{1.37}$$

As each g-division $[u]_\beta \div_g [v]_\beta$ exists for $\beta \in [0, 1]$, $z = u\widetilde{\div}v$ can be considered as a generalization of division of fuzzy numbers, existing for any u, v with $0 \notin [v]_\beta$ for $\beta \in [0, 1]$.

A discretized version of $z = u\widetilde{\div}v$ on a partition $0 = \alpha_0 < \alpha_1 < \cdots < \alpha_N = 1$ of $[0, 1]$ is obtained using $[w_i^-, w_i^+] = [u]_{\alpha_i} \div_g [v]_{\alpha_i}$ and

$$z_N^- = w_N^-, \quad z_N^+ = w_N^+.$$

$$\text{For} \quad k = N-1, \ldots, 0: \begin{cases} z_k^- &= \min\{z_{k+1}^-, w_k^-\} \\ z_k^+ &= \max\{z_{k+1}^+, w_k^+\}. \end{cases}$$

Example 1.4.1 (*see* [75]) For all values of $\alpha \in [0, 1]$ the interval g-divisions $[1 + 0.5\alpha, 5 - 3.5\alpha] \div_g [-4 + 2\alpha, -1 - \alpha] = [(5 - 3.5\alpha)/(-4 + 2\alpha), (1 + 0.5\alpha)/(-1 - \alpha)]$ exist but the resulting intervals are not the α-cuts of a fuzzy number; applying (1.37) we obtain the fuzzy number $[1 + 0.5\alpha, 5 - 3.5\alpha]\widetilde{\div}[-4 + 2\alpha, -1 - \alpha] = [(5 - 3.5\alpha)/(-4 + 2\alpha), -0.75]$. □

1.5 Applications of gH-Difference and g-Division

In the literature [75], the authors present some applications of gH-difference and g-division.

Two important applications of the generalized difference and division are in the field of fuzzy linear equations and fuzzy derivative.

Solving a interval or a fuzzy linear equation $AX + B = C$ has been addressed by various authors and the recent literature has concentrated on different approaches; essentially, the problem cannot be solved from a strict algebraic point of view as it is well known that in general no X (interval or fuzzy number) exists such that equality $AX + B = C$ is satisfied. For example, the solution based on the application of the extension principle (i.e., $X = \{x|ax + b = c \text{ for } a \in A, b \in B \text{ and } c \in C\}$) will not satisfy the equation algebraically.

For the fuzzy case, some authors (see [19]) have proposed a procedure where the possible solutions are ordered by preference, like:

1. try to find the interval-arithmetic-based solution X_1 (using the α-cuts) satisfying exactly our equation; if such a solution exists (i.e., the resulting α-cuts produce a fuzzy number) then accept it;
2. otherwise, apply the extension principle or, depending on the problem at hand,
3. apply approximated interval arithmetic.

In the following Sect. 1.5.1, we illustrate that the proposed generalized difference and division can be used to produce a solution of interval and algebraic equations.

The second Sect. 1.5.2 addresses the definition of the derivative for interval-valued and fuzzy-valued functions and its application to interval and fuzzy differential equations. Also this problem has received great attention in the recent literature. Using the gH-difference in this chapter, in [74] the advantages of this new definition are illustrated for the interval case and it seems to be a useful tool to overcome several difficulties posed by the standard Hukuhara derivative, e.g. its restricted range of existence and the possibilistic irreversibility effect described e.g. in [28, 30].

1.5.1 Interval and Fuzzy Algebraic Equations

The gH-difference can be used to solve the simple interval equation

$$A + X = B, \tag{1.38}$$

where $A = [a^-, a^+]$, $B = [b^-, b^+]$ are given intervals and $X = [x^-, x^+]$ is an interval to be determined satisfying (1.38). We have seen that, for unidimensional intervals, the gH-difference always exists. Denote by $len(A) = a^+ - a^-$ the length of interval A. It is well known from classical interval arithmetic that an interval X satisfying (1.38) exists only if $len(B) \geq len(A)$ (in Minkowski arithmetic we have $len(A + X) \geq \max\{len(A), len(X)\}$); in fact, no such X exists with $x^- \leq x^+$ if $len(B) < len(A)$ and we cannot solve (1.38) unless we interpret it as $B - X = A$.

If we do so, we get

$$case\, len(B) \leq len(A) : \begin{cases} a^- + x^- = b^-, \\ a^+ + x^+ = b^+, \end{cases} \quad i.e., \quad \begin{cases} x^- = b^- - a^-, \\ x^+ = b^+ - a^+, \end{cases}$$

$$case\, len(B) \geq len(A) : \begin{cases} b^- - x^+ = a^-, \\ b^+ - x^- = a^+, \end{cases} \quad i.e., \quad \begin{cases} x^- = b^+ - a^+, \\ x^+ = b^- - a^-. \end{cases}$$

We then obtain that $X = B \ominus_{gH} A$ is the unique solution to (1.38) and it always exists, i.e.,

Proposition 1.16 (see [75]) *Let $A, B \in \mathcal{K}_C(\mathbb{R})$; the gH-difference $X = B \ominus_{gH} A$ always exists and either $A + X = B$ or $A = B - X$.*

To help interpretation of $B \ominus_{gH} A$ as the solution of equation $A + X = B$ we suggest the following definitions:

1. if equality $A + X = B$ is satisfied, in the standard addition sense, we say that X is an additive solution of (1.38);

2. if equality $A = B - X$ is satisfied, in the standard subtraction sense, we say that X is a subtractive solution of (1.38).

From Proposition 1.2, a similar result is true for equation $A + X = B$ with $A, B \in \mathcal{K}_C(\mathbb{R})$ but for $n > 1$ the gH-difference may not exist.

In a similar way, Proposition 1.14 is a motivation to consider $X = B \div A$ (if $0 \notin A$) as the solution to the interval equation

$$AX = B. \tag{1.39}$$

Proposition 1.17 (see [75]) *Let $A, B \in \mathcal{K}_C(\mathbb{R})$; the g-division $X = B \div_g A$ always exists (if $0 \notin A$) and (at least) one of the following equalities is satisfied:*

1. $AX = B$;
2. $B/X = A$.

We say that $X = B \div_g A$ is the solution of (1.39) in the multiplicative sense if $AX = B$ and in the divisive sense if $BX^{-1} = A$.

More generally, given $A, B, C \in \mathcal{K}_C(\mathbb{R})$ with $0 \notin A$, consider the interval equation

$$AX + B = C. \tag{1.40}$$

Using gH-difference and g-division, Eq. (1.40) has the solution $X = (C \ominus_{gH} B) \div_g A$ and it always exists, i.e., (the proof is immediate):

Proposition 1.18 (see [75]) *Let $A, B, C \in \mathcal{K}_C(\mathbb{R})$ with $0 \notin A$; the interval $X = (C \ominus_{gH} B) \div_g A$ always exists and (at least) one of the four equalities is satisfied:*

1. $AX + B = C$ *(X is additive and multiplicative);*
2. $B = C - AX$ *(X is subtractive and multiplicative);*
3. $A \div_g X^{-1} + B = C$ *(X is additive and divisive);*
4. $B = C - A \div_g X^{-1}$ *(X is subtractive and divisive).*

The same results can be applied to Eqs. (1.38) and (1.40) in the case where A, B and C are fuzzy numbers, i.e., to equations

$$ux = v, \tag{1.41}$$

$$ux + v = w. \tag{1.42}$$

We obtain the fuzzy solutions $x = v \div_g u$ and $x = (w \ominus_{gH} v) \div_g u$ or, more generally, using the approximated gH-difference and g-division of Sects. 1.3.3 and 1.4.2, $x = v \tilde{\div} u$ and $x = (w \tilde{\ominus} v) \tilde{\div} u$.

We illustrate the application of gH-difference and g-division to Eqs. (1.41) and (1.42) with some examples.

The data are triangular fuzzy numbers of the form $\langle a, b, c \rangle$ with linear left and right membership sides. The following examples can be found in [75].

Example 1.5.1 $u = \langle 0.2, 0.4, 0.6 \rangle$, $v = \langle 0.1, 0.3, 0.5 \rangle$; the solution $x = v \div_g u$ has α-cuts

$$[x]_\alpha = \left[\frac{0.1 + 0.2\alpha}{0.2 + 0.2\alpha}, \frac{0.5 - 0.2\alpha}{0.6 - 0.2\alpha} \right] \text{ for } \alpha \in [0, 1]$$

and satisfies case 1 of Proposition 1.17. □

Example 1.5.2 $u = \langle 0.2, 0.4, 0.6 \rangle$, $v = \langle 0.6, 0.8, 1 \rangle$; the solution $x = v \div_g u$ has α-cuts

$$[x]_\alpha = \left[\frac{1 - 0.2\alpha}{0.6 - 0.2\alpha}, \frac{0.6 + 0.2\alpha}{0.2 - 0.2\alpha} \right] \text{ for } \alpha \in [0, 1]$$

and satisfies case 2 of Proposition 1.17. □

Example 1.5.3 $u = \langle 0.5, 0.8, 1 \rangle$, $v = \langle 0.3, 0.5, 0.6 \rangle$; the solution $x = v \div_g u$ does not exist, but $x = v \tilde{\div} u$ exists and its α-cuts are

$$[x]_\alpha = \left[\frac{0.6 - 0.1\alpha}{1.0 - 0.2\alpha}, \frac{5}{8} \right] \text{ for } \alpha \in [0, 1].$$

Example 1.5.4 (*from* [19, Chap.10]): $u = \langle 1, 2, 3 \rangle$, $v = \langle -3, -2, -1 \rangle$, $w = \langle 3, 4, 5 \rangle$; note that $w = v + \{6\}$ but it is impossible to find x such that $ux = \{6\}$. Instead there exists x such that $u = 6x^{-1}$ i.e., $x = 6u^{-1}$ and cases 3 and 4 of Proposition 1.18 are both satisfied. □

Example 1.5.5 (*from* [19, Chap.10]): $u = \langle 8, 9, 10 \rangle$, $v = \langle -3, -2, -1 \rangle$, $w = \langle 3, 5, 7 \rangle$; note that $w = 2v + \{9\}$ and it is possible to find x such that $ux = v + \{9\}$. The solution $x = (w \ominus_{gH} v) \div_g u$ has α-cuts

$$[x]_\alpha = \left[\frac{6 + \alpha}{8 + \alpha}, \frac{8 - \alpha}{10 - \alpha} \right] \text{ for } \alpha \in [0, 1]$$

and satisfies cases 1 of Proposition 1.18. □

1.5.2 Interval and Fuzzy Differential Equations

Equations with fuzzy numbers was discussed in [17]. In the literature [75], Stefanini summarized some results of some literatures related to interval and fuzzy differential equations based on Hukuhara difference.

The generalized Hukuhara difference is (implicitly) used by Bede and Gal (see [10]) in their definition of generalized differentiability of a fuzzy-valued function. The gH-difference allows a new definition of the derivative for interval (or fuzzy) valued functions $f : [a, b] \to K_C(\mathbb{R})$ (or $f : [a, b] \to \mathscr{F}$). If $x \in]a, b[$ and $x + h \in]a, b[$, the gH-derivative of f at x is defined as the limit

$$\lim_{h\to 0} \frac{f(x+h) \ominus_g f(x)}{h} = f'(x)$$

provided that the gH-differences exist for small h. For interval and fuzzy-valued functions, using the results of the preceding sections, the gH-differences are very easy to be computed in terms of the lower and upper bounds of the intervals $f(x) = [f^-(x), f^+(x)]$ (or of the α-cuts $[f(x)]_\alpha = [f_\alpha^-(x), f_\alpha^+(x)]$, $\alpha \in [0, 1]$ in the fuzzy case) the interval derivative is

$$f'(x) = \left[\min\left\{\frac{\mathrm{d}f^-(x)}{\mathrm{d}x}, \frac{\mathrm{d}f^+(x)}{\mathrm{d}x}\right\}, \max\left\{\frac{\mathrm{d}f^-(x)}{\mathrm{d}x}, \frac{\mathrm{d}f^+(x)}{\mathrm{d}x}\right\}\right]$$

and the fuzzy derivative, in terms of level-cuts, is

$$[f'(x)]_\alpha = \left[\min\left\{\frac{\mathrm{d}f_\alpha^-(x)}{\mathrm{d}x}, \frac{\mathrm{d}f_\alpha^+(x)}{\mathrm{d}x}\right\}, \max\left\{\frac{\mathrm{d}f_\alpha^-(x)}{\mathrm{d}x}, \frac{\mathrm{d}f_\alpha^+(x)}{\mathrm{d}x}\right\}\right].$$

This definition of the derivative allows to characterize a new solution concept in interval and fuzzy differential equations and, in particular, to design new computational procedures based on classical ODE methods (see [73]).

A study by Bede and Stefanini (see [74]) presented a new approach to model interval uncertainty in dynamical systems and interval differential equations, related but more general than the differential inclusion model (see [45, 55]). In particular, the derivative of an interval-valued function, based on the new gH-difference, allows the possibility of solutions without restrictions imposed by the classical Hukuhara derivative, i.e., the increasing uncertainty effect. See [29–32] for a critic of the use of classical Hukuhara derivative in fuzzy differential equations; see also [11] for an extended approach to fuzzy differential equations.

An Embedding Theorem and Multiplication of Fuzzy Vectors

2

In this chapter, an embedding theorem of fuzzy vectors is established. Furthermore, two new types of multiplication of fuzzy vectors are introduced and some of their operation regulations are established.

Let x be a point in $\mathbb{R}^n$ and A a nonempty subset of $\mathbb{R}^n$. The distance $d(x, A)$ from x to A is defined by

$$d(x, A) = \inf\{\|x - a\| : a \in A\}.$$

Thus $d(x, A) = d(x, \bar{A}) \geq 0$ and $d(x, A) = 0$ if and only if $x \in \bar{A}$, the closure of A in $\mathbb{R}^n$. We shall call the subset $S_\varepsilon(A) = \{x \in \mathbb{R}^n : d(x, A) < \varepsilon\}$ an ε-neighborhood of A. Its closure is the subset $\bar{S}_\varepsilon(A) = \{x \in \mathbb{R}^n : d(x, A) \leq \varepsilon\}$. In particular, we shall denote the closed unit ball in $\mathbb{R}^n$ by $\bar{S}_1^n$, that is, $\bar{S}_1^n = \bar{S}_1(\{0\})$. For more details of this knowledge, we refer the readers to the book [28]. So if we let $d''_H(B, A) = \inf\{\varepsilon > 0 : B \subseteq A + \varepsilon \bar{S}_1^n\}$, then the Hausdorff distance between nonempty subsets A and B of $\mathbb{R}^n$ is defined by

$$d'_H(A, B) = \max\{d''_H(A, B), d''_H(B, A)\}. \tag{2.1}$$

Now, we consider a circumscribed cube $\bar{V}_1^n$ of $\bar{S}_1^n$, i.e.,

$$V_1^n = (-1, 1) \times (-1, 1) \times \cdots \times (-1, 1) := \times_{i=1}^n (-1, 1),$$

where $\times_{i=1}^n$ denotes the Cartesian product. Now $\bar{V}_1^n$ is obviously a compact subset of $\mathbb{R}^n$. We say

$$V_\varepsilon(A) := A + \varepsilon V_1^n$$

is the ε-cube neighborhood of A. Obviously, $S_\varepsilon(A) \subset V_\varepsilon(A)$. Let S^{n-1} and V^{n-1} be the surfaces of $\bar{S}_1^n$ and $\bar{V}_1^n$, then the tangential points set between the surfaces of $\bar{S}_1^n$ (i.e., S^{n-1}) and $\bar{V}_1^n$ (i.e., V^{n-1}) is

C. Wang and R. P. Agarwal, *Dynamic Equations and Almost Periodic Fuzzy Functions on Time Scales*, Synthesis Lectures on Mathematics & Statistics,
https://doi.org/10.1007/978-3-031-11236-2_2

$$S^{n-1} \cap V^{n-1} = \{P_1, P_1^*, P_2, P_2^*, \dots, P_n, P_n^*\},$$

where

$$P_i = (0, \dots, 0, \underbrace{1}_{\text{the } i\text{th element}}, 0, \dots, 0),$$

$$P_i^* = (0, \dots, 0, \underbrace{-1}_{\text{the } i\text{th element}}, 0, \dots, 0),$$

and $S^{n-1} = \{p \in \mathbb{R}^n : \|p\| = 1\}$.

Now let A and B be nonempty subsets of $\mathbb{R}^n$. We define the Hausdorff separation of B from A by

$$d_H^*(B, A) = \inf\{\varepsilon > 0 : B \subseteq A + \varepsilon \bar{V}_1^n\},$$

where $\varepsilon \bar{V}_1^n = [-\varepsilon, \varepsilon] \times [-\varepsilon, \varepsilon] \times \cdots \times [-\varepsilon, \varepsilon] := \times_{i=1}^n [-\varepsilon, \varepsilon]$.

We define the Hausdorff distance between nonempty subsets A and B of $\mathbb{R}^n$ by

$$d_H(A, B) = \max\left\{d_H^*(A, B), d_H^*(B, A)\right\}. \tag{2.2}$$

If we restrict our attention to nonempty closed subsets of $\mathbb{R}^n$, we find that the distance (2.2) is a metric which is strongly equivalent to the Pompeiu-Hausdorff metric [40] when the ε-cube neighborhood is applied.

Following Diamond and Kloeden (see [28]), denote by $\mathscr{K}(\mathbb{R}^n)$ and $\mathscr{K}_C(\mathbb{R}^n)$ the spaces of nonempty compact and compact convex sets of $\mathbb{R}^n$. For a vector $a \in \mathbb{R}^n$ and convex set $\mathbb{C} \subseteq \mathbb{R}^n$, denote by $s(\mathbb{C}, a) := \sup\{\langle a, x\rangle, x \in \mathbb{C}\}$ the support function of $\mathbb{C}$ in direction a. In addition, for a convex function $h : \mathbb{C} \to \mathbb{R}$ on some convex set $\mathbb{C} \subseteq \mathbb{R}^n$ and $x \in \mathbb{C}$, we write ∂h for the subdifferential of h in x and $N_{\mathbb{C}}(x) := \left\{a \in \mathbb{R}^n : s(\mathbb{C}, a) = \langle a, x\rangle\right\}$ the normal cone of $\mathbb{C}$ at x (for more knowledge of the normal cone, tangential cone and computational geometry, one may consult [15, 38, 54]).

Theorem 2.1 *Let* $A, B \in \mathscr{K}_C(\mathbb{R}^n)$. *Then there exists a number* $\mathscr{N}_n$ *with* $1 \geq \mathscr{N}_n \geq \frac{1}{2\sqrt{n}}$ *such that*

$$d_H(A, B) = \mathscr{N}_n \sup_{u \in V^{n-1}} |s(A, u) - s(B, u)|. \tag{2.3}$$

Proof Let $\varepsilon := d_H(A, B)$ and $u \in \bar{V}_1^n$. Since $A \subset B + \varepsilon \bar{V}_1^n$, it follows that

$$s(A, u) \leq s(B + \varepsilon \bar{V}_1^n, u) \leq s(B, u) + \varepsilon s(\bar{V}_1^n, u) \leq s(B, u) + \mathscr{N}_n^* \varepsilon,$$

where $\mathscr{N}_n^*$ is a positive number such that $2\sqrt{n} \geq \mathscr{N}_n^* \geq s(\bar{V}_1^n, u)$. Similarly, we can also obtain $s(B, u) \leq s(A, u) + \mathscr{N}_n^* \varepsilon$. Since $u \in \bar{V}_1^n$ is arbitrary, this yields that

$$\frac{1}{\mathscr{N}_n^*} \sup_{u \in \bar{V}_1^n} |s(A, u) - s(B, u)| \leq \varepsilon. \tag{2.4}$$

For another inequality, let $a^* \in A$ and $b^* \in B$ such that $\varepsilon = \|a^* - b^*\|$. Let $h(b) := \|b - a^*\|$. Since $h(b^*) \leq h(b)$ for all $b \in B$, there exists

$$u^{'} \in \partial h(b^*) = \left\{u \in \bar{V}_1^n : \langle b^* - a^*, u\rangle = h(b^*)\right\}$$

such that $-u^{'} \in N_B(b^*)$. Thus, for $u^* = -u^{'}$, we have

$$\begin{aligned} \varepsilon = h(b^*) &= \langle a^* - b^*, u^*\rangle = \langle a^*, u^*\rangle - \max_{b\in B}\langle b, u^*\rangle \\ &\leq |s(A, u^*) - s(B, u^*)| \leq \sup_{u\in \bar{V}_1^n} |s(A, u) - s(B, u)|. \end{aligned} \tag{2.5}$$

Because of the homogeneity of $s(A, \cdot)$, $s(B, \cdot)$ and $|\cdot|$, the supremum in (2.3) is attained for some vector $u^* \in V^{n-1}$. Hence, by (2.4) and (2.5), there exists $1 \geq \mathcal{N}_n \geq \frac{1}{2\sqrt{n}}$ such that (2.3) holds. This completes the proof. □

Example 2.1 Consider $A = [0, 1] \times \{1\}$ and $B = [-3, 1] \times \{0\}$ two compact and convex sets of $\mathbb{R}^2$. Then

$$d_H^*(A, B) = 2, \ d_H^*(B, A) = 3,$$

and so $d_H(A, B) = 3$. Now for $(q_1, q_2) \in V^1$, we have

$$\begin{aligned} s(q, A) &= \sup\{\langle q, a\rangle, a \in A\} = \sup\langle (q_1, q_2), (a_1, 1)\rangle : a_1 \in [0, 1]\} \\ &= \sup\{q_1 a_1 + q_2, a_1 \in [0, 1]\} \\ &= \begin{cases} q_1 + q_2 \text{ if } q_1 \geq 0, \\ q_2 \ \text{ if } q_1 < 0 \end{cases} \end{aligned}$$

and

$$\begin{aligned} s(q, B) &= \sup\{\langle q, b\rangle, b \in B\} = \sup\langle (q_1, q_2), (b_1, 0)\rangle : b_1 \in [-3, -1]\} \\ &= \sup\{q_1 b_1, b_1 \in [-3, -1]\} \\ &= \begin{cases} -q_1 \text{ if } q_1 \geq 0, \\ -3q_1 \ \text{ if } q_1 < 0. \end{cases} \end{aligned}$$

Then we have

$$\begin{aligned} |s(q, A) - s(q, B)| &= \begin{cases} |q_1 + q_2 - (-q_1)|, \ \text{ if } q_1 \geq 0, \\ |q_2 - (-3q_1)| \text{ if } q_1 < 0 \end{cases} \\ &= \begin{cases} |2q_1 + q_2|, \ \text{ if } q_1 \geq 0, \\ |3q_1 + q_2| \text{ if } q_1 < 0. \end{cases} \end{aligned}$$

Thus by Lemma 2.1, there exists $\mathcal{N}_n = \frac{3}{4}$ such that

$$\frac{3}{4} \sup_{u \in V^1} |s(A, u) - s(B, u)| = \frac{3}{4} \cdot 4 = 3.$$

Remark 2.1 As is well known that if the unit **ball neighborhood** is adopted (see [15, 28, 38, 54]), the Hausdorff distance via support functions can be represented by $d_H^{'}(A, B) = \sup_{u \in S^{n-1}} |s(A, u) - s(B, u)|$. Since $\bar{V}_1^n$ is the circumscribed cube of $\bar{S}_1^n$, then from Theorem 2.1, we can obtain $d_H(A, B)$ and $d_H^{'}(A, B)$ are strongly equivalent, i.e., there exist positive constants $\tilde{\alpha}$ and $\tilde{\beta}$ such that for any $A, B \in \mathscr{K}_C(\mathbb{R}^n)$ we have $\tilde{\alpha} d_H^{'}(A, B) \leq d_H(A, B) \leq \tilde{\beta} d_H^{'}(A, B)$. Hence, the metric $d_H(A, B)$ is strongly equivalent to classical Pompeiu-Hausdorff metric defined by (2.1).

In the sequel, we recall briefly the necessary knowledge of fuzzy sets.

Definition 2.1 (*see* [13, 75]) Let X be a nonempty set. A fuzzy set $u \in X$ is characterized by its membership function $u : X \to [0, 1]$. Then $u(x)$ is interpreted as the degree of membership of a element x in the fuzzy set u for each $x \in X$.

Let us denote by $\mathbb{R}^n_{\mathscr{F}}$ the class of fuzzy subsets u of $\mathbb{R}^n$ (i.e., $u : \mathbb{R}^n \to [0, 1]$), satisfying the following properties:

(i) u is normal, i.e., there exists $x_0 \in \mathbb{R}^n$ with $u(x_0) = 1$;
(ii) u is convex fuzzy set (i.e., $u(tx + (1 - t)y) \geq \min\{u(x), u(y)\}$, $\forall t \in [0, 1]$, $x, y \in \mathbb{R}^n$);
(iii) u is upper semicontinuous on $\mathbb{R}^n$;
(iv) $\mathrm{cl}\{x \in \mathbb{R}^n, u(x) > 0\}$ is compact, where cl denotes the closure of a subset.

Then, $\mathbb{R}^n_{\mathscr{F}}$ is called the space of fuzzy vectors . Obviously, $\mathbb{R}^n \subset \mathbb{R}^n_{\mathscr{F}}$. Here $\mathbb{R}^n \subset \mathbb{R}^n_{\mathscr{F}}$ is understood as $\mathbb{R}^n = \{\chi_{\{x\}} : x \in \mathbb{R}^n \text{is a usual real-vector}\}$, where $\chi_A = 1$ if $x \in A$, $\chi_A = 0$ if $x \notin A$. For $0 < \alpha \leq 1$, denote $[u]^\alpha = \{x \in \mathbb{R}^n : u(x) \geq \alpha\}$ (the α-level set) and $[u]^0 = \mathrm{cl}\{x \in \mathbb{R}^n : u(x) > 0\}$.

For $u, v \in \mathbb{R}^n_{\mathscr{F}}$, and $\lambda \in \mathbb{R}$, the sum $u \tilde{+} v$ and the product $\lambda \cdot u$ are defined by $[u \tilde{+} v]^\alpha = [u]^\alpha + [v]^\alpha$, $[\lambda \cdot u]^\alpha = \lambda [u]^\alpha$, $\forall \alpha \in [0, 1]$, where $[u]^\alpha + [v]^\alpha = \{x + y,\ x \in [u]^\alpha, y \in [v]^\alpha\}$ means the usual addition of two sets in $\mathbb{R}^n$ and $[\lambda \cdot u]^\alpha = \{\lambda x : x \in [u]^\alpha\}$ means the usual product between a scalar and a subset of $\mathbb{R}^n$.

Definition 2.2 (*see* [13, 75]) The Hausdorff distance between two fuzzy vectors is the function $D_\infty : \mathbb{R}^n_{\mathscr{F}} \times \mathbb{R}^n_{\mathscr{F}} \to \mathbb{R}^+ \cup \{0\}$, defined in terms of the Hausdorff distance between their level sets, that is,

$$D_\infty(u,v) = \sup\left\{d_H([u]^\alpha,[v]^\alpha) : \alpha \in [0,1]\right\}.$$

Remark 2.2 Since $d_H([u]^\alpha,[v]^\alpha)$ and $d_H^{'}([u]^\alpha,[v]^\alpha)$ are strongly equivalent, then $D_\infty(u,v)$ and $D_\infty^{'}(u,v) := \sup\left\{d_H^{'}([u]^\alpha,[v]^\alpha) : \alpha \in [0,1]\right\}$ are also strongly equivalent, which indicates that they only exist numerically different.

For convenience, let us denote by $\|\cdot\|_{\mathcal{F}}$ the function $\|u\|_{\mathcal{F}} = D_\infty(u,\tilde{\mathbf{0}})$ defined for all $u \in \mathbb{R}^n_{\mathcal{F}}$, where $\tilde{\mathbf{0}} = (\tilde{0},\tilde{0},\ldots,\tilde{0})$ and $\tilde{0}$ is a zero element of $\mathbb{R}_{\mathcal{F}}$. The next lemma asserts that $\|\cdot\|_{\mathcal{F}}$ has properties similar to the properties of a norm in the usual crisp sense, without being a norm. It is not a norm because $\mathbb{R}^n_{\mathcal{F}}$ is not a linear space and, consequently, $(\mathbb{R}^n_{\mathcal{F}}, \|\cdot\|_{\mathcal{F}})$ is not a normed space.

Due to the strong equivalence of the metrics $D_\infty(u,v)$ and $D_\infty^{'}(u,v)$, the following lemma can be obtained immediately.

Lemma 2.1 (see [13, 75]) *The function* $\|\cdot\|_{\mathcal{F}}$ *has the following properties:*

(*i*) $\|u\|_{\mathcal{F}} = 0$ *if and only if* $u = \tilde{0}$;
(*ii*) $\|\lambda \cdot u\|_{\mathcal{F}} = |\lambda| \cdot \|u\|_{\mathcal{F}}$ *for all* $u \in \mathbb{R}^n_{\mathcal{F}}$ *and* $\lambda \in \mathbb{R}$;
(*iii*) $\|u\tilde{+}v\|_{\mathcal{F}} \le \|u\|_{\mathcal{F}} + \|v\|_{\mathcal{F}}$ *for all* $u, v \in \mathbb{R}^n_{\mathcal{F}}$.

Lemma 2.2 (see [13, 75])

(*i*) *If we denote* $\tilde{\mathbf{0}} = \chi_{\{\mathbf{0}\}}$ *then* $\tilde{\mathbf{0}} \in \mathbb{R}^n_{\mathcal{F}}$ *is neutral element with respect to* $\tilde{+}$, *i.e.,* $u\tilde{+}\tilde{\mathbf{0}} = \tilde{\mathbf{0}}\tilde{+}u = u$, *for all* $u \in \mathbb{R}^n_{\mathcal{F}}$.
(*ii*) *For any* $a, b \in \mathbb{R}$ *with* $a, b \le 0$ *or* $a, b \ge 0$ *and any* $u \in \mathbb{R}^n_{\mathcal{F}}$, *we have* $(a+b)\cdot u = a\cdot u\tilde{+}b\cdot u$; *for any general* $a, b \in \mathbb{R}$, *the above property does not hold.*
(*iii*) *For any* $\lambda \in \mathbb{R}$ *and any* $u, v \in \mathbb{R}^n_{\mathcal{F}}$, *we have* $\lambda\cdot(u\tilde{+}v) = \lambda\cdot u\tilde{+}\lambda\cdot v$.
(*iv*) *For any* $\lambda, \mu \in \mathbb{R}$ *and any* $u \in \mathbb{R}^n_{\mathcal{F}}$, *we have* $\lambda\cdot(\mu\cdot u) = (\lambda\mu)\cdot u$.

Definition 2.3 (*see* [75]) Given $u, v \in \mathbb{R}^n_{\mathcal{F}}$, the gH-difference is the fuzzy vector w, if it exists, such that

$$u\tilde{-}_{gH}v = w \Leftrightarrow \begin{cases} (I)\ u = v\tilde{+}w \text{ or} \\ (II)\ v = u\tilde{+}(-1)\cdot w. \end{cases}$$

Remark 2.3 In the above definition, we also introduce the following notations:

$$u\tilde{-}_{gH_{(I)}}v = w \Leftrightarrow (I)\ u = v\tilde{+}w$$

$$u\tilde{-}_{gH_{(II)}}v = w \Leftrightarrow (II)\ v = u\tilde{+}(-1)\cdot w.$$

Remark 2.4 Note that it is possible that the gH-difference of two fuzzy vectors does not exist. If $u\tilde{-}_{gH}v$ exists, then $v\tilde{-}_{gH}u$ exists and $v\tilde{-}_{gH}u = -(u\tilde{-}_{gH}v)$.

Lemma 2.3 (see [75])

(*i*) *If the gH-difference exists, it is unique;*
(*ii*) $u\tilde{-}_{gH}v = u\tilde{-}v$ *or* $u\tilde{-}_{gH}v = -(u\tilde{-}v)$ *whenever the expressions on the right exist; in particular,* $u\tilde{-}_{gH}u = u\tilde{-}u = \tilde{\mathbf{0}}$*, where* $u\tilde{-}v$ *denotes the Hukuhara difference of* u, v*;*
(*iii*) *if* $u\tilde{-}_{gH}v$ *exists in the sense* (I)*, then* $v\tilde{-}_{gH}u$ *exists in the sense* (II) *and vice versa;*
(*iv*) $(u\tilde{+}v)\tilde{-}_{gH}v = u$*;*
(*v*) $u\tilde{-}_{gH}v = v\tilde{-}_{gH}u = w$ *if and only if* $w = -w$ *; furthermore,* $w = \tilde{\mathbf{0}}$ *if and only if* $u = v$*.*

Definition 2.4 (*see* [13, 75]) Let $\mathscr{K}_C^n$ be the space of nonempty compact convex set of $\mathbb{R}^n$, $A, B \in \mathscr{K}_C^n$, we define the generalized Hukuhara difference of A and B as the set $C \in \mathscr{K}_C^n$ such that

$$A \boxminus_{gH} B = C \Leftrightarrow \begin{cases} (I)\ A = B + C \text{ or} \\ (II)\ B = A + (-1)\cdot C. \end{cases} \tag{2.6}$$

Remark 2.5 In the above definition, we also introduce the following notations:

$$A \boxminus_{gH_{(I)}} B = C \Leftrightarrow (I)\ A = B + C$$

$$A \boxminus_{gH_{(II)}} B = C \Leftrightarrow (II)\ B = A + (-1)\cdot C.$$

Remark 2.6 In the Sect. 3 of the literature [75], Stefanini established some important basic results of gH-difference between compact intervals in $\mathbb{R}^n$, including the special case of unidimensional and multidimensional intervals (boxes). For a compact subset A of $\mathbb{R}^n$, we will consider its ε-cube neighborhood of A, then the gH-difference between multidimensional boxes can be applied. From [75], for any multidimensional boxes A, B of $\mathbb{R}^n$, if $A \boxminus_{gH} B$ exists, then $A \boxminus_{gH} B = \times_{i=1}^n (A_i \boxminus_{gH} B_i)$, where $A = \times_{i=1}^n A_i = \times_{i=1}^n [a_i^-, a_i^+]$ and $B = \times_{i=1}^n B_i = \times_{i=1}^n [b_i^-, b_i^+]$.

Lemma 2.4 (see [75]) *Let* A, B *be two multidimensional boxes of* $\mathbb{R}^n$*. A necessary and sufficient condition for* $A \boxminus_{gH} B$ *to exist is that either* A *contains a translate of* B *or* B *contains a translate of* A*.*

Remark 2.7 Given $A = ([1, 2], [-2, 1])$ and $B = ([-2, 1], [1, 2])$, let $\chi_A, \chi_B : \mathbb{R}^2 \to [0, 1]$ be given, respectively, by $\chi_A(x) = 1$ if $x \in A$; $\chi_A = 0$ if $x \notin A$, and $\chi_B(x) = 1$ if $x \in B$; $\chi_B = 0$ if $x \notin B$. Then $\chi_A, \chi_B \in \mathbb{R}^2_{\mathscr{F}}$ and $[\chi_A]^\alpha = A, [\chi_B]^\alpha = B$ for all $\alpha \in [0, 1]$.

However, since A is not a translate of B and B is not a translate of A, we obtain that $[\chi_A]^\alpha \boxminus_{gH} [\chi_B]^\alpha$ does not exist. Hence, Lemma 2.4 provides the significant condition to guarantee the existence of the gH-difference between two multidimensional boxes.

Now, let $u \in \mathbb{R}^n_{\mathscr{F}}$ and define $s_u : [0, 1] \times V^{n-1} \to \mathbb{R}$ by

$$s_u(\alpha, p) = s(p, [u]^\alpha) = \sup\{\langle p, a\rangle : a \in [u]^\alpha\} \tag{2.7}$$

for $(\alpha, p) \in [0, 1] \times V^{n-1}$, where $s(\cdot, [u]^\alpha)$ is the support function of $[u]^\alpha$. We shall call s_u the support function of the fuzzy set u. Note that the supremum in (2.7) is actually attained since the level set $[u]^\alpha$ is compact and so can be replaced by the maximum.

Lemma 2.5 *The support function s_u has the following properties:*

(i) $u = v$ *if and only if* $s_u = s_v$.
(ii) *uniformly bounded on* $[0, 1] \times V^{n-1}$.
(iii) *Lipschitz in* $p \in V^{n-1}$ *uniformly on* $[0, 1]$.
(iv) *For each* $\alpha \in [0, 1]$, $d_H([u]^\alpha, [v]^\alpha) = \frac{1}{\sqrt{n}} \sup\{|s_u(\alpha, p) - s_v(\alpha, p)| : p \in V^{n-1}\}$.
(v) $s_u(\cdot, p)$ *is nonincreasing and left continuous in* $\alpha \in [0, 1]$ *for each* $p \in V^{n-1}$.
(vi) *A fuzzy set* $u \in \mathbb{R}^n_{\mathscr{F}}$ *is called a Lipschitzian fuzzy set if it is a Lipschitz function of its membership grade in the sense that* $d_H([u]^\alpha, [u]^\beta) \le K|\alpha - \beta|$ *for all* $\alpha, \beta \in [0, 1]$ *and some fixed finite constant* K. *The support function* $s_u(\cdot, p)$ *is Lipschitz uniformly in* $p \in V^{n-1}$ *if and only if* u *is a Lipschitzian fuzzy set.*

Proof (i). Since the support function on $\mathscr{K}_C(\mathbb{R}^n)$ uniquely characterizes the elements of $\mathscr{K}_C(\mathbb{R}^n)$, the result is obvious.

(ii). Since $[u]^\alpha \subseteq [u]^0$, so $|s_u(\alpha, p)| = |s(p, [u]^\alpha)| \le \|[u]^\alpha\|\|p\| \le \sqrt{n}\|[u]^0\|$, then the result follows.

(iii). From the following inequality,

$$|s_u(\alpha, p) - s_u(\alpha, q)| = |s(p, [u]^\alpha) - s(q, [u]^\alpha)| \le \|[u]^\alpha\|\|p - q\| \le \|[u]^0\|\|p - q\|.$$

Then we can obtain the results.

(iv). By Theorem 2.1, (iv) is the restatement of (2.3).

(v). Since $[u]^\beta \subseteq [u]^\alpha$ for $0 \le \alpha \le \beta$,

$$s_u(\beta, p) = s(p, [u]^\beta) \le s(p, [u]^\alpha) = s_u(\alpha, p)$$

so $s_u(\cdot, p)$ is nonincreasing for each $p \in V^{n-1}$. Moreover, for a nondecreasing sequence $\alpha_{\tilde{n}} \uparrow \alpha$ in $[0, 1]$,

$$|s_u(\alpha_{\tilde{n}}, p) - s_u(\alpha, p)| \le \sqrt{n} d_H([u]^{\alpha_{\tilde{n}}}, [u]^\alpha) \to 0 \text{ as} \tilde{n} \to \infty.$$

(vi). By (iii) and (iv), we can obtain the desired result immediately. This completes the proof. □

From Definitions 2.2, 2.4 and Lemma 2.4, the following theorem follows immediately.

Theorem 2.2 *For any $u, v \in \mathbb{R}^n_{\mathscr{F}}$ and $0 \leq \alpha \leq 1$, if $u\tilde{-}_{gH}v$ exists, then*

$$D_\infty(u, v) = \sup_{\alpha\in[0,1]} \{\|[u]^\alpha \boxminus_{gH} [v]^\alpha\|_*\} = \|u\tilde{-}_{gH}v\|_{\mathscr{F}}, \tag{2.8}$$

where $\|[u]^\alpha \boxminus_{gH} [v]^\alpha\|_* = d_H([u]^\alpha \boxminus_{gH} [v]^\alpha, \{0\})$ and

$$[u]^\alpha \boxminus_{gH} [v]^\alpha = [w]^\alpha \Leftrightarrow \begin{cases} (I)\ [u]^\alpha = [v]^\alpha + [w]^\alpha \text{ or} \\ (II)\ [v]^\alpha = [u]^\alpha + (-1)\cdot[w]^\alpha. \end{cases}$$

Remark 2.8 The existence and nonexistence of $u\tilde{-}_{gH}v$ for $u, v \in \mathbb{R}^n_{\mathscr{F}}$ via support functions were established in Sect. 4.1 of the literature [75] (see pp. 1571-1572) by Stefanini. From the Proposition 20 in [75], Stefanini provided a necessary and sufficient condition to guarantee the existence of $u\tilde{-}_{gH}v$. Hence, when we suppose that $s_u - s_v$, $s_{-v} - s_{-u}$ is a support function and is nonincreasing with α for all p in Theorem 2.2, then $u\tilde{-}_{gH}v \in \mathbb{R}^n_{\mathscr{F}}$ exists. The monotonicity of $s_u - s_v$, $s_{-v} - s_{-u}$ is an important condition for the existence $u\tilde{-}_{gH}v$ and it may not be satisfied. Consider $u, v \in \mathbb{R}_{\mathscr{F}}$ such that $[u]^\alpha = [u^-_\alpha, u^+_\alpha] = [5 + 4\alpha, 14 - \alpha]$ and $[v]^\alpha = [v^-_\alpha, v^+_\alpha] = [9 + \alpha, 12 - 2\alpha]$ for all $\alpha \in [0, 1]$, through calculation, we have $s_u(\alpha, 1) = s_{-u}(\alpha, -1) = 14 - \alpha$ and $s_v(\alpha, 1) = s_{-v}(\alpha, -1) = 12 - 2\alpha$, then we obtain $s_u(\alpha, 1) - s_v(\alpha, 1) = s_{-u}(\alpha, -1) - s_{-v}(\alpha, -1) = 2 + \alpha$ is increasing with α. For the unidimensional case, the monotonicity of $u^-_\alpha - v^-_\alpha$ and $u^+_\alpha - v^+_\alpha$ must be guaranteed for the existence of $u\tilde{-}_{gH}v \in \mathbb{R}_{\mathscr{F}}$ (see Proposition 21 in [75] that is a particular case of Proposition 20).

2.1 An Embedding Theorem for Fuzzy Multidimensional Space

In this subsection, we will establish an embedding theorem for fuzzy multidimensional space.

Definition 2.5 Let $u_i \in \mathbb{R}_{\mathscr{F}}$ for each $i = 1, 2, \ldots, n$. We say $u = (u_1, u_2, \ldots, u_n) \in \underbrace{\mathbb{R}_{\mathscr{F}} \times \mathbb{R}_{\mathscr{F}} \times \cdots \times \mathbb{R}_{\mathscr{F}}}_{\text{n terms}} = \times^n_{i=1}\{\mathbb{R}_{\mathscr{F}}\} := [\mathbb{R}^n_{\mathscr{F}}]$ is a fuzzy (box) vector , where $\times^n_{i=1}$ denotes the Cartesian product .

Remark 2.9 Let $u = (u_1, u_2, \ldots, u_n) \in [\mathbb{R}^n_{\mathscr{F}}]$, then the α-level of u are multidimensional intervals (box) of $\mathbb{R}^n$ (see Sect. 3 from Stefanini [75]). In fact, a multidimensional intervals (box) of $\mathbb{R}^n$ can be regarded as a fuzzy (box) vector.

Let $u = (u_1, u_2, \ldots, u_n)$ and $v = (v_1, v_2, \ldots, v_n)$ be two fuzzy vectors with (box) α-levels:

$$[u]^\alpha = [u^-_{1,\alpha}, u^+_{1,\alpha}] \times [u^-_{2,\alpha}, u^+_{2,\alpha}] \times \cdots \times [u^-_{n,\alpha}, u^+_{n,\alpha}] := \times_{i=1}^n [u^-_{i,\alpha}, u^+_{i,\alpha}],$$

$$[v]^\alpha = [v^-_{1,\alpha}, v^+_{1,\alpha}] \times [v^-_{2,\alpha}, v^+_{2,\alpha}] \times \cdots \times [v^-_{n,\alpha}, v^+_{n,\alpha}] := \times_{i=1}^n [v^-_{i,\alpha}, v^+_{i,\alpha}].$$

The distance is defined by

$$\begin{aligned} &D_\infty(u, v) \\ &= \sup_{\alpha\in[0,1]} \max \left\{ \left[\sum_{i=1}^n |s_u(\alpha, P_i) - s_v(\alpha, P_i)|^2 \right]^{\frac{1}{2}}, \left[\sum_{i=1}^n |s_u(\alpha, P_i^*) - s_v(\alpha, P_i^*)|^2 \right]^{\frac{1}{2}} \right. \\ &\left. : \alpha \in [0, 1],\ P_i, P_i^* \in S^{n-1} \cap V^{n-1},\ i = 1, 2, \ldots, n \right\}, \end{aligned} \tag{2.9}$$

the distance $D_\infty(\cdot, \cdot)$ induces $\|\cdot\|_{\mathscr{F}}$ on $[\mathbb{R}^n_{\mathscr{F}}]$ defined by $\|u\|_{\mathscr{F}} = D_\infty(u, \tilde{\mathbf{0}})$, where $\tilde{\mathbf{0}} = (\tilde{0}, \tilde{0}, \ldots, \tilde{0})$ and $\tilde{0}$ is a zero element of $\mathbb{R}_{\mathscr{F}}$. In fact, because

$$\big[-s_u(\alpha, P_i^*), s_u(\alpha, P_i) \big] = [u^-_{i,\alpha}, u^+_{i,\alpha}],\ i = 1, 2, \ldots, n,$$

$$\big[-s_v(\alpha, P_i^*), s_v(\alpha, P_i) \big] = [v^-_{i,\alpha}, v^+_{i,\alpha}],\ i = 1, 2, \ldots, n,$$

then

$$\begin{aligned} [u \tilde{-}_{gH} v]^\alpha &= [u]^\alpha \boxminus_{gH} [v]^\alpha \\ &= \begin{cases} (i)\ \times_{i=1}^n [s_v(\alpha, P_i^*) - s_u(\alpha, P_i^*), s_u(\alpha, P_i) - s_v(\alpha, P_i)] \text{ or} \\ (ii)\ \times_{i=1}^n [s_u(\alpha, P_i) - s_v(\alpha, P_i), s_v(\alpha, P_i^*) - s_u(\alpha, P_i^*)], \end{cases} \end{aligned}$$

so from (2.8), we have

$$
\begin{aligned}
&D_\infty(u, v) \\
&= \sup_{\alpha\in[0,1]} \{\|[u]^\alpha \boxminus_{gH} [v]^\alpha\|_*\} = \|u \tilde{-}_{gH} v\|_{\mathscr{F}} \\
&= \sup_{\alpha\in[0,1]} \max\left\{\left[\sum_{i=1}^{n} |s_u(\alpha, P_i) - s_v(\alpha, P_i)|^2\right]^{\frac{1}{2}}, \left[\sum_{i=1}^{n} |s_u(\alpha, P_i^*) - s_v(\alpha, P_i^*)|^2\right]^{\frac{1}{2}}\right. \\
&\quad \left. : \alpha \in [0,1],\ P_i, P_i^* \in S^{n-1} \cap V^{n-1},\ i = 1, 2, \ldots, n\right\}.
\end{aligned}
$$

Remark 2.10 For each $i = 1, 2, \ldots, n$, if we introduce the distance

$$
\begin{aligned}
D_\infty^{(i)}(u_i, v_i) = \sup_{\alpha\in[0,1]} \max\Big\{&|s_u(\alpha, P_i) - s_v(\alpha, P_i)|, |s_u(\alpha, P_i^*) - s_v(\alpha, P_i^*)| : \\
&\alpha \in [0,1],\ P_i, P_i^* \in S^{n-1} \cap V^{n-1}\Big\},
\end{aligned}
$$

the distance $D_\infty^{(i)}(\cdot, \cdot)$ induces $\|\cdot\|_{\mathscr{F}_0}$ on $\mathbb{R}_{\mathscr{F}}$ defined by $\|u_i\|_{\mathscr{F}_0} = D_\infty(u_i, \tilde{0})$, and then it follows that

$$
D_\infty(u, v) = \|u \tilde{-}_{gH} v\|_{\mathscr{F}} = \left(\sum_{i=1}^{n} D_\infty^{(i)}(u_i, v_i)\right)^{\frac{1}{2}} = \left(\sum_{i=1}^{n} \|u_i - v_i\|_{\mathscr{F}_0}^2\right)^{\frac{1}{2}}.
$$

Theorem 2.3 *The metric space* $([\mathbb{R}^n_{\mathscr{F}}], D_\infty)$ *is complete .*

Proof Let $\{u_k\} = \{u_1^{(k)}, u_2^{(k)}, \ldots, u_n^{(k)}) \in [\mathbb{R}^n_{\mathscr{F}}]$ be a Cauchy sequence such that $D_\infty(u_m, u_n) \to 0$ as $m, n \to \infty$. For all $\alpha \in [0, 1]$, we obtain the following α-levels for each $k \in \mathbb{Z}^+$:

$$
[u_k]^\alpha = [u_{1,\alpha}^{(k)-}, u_{1,\alpha}^{(k)+}] \times \cdots \times [u_{n,\alpha}^{(k)-}, u_{n,\alpha}^{(k)+}] := \times_{i=1}^{n} [u_{i,\alpha}^{(k)-}, u_{i,\alpha}^{(k)+}].
$$

Then we obtain

$$
\begin{aligned}
D_\infty(u_m, u_n) &= \|u_m - u_n\|_{\mathscr{F}} = \left(\sum_{i=1}^{n} \|u_i^{(m)} - u_i^{(n)}\|_{\mathscr{F}_0}\right)^{\frac{1}{2}} \\
&= \sup_{\alpha\in[0,1]} \max\left\{\left[\sum_{i=1}^{n} \left(u_{i,\alpha}^{(m)-} - u_{i,\alpha}^{(n)-}\right)^2\right]^{\frac{1}{2}}, \left[\sum_{i=1}^{n} \left(u_{i,\alpha}^{(m)+} - u_{i,\alpha}^{(n)+}\right)^2\right]^{\frac{1}{2}}\right\} \\
&\to 0 \text{ as } n, m \to \infty. \qquad (2.10)
\end{aligned}
$$

So for each $i = 1, 2, \ldots, n$, (2.10) yields that

$$
|u_{i,\alpha}^{(m)-} - u_{i,\alpha}^{(n)-}| \to 0 \text{ and } |u_{i,\alpha}^{(m)+} - u_{i,\alpha}^{(n)+}| \to 0 \text{ as } n, m \to \infty.
$$

Hence, there exist $u_{i,\alpha}^{(0)-}$ and $u_{i,\alpha}^{(0)+}$ such that $u_{i,\alpha}^{(k)-} \to u_{i,\alpha}^{(0)-}$ and $u_{i,\alpha}^{(k)+} \to u_{i,\alpha}^{(0)+}$ as $k \to \infty$. Therefore, there exists some $u_0 \in [\mathbb{R}_{\mathscr{F}}^n]$ with the α-level which is also a multidimensional intervals (box):

$$[u_0]^\alpha = [u_{1,\alpha}^{(0)-}, u_{1,\alpha}^{(0)+}] \times \cdots \times [u_{n,\alpha}^{(0)-}, u_{n,\alpha}^{(0)+}] = \times_{i=1}^n [u_{i,\alpha}^{(0)-}, u_{i,\alpha}^{(0)+}] \in [\mathbb{R}_{\mathscr{F}}^n]$$

such that $D_\infty(u_k, u_0) \to 0$ as $k \to \infty$. The proof is completed. $\qquad\square$

The space $\bar{C}[0, 1]$ on interval $[0, 1]$ is the class of all real-valued bounded functions f on $[0, 1]$ such that f is left continuous for any $t \in (0, 1]$ and f has right limit for any $t \in [0, 1)$, especially f is right continuous at 0.

Lemma 2.6 (see [16]) $(\bar{C}[0, 1], \|\cdot\|_{\bar{C}})$ *is a Banach space with the norm* $\|f\|_{\bar{C}} = \sup_{t \in [0,1]} |f(t)|$.

In addition, it is easy to verify:

Theorem 2.4 $\times_{i=1}^n (\bar{C}[0, 1] \times \bar{C}[0, 1])$, *with the norm defined by*

$$\left\|((f_1, g_1), (f_2, g_2), \ldots, (f_n, g_n))\right\|_{\times_{i=1}^n (\bar{C} \times \bar{C})}$$

$$= \sup_{x \in [0,1]} \max \left\{ \left(\sum_{i=1}^n f_i^2(x) \right)^{\frac{1}{2}}, \left(\sum_{i=1}^n g_i^2(x) \right)^{\frac{1}{2}} \right\}$$

is a Banach space.

Proof For any Cauchy sequence $\{h_{n_0}\}_{n_0 \in \mathbb{N}} \subset \times_{i=1}^n (\bar{C}[0, 1] \times \bar{C}[0, 1])$, denote

$$h_{n_0} = ((f_1^{(n_0)}, g_1^{(n_0)}), (f_2^{(n_0)}, g_2^{(n_0)}), \ldots, (f_n^{(n_0)}, g_n^{(n_0)})),$$

and then for any $\varepsilon > 0$, there exists $N > 0$ such that $n_0, m_0 > N$ implies $\|h_{m_0} - h_{n_0}\|_{\times_{i=1}^n (\bar{C} \times \bar{C})} < \varepsilon$, that is

$$\|h_{m_0} - h_{n_0}\|_{\times_{i=1}^n (\bar{C} \times \bar{C})}$$

$$= \left\|((f_1^{(m_0)}, g_1^{(m_0)}), \ldots, (f_n^{(m_0)}, g_n^{(m_0)})) - ((f_1^{(n_0)}, g_1^{(n_0)}), \ldots, (f_n^{(n_0)}, g_n^{(n_0)}))\right\|_{\times_{i=1}^n (\bar{C} \times \bar{C})}$$

$$= \left\|((f_1^{(m_0)} - f_1^{(n_0)}, g_1^{(m_0)} - g_1^{(n_0)}), \ldots, (f_n^{(m_0)} - f_n^{(n_0)}, g_n^{(m_0)} - g_n^{(n_0)}))\right\|_{\times_{i=1}^n (\bar{C} \times \bar{C})}$$

$$= \sup_{x \in [0,1]} \max \left\{ \left(\sum_{i=1}^n (f_i^{(m_0)}(x) - f_i^{(n_0)}(x))^2 \right)^{\frac{1}{2}}, \left(\sum_{i=1}^n (g_i^{(m_0)}(x) - g_i^{(n_0)}(x))^2 \right)^{\frac{1}{2}} \right\},$$

which yields that

$$f_i^{(n_0)}(x) \to f_i(x) \text{ and } g_i^{(n_0)}(x) \to g_i(x) \text{ as } n_0 \to +\infty$$

for each $i = 1, 2, \dots, n$. According to Lemma 2.6, we have $f_i(x), g_i(x) \in \bar{C}[0, 1]$ for each $i = 1, 2, \dots, n$. Therefore, for any $\varepsilon > 0$, there is a $N > 0$ such that $n_0 > N$ implies

$$\begin{aligned} h_{n_0} &= \big((f_1^{(n_0)}, g_1^{(n_0)}), \dots, (f_n^{(n_0)}, g_n^{(n_0)})\big) \to \big((f_1, g_1), \dots, (f_n, g_n)\big) \\ &:= h \in \times_{i=1}^n \big(\bar{C}[0, 1] \times \bar{C}[0, 1]\big). \end{aligned}$$

This completes the proof. □

The embedding theorem is established as follows.

Theorem 2.5 (Embedding theorem of fuzzy multidimensional space) *For all $u \in [\mathbb{R}^n_{\mathscr{F}}]$, denote $j(u) = \times_{i=1}^n(u_i^-, u_i^+)$. Then $j([\mathbb{R}^n_{\mathscr{F}}])$ is a closed convex cone with vertex $\mathbf{0}$ in $\times_{i=1}^n\big(\bar{C}[0, 1] \times \bar{C}[0, 1]\big)$ and $j : [\mathbb{R}^n_{\mathscr{F}}] \to \times_{i=1}^n\big(\bar{C}[0, 1] \times \bar{C}[0, 1]\big)$ satisfies:*

(i) for all $u, v \in [\mathbb{R}^n_{\mathscr{F}}]$, $\hat{s}, t \geq 0$, $j(\hat{s} \cdot u \tilde{+} t \cdot v) = \hat{s} j(u) + t j(v)$;
(ii) $D_\infty(u, v) = \|j(u) - j(v)\|_{\times_{i=1}^n(\bar{C}\times\bar{C})}$;

i.e., j embeds $[\mathbb{R}^n_{\mathscr{F}}]$ into $\times_{i=1}^n\big(\bar{C}[0, 1] \times \bar{C}[0, 1]\big)$ isometrically and isomorphically.

Proof From Lemma 2.5, j is meaningful and $j([\mathbb{R}^n_{\mathscr{F}}]) \subset \times_{i=1}^n\big(\bar{C}[0, 1] \times \bar{C}[0, 1]\big)$. It is easy to see that j is faithful (in fact, from (i) in Lemma 2.5, one can see j is absolutely a one-to-one mapping). In order to show that $j([\mathbb{R}^n_{\mathscr{F}}])$ is a convex cone, it suffices to show (i), (ii) in Theorem 2.5.

By the equations $[u \tilde{+} v]^\alpha = [u]^\alpha + [v]^\alpha$, $[k \cdot u]^\alpha = k[u]^\alpha$ $(k \geq 0)$, for $i = 1, 2, \dots, n$, we have $(u_i \tilde{+} v_i)^- = u_i^- + v_i^-$, $(u_i \tilde{+} v_i)^+ = u_i^+ + v_i^+$, $(k \cdot u_i)^- = k u_i^-$, $(k \cdot u_i)^+ = k u_i^+$ whenever $k \geq 0$. This implies that

$$\begin{aligned} j(\hat{s} \cdot u \tilde{+} t \cdot v) &= \times_{i=1}^n\big((\hat{s} \cdot u_i \tilde{+} t \cdot v_i)^-, (\hat{s} \cdot u_i \tilde{+} t \cdot v_i)^+\big) \\ &= \times_{i=1}^n\big((\hat{s} \cdot u_i)^- + (t \cdot v_i)^-, (\hat{s} \cdot u_i)^+ + (t \cdot v_i)^+\big) \\ &= \times_{i=1}^n(\hat{s} u_i^- + t v_i^-, \hat{s} u_i^+ + t v_i^+) \\ &= \hat{s}\big[\times_{i=1}^n (u_i^-, u_i^+)\big] \\ &\quad + t\big[\times_{i=1}^n (v_i^-, v_i^+)\big] \\ &= \hat{s} j(u) + t j(v) \end{aligned}$$

for any $u, v \in [\mathbb{R}^n_{\mathscr{F}}]$, $\hat{s} \geq 0$, $t \geq 0$. Hence we complete the proof of (i).

Now, we prove (ii). Since

$$\begin{aligned}
&D_\infty(u, v)\\
&= \sup_{\alpha\in[0,1]} d_H([u]^\alpha, [v]^\alpha)\\
&= \sup_{\alpha\in[0,1]} \max\left\{\left[\sum_{i=1}^{n} |s_u(\alpha, P_i) - s_v(\alpha, P_i)|^2\right]^{\frac{1}{2}}, \left[\sum_{i=1}^{n} |s_u(\alpha, P_i^*) - s_v(\alpha, P_i^*)|^2\right]^{\frac{1}{2}}\right.\\
&\quad \left. : \alpha \in [0, 1],\ P_i, P_i^* \in S^{n-1} \cap V^{n-1},\ i = 1, 2, \ldots, n\right\}\\
&= \Big\{\big\| \times_{i=1}^{n} \big[\big(-s_u(\alpha, P_i^*), s_u(\alpha, P_i)\big) - \big(-s_v(\alpha, P_i^*), s_v(\alpha, P_i)\big)\big]\big\|_{\times_{i=1}^{n}(\bar{C}\times\bar{C})} :\\
&\quad \alpha \in [0, 1],\ P_i, P_i^* \in S^{n-1} \cap V^{n-1},\ i = 1, 2, \ldots, n\Big\}\\
&= \big\| \times_{i=1}^{n} \big[(u_i^-, u_i^+) - (v_i^-, v_i^+)\big]\big\|_{\times_{i=1}^{n}(\bar{C}\times\bar{C})}\\
&= \big\| \times_{i=1}^{n} (u_i^-, u_i^+) - \times_{i=1}^{n}(v_i^-, v_i^+)\big\|_{\times_{i=1}^{n}(\bar{C}\times\bar{C})}\\
&= \|j(u) - j(v)\|_{\times_{i=1}^{n}(\bar{C}\times\bar{C})}
\end{aligned}$$

and noting that $([\mathbb{R}^n_{\mathscr{F}}], D_\infty)$ is complete, we obtain that $j([\mathbb{R}^n_{\mathscr{F}}])$ is a closed convex cone in $\times_{i=1}^{n}\big(\bar{C}[0, 1] \times \bar{C}[0, 1]\big)$. This completes the proof. □

Similar to Theorem 2.5, we introduce another embedding by

$$\widetilde{j} : [\mathbb{R}^n_{\mathscr{F}}] \to \times_{i=1}^{n}\big(\bar{C}[0, 1] \times \bar{C}[0, 1]\big),$$

$\widetilde{j}(u) = j(-u)$, $u \in [\mathbb{R}^n_{\mathscr{F}}]$. It is easy to prove the following properties:

(*i*) for all $u, v \in [\mathbb{R}^n_{\mathscr{F}}]$, $\hat{s}, t \geq 0$, $\widetilde{j}(\hat{s} \cdot u \widetilde{+} t \cdot v) = \hat{s}\widetilde{j}(u) + t\widetilde{j}(v)$;
(*ii*) $D_\infty(u, v) = \|\widetilde{j}(u) - \widetilde{j}(v)\|_{\times_{i=1}^{n}(\bar{C}\times\bar{C})}$, $\widetilde{j}([\mathbb{R}^n_{\mathscr{F}}]) = j([\mathbb{R}^n_{\mathscr{F}}])$.

Remark 2.11 In Theorem 2.5, let $n = 1$, then $j(u) = (u^-, u^+)$, so one can obtain the Embedding Theorem from [119, 120] immediately.

Remark 2.12 From the definitions of j and $\widetilde{j}$, one can easily see that $j(t \cdot u) = tj(u)$ and $\widetilde{j}(t \cdot u) = t\widetilde{j}(u)$ for all $t \in \mathbb{R}$.

2.2 Multiplication of Fuzzy Vectors in Fuzzy Multidimensional Space

In this section, we will introduce six new types of multiplication of two compact intervals. Let $[u^-, u^+]$ and $[v^-, v^+]$ be two compact intervals and ab denote the ordinary product of real numbers a, b. For convenience, we introduce the following notations:

$$I_{u,v}^{(I)} = \begin{vmatrix} u^- & u^+ \\ v^- & v^+ \end{vmatrix}, \quad I_{u,v}^{(II)} = \begin{vmatrix} u^+ & u^- \\ v^- & v^+ \end{vmatrix}, \quad I_{u,v}^{(III)} = \begin{vmatrix} u^- & u^- \\ v^- & v^+ \end{vmatrix},$$

$$I_{u,v}^{(IV)} = \begin{vmatrix} u^+ & u^+ \\ v^- & v^+ \end{vmatrix}, \quad I_{u,v}^{(V)} = \begin{vmatrix} u^- & u^+ \\ v^- & v^- \end{vmatrix}, \quad I_{u,v}^{(VI)} = \begin{vmatrix} u^- & u^+ \\ v^+ & v^+ \end{vmatrix}.$$

For any $[a^-, a^+] \subseteq [u^-, u^+]$ and $[b^-, b^+] \subseteq [v^-, v^+]$, we define the following multiplications:

$$\textbf{Type I.} \quad [a^-, a^+] \circ [b^-, b^+] = \{a \ast b : a \in [a^-, a^+],\ b \in [b^-, b^+]\}, \tag{2.11}$$

where if $I_{u,v}^{(I)} \le 0$, then

$$a \ast b = \begin{cases} ab, & ab \in [u^-v^+, u^+v^-], \\ u^-v^+, & ab < u^-v^+, \\ u^+v^-, & ab > u^+v^-; \end{cases}$$

if $I_{u,v}^{(I)} \ge 0$, then

$$a \ast b = \begin{cases} ab, & ab \in [u^+v^-, u^-v^+], \\ u^+v^-, & ab < u^+v^-, \\ u^-v^+, & ab > u^-v^+. \end{cases}$$

$$\textbf{Type II.} \quad [a^-, a^+] \circledcirc [b^-, b^+] = \{a \ast b : a \in [a^-, a^+],\ b \in [b^-, b^+]\}, \tag{2.12}$$

where if $I_{u,v}^{(II)} \le 0$, then

$$a \ast b = \begin{cases} ab, & ab \in [u^+v^+, u^-v^-], \\ u^+v^+, & ab < u^+v^+, \\ u^-v^-, & ab > u^-v^-; \end{cases}$$

if $I_{u,v}^{(II)} \ge 0$, then

$$a \ast b = \begin{cases} ab, & ab \in [u^-v^-, u^+v^+], \\ u^-v^-, & ab < u^-v^-, \\ u^+v^+, & ab > u^+v^+. \end{cases}$$

$$\textbf{Type III.} \quad [a^-, a^+] \boxtimes [b^-, b^+] = \{a \ast b : a \in [a^-, a^+],\ b \in [b^-, b^+]\}, \tag{2.13}$$

where if $I_{u,v}^{(III)} \le 0$, then

$$a \circledast b = \begin{cases} ab, & ab \in [u^-v^+, u^-v^-], \\ u^-v^+, & ab < u^-v^+, \\ u^-v^-, & ab > u^-v^-; \end{cases}$$

if $I_{u,v}^{(III)} \geq 0$, then

$$a \circledast b = \begin{cases} ab, & ab \in [u^-v^-, u^-v^+], \\ u^-v^-, & ab < u^-v^-, \\ u^-v^+, & ab > u^-v^+. \end{cases}$$

$$\textbf{Type IV.} \quad [a^-, a^+] \boxdot [b^-, b^+] = \big\{a \circledast b : a \in [a^-, a^+],\ b \in [b^-, b^+]\big\}, \tag{2.14}$$

where if $I_{u,v}^{(IV)} \leq 0$, then

$$a \circledast b = \begin{cases} ab, & ab \in [u^+v^-, u^+v^+], \\ u^+v^-, & ab < u^+v^-, \\ u^+v^+, & ab > u^+v^+; \end{cases}$$

if $I_{u,v}^{(IV)} \geq 0$, then

$$a \circledast b = \begin{cases} ab, & ab \in [u^+v^+, u^+v^-], \\ u^+v^+, & ab < u^+v^+, \\ u^+v^-, & ab > u^+v^-. \end{cases}$$

$$\textbf{Type V.} \quad [a^-, a^+] \otimes [b^-, b^+] = \big\{a \circledast b : a \in [a^-, a^+],\ b \in [b^-, b^+]\big\}, \tag{2.15}$$

where if $I_{u,v}^{(V)} \leq 0$, then

$$a \circledast b = \begin{cases} ab, & ab \in [u^-v^-, u^+v^-], \\ u^-v^-, & ab < u^-v^-, \\ u^+v^-, & ab > u^+v^-; \end{cases}$$

if $I_{u,v}^{(V)} \geq 0$, then

$$a \circledast b = \begin{cases} ab, & ab \in [u^+v^-, u^-v^-], \\ u^+v^-, & ab < u^+v^-, \\ u^-v^-, & ab > u^-v^-. \end{cases}$$

$$\textbf{Type VI.} \quad [a^-, a^+] \odot [b^-, b^+] = \big\{a \circledast b : a \in [a^-, a^+],\ b \in [b^-, b^+]\big\}, \tag{2.16}$$

where if $I_{u,v}^{(VI)} \leq 0$, then

$$a \circledast b = \begin{cases} ab, & ab \in [u^- v^+, u^+ v^+], \\ u^- v^+, & ab < u^- v^+, \\ u^+ v^+, & ab > u^+ v^+; \end{cases}$$

if $I_{u,v}^{(VI)} \geq 0$, then

$$a \circledast b = \begin{cases} ab, & ab \in [u^+ v^+, u^- v^+], \\ u^+ v^+, & ab < u^+ v^+, \\ u^- v^+, & ab > u^- v^+. \end{cases}$$

Remark 2.13 From the above definition of multiplications, for any $[a^-, a^+] \subseteq [u^-, u^+]$ and $[b^-, b^+] \subseteq [v^-, v^+]$, we have $[a^-, a^+] \circ [b^-, b^+] \subseteq [u^-, u^+] \circ [v^-, v^+]$ and $[a^-, a^+] \circledcirc [b^-, b^+] \subseteq [u^-, u^+] \circledcirc [v^-, v^+]$, similarly, the multiplications $\boxtimes, \boxdot, \otimes, \odot$ between two compact intervals also have the same inclusion isotonicity .

Now, we introduce six types of the multiplication of fuzzy vectors induced by the multiplications of compact intervals defined by (2.11)–(2.16). For any $\alpha \in [0, 1]$ and $i = 1, 2, \ldots, n$, we introduce the notations:

$$I_{u_i,v_i}^{\alpha,(I)} = \begin{vmatrix} u_{i,\alpha}^- & u_{i,\alpha}^+ \\ v_{i,\alpha}^- & v_{i,\alpha}^+ \end{vmatrix}, \quad I_{u_i,v_i}^{\alpha,(II)} = \begin{vmatrix} u_{i,\alpha}^+ & u_{i,\alpha}^- \\ v_{i,\alpha}^- & v_{i,\alpha}^+ \end{vmatrix}, \quad I_{u_i,v_i}^{\alpha,(III)} = \begin{vmatrix} u_{i,\alpha}^- & u_{i,\alpha}^- \\ v_{i,\alpha}^- & v_{i,\alpha}^+ \end{vmatrix},$$

$$I_{u_i,v_i}^{\alpha,(IV)} = \begin{vmatrix} u_{i,\alpha}^+ & u_{i,\alpha}^+ \\ v_{i,\alpha}^- & v_{i,\alpha}^+ \end{vmatrix}, \quad I_{u_i,v_i}^{\alpha,(V)} = \begin{vmatrix} u_{i,\alpha}^- & u_{i,\alpha}^+ \\ v_{i,\alpha}^- & v_{i,\alpha}^- \end{vmatrix}, \quad I_{u_i,v_i}^{\alpha,(VI)} = \begin{vmatrix} u_{i,\alpha}^- & u_{i,\alpha}^+ \\ v_{i,\alpha}^+ & v_{i,\alpha}^+ \end{vmatrix},$$

then we define the following types $I - VI$ with the (compact box) α-level set:

$$\textbf{Type I.} \quad [u * v]^\alpha = \times_{i=1}^n \big([u_{i,\alpha}^-, v_{i,\alpha}^+] \circ [u_{i,\alpha}^-, v_{i,\alpha}^+]\big),$$

$$\text{where} \quad [u_{i,\alpha}^-, u_{i,\alpha}^+] \circ [v_{i,\alpha}^-, v_{i,\alpha}^+] = \begin{cases} [u_{i,\alpha}^- v_{i,\alpha}^+, u_{i,\alpha}^+ v_{i,\alpha}^-] \text{ if } I_{u_i,v_i}^{\alpha,(I)} \leq 0, \\ [u_{i,\alpha}^+ v_{i,\alpha}^-, u_{i,\alpha}^- v_{i,\alpha}^+] \text{ if } I_{u_i,v_i}^{\alpha,(I)} \geq 0; \end{cases} \tag{2.17}$$

$$\textbf{Type II.} \quad [u \circledast v]^\alpha = \times_{i=1}^n \big([u_{i,\alpha}^-, u_{i,\alpha}^+] \circledcirc [v_{i,\alpha}^-, v_{i,\alpha}^+]\big),$$

$$\text{where} \quad [u_{i,\alpha}^-, u_{i,\alpha}^+] \circledcirc [v_{i,\alpha}^-, v_{i,\alpha}^+] = \begin{cases} [u_{i,\alpha}^+ v_{i,\alpha}^+, u_{i,\alpha}^- v_{i,\alpha}^-] \text{ if } I_{u_i,v_i}^{\alpha,(II)} \leq 0, \\ [u_{i,\alpha}^- v_{i,\alpha}^-, u_{i,\alpha}^+ v_{i,\alpha}^+] \text{ if } I_{u_i,v_i}^{\alpha,(II)} \geq 0; \end{cases} \tag{2.18}$$

$$\textbf{Type III.} \quad [u \hat{*} v]^\alpha = \times_{i=1}^n \big([u_{i,\alpha}^-, v_{i,\alpha}^+] \boxtimes [u_{i,\alpha}^-, v_{i,\alpha}^+]\big),$$

$$\text{where } [u_{i,\alpha}^-, u_{i,\alpha}^+] \boxtimes [v_{i,\alpha}^-, v_{i,\alpha}^+] = \begin{cases} [u_{i,\alpha}^- v_{i,\alpha}^+, u_{i,\alpha}^- v_{i,\alpha}^-] \text{ if } I_{u_i,v_i}^{\alpha,(III)} \leq 0, \\ [u_{i,\alpha}^- v_{i,\alpha}^-, u_{i,\alpha}^- v_{i,\alpha}^+] \text{ if } I_{u_i,v_i}^{\alpha,(III)} \geq 0; \end{cases} \tag{2.19}$$

$$\textbf{Type IV.}\ [u \hat{\circledast} v]^\alpha = \times_{i=1}^n \big([u_{i,\alpha}^-, v_{i,\alpha}^+] \boxdot [u_{i,\alpha}^-, v_{i,\alpha}^+]\big),$$

$$\text{where } [u_{i,\alpha}^-, u_{i,\alpha}^+] \boxdot [v_{i,\alpha}^-, v_{i,\alpha}^+] = \begin{cases} [u_{i,\alpha}^+ v_{i,\alpha}^-, u_{i,\alpha}^+ v_{i,\alpha}^+] \text{ if } I_{u_i,v_i}^{\alpha,(IV)} \leq 0, \\ [u_{i,\alpha}^+ v_{i,\alpha}^+, u_{i,\alpha}^+ v_{i,\alpha}^-] \text{ if } I_{u_i,v_i}^{\alpha,(IV)} \geq 0; \end{cases} \tag{2.20}$$

$$\textbf{Type V.}\ [u \tilde{*} v]^\alpha = \times_{i=1}^n \big([u_{i,\alpha}^-, v_{i,\alpha}^+] \otimes [u_{i,\alpha}^-, v_{i,\alpha}^+]\big),$$

$$\text{where } [u_{i,\alpha}^-, u_{i,\alpha}^+] \otimes [v_{i,\alpha}^-, v_{i,\alpha}^+] = \begin{cases} [u_{i,\alpha}^- v_{i,\alpha}^-, u_{i,\alpha}^+ v_{i,\alpha}^-] \text{ if } I_{u_i,v_i}^{\alpha,(V)} \leq 0, \\ [u_{i,\alpha}^+ v_{i,\alpha}^-, u_{i,\alpha}^- v_{i,\alpha}^-] \text{ if } I_{u_i,v_i}^{\alpha,(V)} \geq 0; \end{cases} \tag{2.21}$$

$$\textbf{Type VI.}\ [u \tilde{\circledast} v]^\alpha = \times_{i=1}^n \big([u_{i,\alpha}^-, v_{i,\alpha}^+] \odot [u_{i,\alpha}^-, v_{i,\alpha}^+]\big),$$

$$\text{where } [u_{i,\alpha}^-, u_{i,\alpha}^+] \odot [v_{i,\alpha}^-, v_{i,\alpha}^+] = \begin{cases} [u_{i,\alpha}^- v_{i,\alpha}^+, u_{i,\alpha}^+ v_{i,\alpha}^+] \text{ if } I_{u_i,v_i}^{\alpha,(VI)} \leq 0, \\ [u_{i,\alpha}^+ v_{i,\alpha}^+, u_{i,\alpha}^- v_{i,\alpha}^+] \text{ if } I_{u_i,v_i}^{\alpha,(VI)} \geq 0. \end{cases} \tag{2.22}$$

Remark 2.14 For $I_{u_i,v_i}^{\alpha,(I)} = 0$ for all $i = 1, 2, \ldots, n$, from (2.17), we have $u_{i,\alpha}^- v_{i,\alpha}^+ = u_{i,\alpha}^+ v_{i,\alpha}^-$, then

$$[u * v]^\alpha = \times_{i=1}^n [u_{i,\alpha}^-, u_{i,\alpha}^+] \circ [v_{i,\alpha}^-, v_{i,\alpha}^+] = \times_{i=1}^n \{u_{i,\alpha}^- v_{i,\alpha}^+\} = \times_{i=1}^n \{u_{i,\alpha}^+ v_{i,\alpha}^-\}.$$

Similarly, For $I_{u_i,v_i}^{\alpha,(II)} = 0$ for all $i = 1, 2, \ldots, n$, from (2.18), we have

$$[u \circledast v]^\alpha = \times_{i=1}^n \big([u_{i,\alpha}^-, u_{i,\alpha}^+] \circledcirc [v_{i,\alpha}^-, v_{i,\alpha}^+]\big) = \times_{i=1}^n \{u_{i,\alpha}^+ v_{i,\alpha}^+\} = \times_{i=1}^n \{u_{i,\alpha}^- v_{i,\alpha}^-\},$$

noticing that $\times_{i=1}^n [a_i, a_i] = \times_{i=1}^n \{a_i\}$ for any $a_i \in \mathbb{R}$. For example, given $u = \chi_{[-a,a]}$ and $v = \chi_{[-b,b]}$ in $\mathbb{R}_{\mathscr{F}}$, where $a, b > 0$, it follows that $[u]^\alpha = [-a, a]$, $[v]^\alpha = [-b, b]$ for all $\alpha \in [0, 1]$. Note that $I_{u,v}^{\alpha,(I)} = I_{u,v}^{\alpha,(II)} = 0$, it indicates that $[u * v]^\alpha = \{-ab\}$ and $[u \circledast v]^\alpha = \{ab\}$, i.e. $u * v = \chi_{\{-ab\}}$ and $u \circledast v = \chi_{\{ab\}}$. In fact, it is easy to see that if there exists some $\hat{I} \in \{I, II, \ldots, VI\}$ such that $I_{u_i,v_i}^{\alpha,(\hat{I})} = 0$, then the corresponding product of α-levels defined by (2.17)–(2.22) is a one-point set for Type $\hat{I}$.

Remark 2.15 Since the interval multiplications defined by (2.11) and (2.16) have a well inclusion isotonicity, then (2.17) and (2.22) also has well inclusion isotonicity naturally. For example, given $u = \chi_{[-1,0]}$ and $v = \chi_{[-1,1]}$, then we have $I_{u,v}^{\alpha,(I)} < 0$ for all $\alpha \in [0, 1]$. Therefore, $u * v$ is given by

$$[u * v]^\alpha = [u_\alpha^-, u_\alpha^+] \circ [v_\alpha^-, v_\alpha^+] = [u_\alpha^- v_\alpha^+, u_\alpha^+ v_\alpha^-] = [-1, 0]$$

for all $\alpha \in [0, 1]$. For any given $a \in [-1, 0] = [u_\alpha^-, u_\alpha^+]$ and $b \in [-1, 1] = [v_\alpha^-, v_\alpha^+]$, it implies that

$$a \circledast b = \begin{cases} ab, & ab \in [-1, 0], \\ -1, & ab < -1, \\ 0, & ab > 0, \end{cases}$$

which indicates that for any $[a, b] \subseteq [u_\alpha^-, u_\alpha^+]$, $[c, d] \subseteq [v_\alpha^-, v_\alpha^+]$, we obtain $[a, b] \circ [c, d] \subseteq [u_\alpha^-, u_\alpha^+] \circ [v_\alpha^-, v_\alpha^+]$.

Remark 2.16 Traditionally, the multiplication of compact intervals is induced by the ordinary multiplication of real numbers, i.e, for the real compact intervals $U = [u^-, u^+]$ and $V = [v^-, v^+]$, the interval $C = [c^-, c^+]$ defining the multiplication $C = UV$ is given by

$$c^- = \min\{u^-v^-, u^-v^+, u^+v^-, u^+v^+\}, \quad c^+ = \max\{u^-v^-, u^-v^+, u^+v^-, u^+v^+\}.$$

In fact, $C = UV = \{ab : a \in U, \ b \in V\}$. However, note that such a multiplication of compact intervals induced by ordinary multiplication of real numbers is completely different from the multiplications of compact intervals induced by $a \circledast b$ above. In the example of Remark 2.15, given $-\frac{1}{2} \in [u_\alpha^-, u_\alpha^+]$, $-\frac{1}{4} \in [v_\alpha^-, v_\alpha^+]$, we have $ab = (-\frac{1}{2})(-\frac{1}{4}) = \frac{1}{8} \notin [-1, 0] = [-1, 0] \circ [-1, 1]$ but $(-\frac{1}{2}) \circledast (-\frac{1}{4}) = 0 \in [-1, 0] = [-1, 0] \circ [-1, 1]$.

Theorem 2.6 *Let $u, v, \omega \in [\mathbb{R}^n_{\mathscr{F}}]$, we have the following related properties with respect to the multiplications $u * v$ and $u \circledast v$. For $(i) - (xi)$ we assume that $u \tilde{-}_{gH_{(I)}} v$ exists.*

- (i) $u * v = v * u;\ u \circledast v = v \circledast u.$
- (ii) *If* $I^{\alpha,(I)}_{u_i,\omega_i}, I^{\alpha,(I)}_{v_i,\omega_i} \geq 0$ *or* $I^{\alpha,(I)}_{u_i,\omega_i}, I^{\alpha,(I)}_{v_i,\omega_i} \leq 0$, *then* $(u \tilde{+} v) * \omega = u * \omega \tilde{+} v * \omega.$
- (iii) *If* $I^{\alpha,(II)}_{u_i,\omega_i}, I^{\alpha,(II)}_{v_i,\omega_i} \geq 0$ *or* $I^{\alpha,(II)}_{u_i,\omega_i}, I^{\alpha,(II)}_{v_i,\omega_i} \leq 0$, *then* $(u \tilde{+} v) \circledast \omega = u \circledast \omega \tilde{+} v \circledast \omega.$
- (iv) *If* $I^{\alpha,(I)}_{u_i \tilde{-}_{gH_{(I)}} v_i,\omega_i} \geq 0$, $I^{\alpha,(I)}_{u_i,\omega_i}, I^{\alpha,(I)}_{v_i,\omega_i} \geq 0$, *then* $(u \tilde{-}_{gH_{(I)}} v) * \omega = u * \omega \tilde{-}_{gH_{(I)}} v * \omega.$
- (v) *If* $I^{\alpha,(I)}_{u_i \tilde{-}_{gH_{(I)}} v_i,\omega_i} \geq 0$, $I^{\alpha,(I)}_{u_i,\omega_i}, I^{\alpha,(I)}_{v_i,\omega_i} \leq 0$, *then* $(u \tilde{-}_{gH_{(I)}} v) * \omega = u * \omega \tilde{-}_{gH_{(II)}} v * \omega.$
- (vi) *If* $I^{\alpha,(I)}_{u_i \tilde{-}_{gH_{(I)}} v_i,\omega_i} \leq 0$, $I^{\alpha,(I)}_{u_i,\omega_i}, I^{\alpha,(I)}_{v_i,\omega_i} \geq 0$, *then* $(u \tilde{-}_{gH_{(I)}} v) * \omega = u * \omega \tilde{-}_{gH_{(II)}} v * \omega.$
- (vii) *If* $I^{\alpha,(I)}_{u_i \tilde{-}_{gH_{(I)}} v_i,\omega_i} \leq 0$, $I^{\alpha,(I)}_{u_i,\omega_i}, I^{\alpha,(I)}_{v_i,\omega_i} \leq 0$, *then* $(u \tilde{-}_{gH_{(I)}} v) * \omega = u * \omega \tilde{-}_{gH_{(I)}} v * \omega.$
- $(viii)$ *If* $I^{\alpha,(II)}_{u_i \tilde{-}_{gH_{(I)}} v_i,\omega_i} \geq 0$, $I^{\alpha,(II)}_{u_i,\omega_i}, I^{\alpha,(II)}_{v_i,\omega_i} \geq 0$, *then* $(u \tilde{-}_{gH_{(I)}} v) \circledast \omega = u \circledast \omega \tilde{-}_{gH_{(I)}} v \circledast \omega.$
- (ix) *If* $I^{\alpha,(II)}_{u_i \tilde{-}_{gH_{(I)}} v_i,\omega_i} \geq 0$, $I^{\alpha,(II)}_{u_i,\omega_i}, I^{\alpha,(II)}_{v_i,\omega_i} \leq 0$, *then* $(u \tilde{-}_{gH_{(I)}} v) \circledast \omega = u \circledast \omega \tilde{-}_{gH_{(II)}} v \circledast \omega.$
- (x) *If* $I^{\alpha,(II)}_{u_i \tilde{-}_{gH_{(I)}} v_i,\omega_i} \leq 0$, $I^{\alpha,(II)}_{u_i,\omega_i}, I^{\alpha,(II)}_{v_i,\omega_i} \geq 0$, *then* $(u \tilde{-}_{gH_{(I)}} v) \circledast \omega = u \circledast \omega \tilde{-}_{gH_{(II)}} v \circledast \omega.$
- (xi) *If* $I^{\alpha,(II)}_{u_i \tilde{-}_{gH_{(I)}} v_i,\omega_i} \leq 0$, $I^{\alpha,(II)}_{u_i,\omega_i}, I^{\alpha,(II)}_{v_i,\omega_i} \leq 0$, *then* $(u \tilde{-}_{gH_{(I)}} v) \circledast \omega = u \circledast \omega \tilde{-}_{gH_{(I)}} v \circledast \omega.$

Proof (i). From (2.17) and (2.18), the results are obvious.

(ii). Case I. If $I^{\alpha,(I)}_{u_i,\omega_i}, I^{\alpha,(I)}_{v_i,\omega_i} \geq 0$, we have

$$[u * \omega]^\alpha = \times_{i=1}^n [u^+_{i,\alpha}\omega^-_{i,\alpha}, u^-_{i,\alpha}\omega^+_{i,\alpha}], \quad [v * \omega]^\alpha = \times_{i=1}^n [v^+_{i,\alpha}\omega^-_{i,\alpha}, v^-_{i,\alpha}\omega^+_{i,\alpha}].$$

Case II. If $I^{\alpha,(I)}_{u_i,\omega_i}, I^{\alpha,(I)}_{v_i,\omega_i} \leq 0$, we have

$$[u * \omega]^\alpha = \times_{i=1}^n [u^-_{i,\alpha}\omega^+_{i,\alpha}, u^+_{i,\alpha}\omega^-_{i,\alpha}], \quad [v * \omega]^\alpha = \times_{i=1}^n [v^-_{i,\alpha}\omega^+_{i,\alpha}, v^+_{i,\alpha}\omega^-_{i,\alpha}].$$

Then for Case I, we have

$$\begin{aligned} I^{\alpha,(I)}_{u_i \tilde{+} v_i,\omega_i} &= \begin{vmatrix} u^-_{i,\alpha} + v^-_{i,\alpha} & u^+_{i,\alpha} + v^+_{i,\alpha} \\ \omega^-_{i,\alpha} & \omega^+_{i,\alpha} \end{vmatrix} \\ &= (u^-_{i,\alpha} + v^-_{i,\alpha})\omega^+_{i,\alpha} - (u^+_{i,\alpha} + v^+_{i,\alpha})\omega^-_{i,\alpha} \geq 0, \end{aligned}$$

and then we obtain

$$\begin{aligned} [(u \tilde{+} v) * \omega]^\alpha &= \times_{i=1}^n [(u^+_{i,\alpha} + v^+_{i,\alpha})\omega^-_{i,\alpha}, (u^-_{i,\alpha} + v^-_{i,\alpha})\omega^+_{i,\alpha}] \\ &= [u * \omega]^\alpha + [v * \omega]^\alpha. \end{aligned}$$

Similarly, we can also have the same results for Case II.

(iii). Case I. If $I^{\alpha,(II)}_{u_i,\omega_i}, I^{\alpha,(II)}_{v_i,\omega_i} \geq 0$, we have

$$[u \circledast \omega]^\alpha = \times_{i=1}^n [u^-_{i,\alpha}\omega^-_{i,\alpha}, u^+_{i,\alpha}\omega^+_{i,\alpha}], \quad [v \circledast \omega]^\alpha = \times_{i=1}^n [v^-_{i,\alpha}\omega^-_{i,\alpha}, v^+_{i,\alpha}\omega^+_{i,\alpha}].$$

Case II. If $I^{\alpha,(II)}_{u_i,\omega_i}, I^{\alpha,(II)}_{v_i,\omega_i} \leq 0$, we have

$$[u \circledast \omega]^\alpha = \times_{i=1}^n [u^+_{i,\alpha}\omega^+_{i,\alpha}, u^-_{i,\alpha}\omega^-_{i,\alpha}], \quad [v \circledast \omega]^\alpha = \times_{i=1}^n [v^+_{i,\alpha}\omega^+_{i,\alpha}, v^-_{i,\alpha}\omega^-_{i,\alpha}].$$

Then for Case I, we have

$$\begin{aligned} I^{\alpha,(II)}_{u_i \tilde{+} v_i,\omega_i} &= \begin{vmatrix} u^+_{i,\alpha} + v^+_{i,\alpha} & u^-_{i,\alpha} + v^-_{i,\alpha} \\ \omega^-_{i,\alpha} & \omega^+_{i,\alpha} \end{vmatrix} \\ &= (u^+_{i,\alpha} + v^+_{i,\alpha})\omega^+_{i,\alpha} - (u^-_{i,\alpha} + v^-_{i,\alpha})\omega^-_{i,\alpha} \geq 0, \end{aligned}$$

and then we obtain

$$\begin{aligned} [(u \tilde{+} v) \circledast \omega]^\alpha &= \times_{i=1}^n [(u^-_{i,\alpha} + v^-_{i,\alpha})\omega^-_{i,\alpha}, (u^+_{i,\alpha} + v^+_{i,\alpha})\omega^+_{i,\alpha}] \\ &= [u \circledast \omega]^\alpha + [v \circledast \omega]^\alpha. \end{aligned}$$

Similarly, we can also have the same results for Case II.

(iv). Since $I^{\alpha,(I)}_{u_i,\omega_i}, I^{\alpha,(I)}_{v_i,\omega_i} \geq 0$, we have

$$[u * \omega]^\alpha = \times_{i=1}^n [u_{i,\alpha}^+ \omega_{i,\alpha}^-, u_{i,\alpha}^- \omega_{i,\alpha}^+], \quad [v * \omega]^\alpha = \times_{i=1}^n [v_{i,\alpha}^+ \omega_{i,\alpha}^-, v_{i,\alpha}^- \omega_{i,\alpha}^+].$$

Moreover, it follows from $I^{\alpha,(I)}_{u_i \tilde{-}_{gH_{(I)}} v_i, \omega_i} \geq 0$ that

$$\begin{aligned} I^{\alpha,(I)}_{u_i \tilde{-}_{gH_{(I)}} v_i, \omega_i} &= \begin{vmatrix} u_{i,\alpha}^- - v_{i,\alpha}^- & u_{i,\alpha}^+ - v_{i,\alpha}^+ \\ \omega_{i,\alpha}^- & \omega_{i,\alpha}^+ \end{vmatrix} \\ &= (u_{i,\alpha}^- - v_{i,\alpha}^-)\omega_{i,\alpha}^+ - (u_{i,\alpha}^+ - v_{i,\alpha}^+)\omega_{i,\alpha}^- \geq 0, \end{aligned}$$

which indicates that $[u * \omega]^\alpha \boxminus_{gH_{(I)}} [v * \omega]^\alpha$ exists, so we obtain

$$\begin{aligned} [(u \tilde{-}_{gH_{(I)}} v) * \omega]^\alpha &= \times_{i=1}^n [(u_{i,\alpha}^+ - v_{i,\alpha}^+)\omega_{i,\alpha}^-, (u_{i,\alpha}^- - v_{i,\alpha}^-)\omega_{i,\alpha}^+] \\ &= [u * \omega]^\alpha \boxminus_{gH_{(I)}} [v * \omega]^\alpha. \end{aligned}$$

(v). Since $I^{\alpha,(I)}_{u_i,\omega_i}, I^{\alpha,(I)}_{v_i,\omega_i} \leq 0$, we have

$$[u * \omega]^\alpha = \times_{i=1}^n [u_{i,\alpha}^- \omega_{i,\alpha}^+, u_{i,\alpha}^+ \omega_{i,\alpha}^-], \quad [v * \omega]^\alpha = \times_{i=1}^n [v_{i,\alpha}^- \omega_{i,\alpha}^+, v_{i,\alpha}^+ \omega_{i,\alpha}^-].$$

Moreover, it follows from $I^{\alpha,(I)}_{u_i \tilde{-}_{gH_{(I)}} v_i, \omega_i} \geq 0$ that

$$\begin{aligned} I^{\alpha,(I)}_{u_i \tilde{-}_{gH_{(I)}} v_i, \omega_i} &= \begin{vmatrix} u_{i,\alpha}^- - v_{i,\alpha}^- & u_{i,\alpha}^+ - v_{i,\alpha}^+ \\ \omega_{i,\alpha}^- & \omega_{i,\alpha}^+ \end{vmatrix} \\ &= (u_{i,\alpha}^- - v_{i,\alpha}^-)\omega_{i,\alpha}^+ - (u_{i,\alpha}^+ - v_{i,\alpha}^+)\omega_{i,\alpha}^- \geq 0, \end{aligned}$$

which indicates that $[u * \omega]^\alpha \boxminus_{gH_{(II)}} [v * \omega]^\alpha$ exists, so we obtain

$$\begin{aligned} [(u \tilde{-}_{gH_{(I)}} v) * \omega]^\alpha &= \times_{i=1}^n [(u_{i,\alpha}^+ - v_{i,\alpha}^+)\omega_{i,\alpha}^+, (u_{i,\alpha}^- - v_{i,\alpha}^-)\omega_{i,\alpha}^-] \\ &= [u * \omega]^\alpha \boxminus_{gH_{(II)}} [v * \omega]^\alpha. \end{aligned}$$

(vi). Since $I^{\alpha,(I)}_{u_i,\omega_i}, I^{\alpha,(I)}_{v_i,\omega_i} \geq 0$, we have

$$[u * \omega]^\alpha = \times_{i=1}^n [u_{i,\alpha}^+ \omega_{i,\alpha}^-, u_{i,\alpha}^- \omega_{i,\alpha}^+], \quad [v * \omega]^\alpha = \times_{i=1}^n [v_{i,\alpha}^+ \omega_{i,\alpha}^-, v_{i,\alpha}^- \omega_{i,\alpha}^+].$$

Moreover, it follows from $I^{\alpha,(I)}_{u_i \tilde{-}_{gH_{(I)}} v_i, \omega_i} \leq 0$ that

$$\begin{aligned} I^{\alpha,(I)}_{u_i \tilde{-}_{gH_{(I)}} v_i, \omega_i} &= \begin{vmatrix} u_{i,\alpha}^- - v_{i,\alpha}^- & u_{i,\alpha}^+ - v_{i,\alpha}^+ \\ \omega_{i,\alpha}^- & \omega_{i,\alpha}^+ \end{vmatrix} \\ &= (u_{i,\alpha}^- - v_{i,\alpha}^-)\omega_{i,\alpha}^+ - (u_{i,\alpha}^+ - v_{i,\alpha}^+)\omega_{i,\alpha}^- \leq 0, \end{aligned}$$

which indicates that $[u * \omega]^\alpha \boxminus_{gH_{(II)}} [v * \omega]^\alpha$ exists, so we obtain

$$\begin{aligned}[(u\widetilde{-}_{gH(I)}v)*\omega]^{\alpha} &= \times_{i=1}^{n}[(u_{i,\alpha}^{-}-v_{i,\alpha}^{-})\omega_{i,\alpha}^{+},(u_{i,\alpha}^{+}-v_{i,\alpha}^{+})\omega_{i,\alpha}^{-}]\\ &= [u*\omega]^{\alpha}\boxminus_{gH(II)}[v*\omega]^{\alpha}.\end{aligned}$$

(vii). Since $I_{u_i,\omega_i}^{\alpha,(I)}, I_{v_i,\omega_i}^{\alpha,(I)} \leq 0$, we have

$$[u*\omega]^{\alpha} = \times_{i=1}^{n}[u_{i,\alpha}^{-}\omega_{i,\alpha}^{+},u_{i,\alpha}^{+}\omega_{i,\alpha}^{-}], \quad [v*\omega]^{\alpha} = \times_{i=1}^{n}[v_{i,\alpha}^{-}\omega_{i,\alpha}^{+},v_{i,\alpha}^{+}\omega_{i,\alpha}^{-}].$$

Moreover, it follows from $I_{u_i\widetilde{-}_{gH(I)}v_i,\omega_i}^{\alpha,(I)} \leq 0$ that

$$\begin{aligned}I_{u_i\widetilde{-}_{gH(I)}v_i,\omega_i}^{\alpha,(I)} &= \begin{vmatrix} u_{i,\alpha}^{-}-v_{i,\alpha}^{-} & u_{i,\alpha}^{+}-v_{i,\alpha}^{+}\\ \omega_{i,\alpha}^{-} & \omega_{i,\alpha}^{+}\end{vmatrix}\\ &= (u_{i,\alpha}^{-}-v_{i,\alpha}^{-})\omega_{i,\alpha}^{+}-(u_{i,\alpha}^{+}-v_{i,\alpha}^{+})\omega_{i,\alpha}^{-}\leq 0,\end{aligned}$$

which indicates that $[u*\omega]^{\alpha}\boxminus_{gH(I)}[v*\omega]^{\alpha}$ exists, so we obtain

$$\begin{aligned}[(u\widetilde{-}_{gH(I)}v)*\omega]^{\alpha} &= \times_{i=1}^{n}[(u_{i,\alpha}^{-}-v_{i,\alpha}^{-})\omega_{i,\alpha}^{+},(u_{i,\alpha}^{+}-v_{i,\alpha}^{+})\omega_{i,\alpha}^{-}]\\ &= [u*\omega]^{\alpha}\boxminus_{gH(I)}[v*\omega]^{\alpha}.\end{aligned}$$

$(viii)$. Since $I_{u_i,\omega_i}^{\alpha,(II)}, I_{v_i,\omega_i}^{\alpha,(II)} \geq 0$, we have

$$[u\circledast\omega]^{\alpha} = \times_{i=1}^{n}[u_{i,\alpha}^{-}\omega_{i,\alpha}^{-},u_{i,\alpha}^{+}\omega_{i,\alpha}^{+}], \quad [v\circledast\omega]^{\alpha} = \times_{i=1}^{n}[v_{i,\alpha}^{-}\omega_{i,\alpha}^{-},v_{i,\alpha}^{+}\omega_{i,\alpha}^{+}].$$

Moreover, it follows from $I_{u_i\widetilde{-}_{gH(I)}v_i,\omega_i}^{\alpha,(II)} \geq 0$ that

$$\begin{aligned}I_{u_i\widetilde{-}_{gH(I)}v_i,\omega_i}^{\alpha,(II)} &= \begin{vmatrix} u_{i,\alpha}^{+}-v_{i,\alpha}^{+} & u_{i,\alpha}^{-}-v_{i,\alpha}^{-}\\ \omega_{i,\alpha}^{-} & \omega_{i,\alpha}^{+}\end{vmatrix}\\ &= (u_{i,\alpha}^{+}-v_{i,\alpha}^{+})\omega_{i,\alpha}^{+}-(u_{i,\alpha}^{-}-v_{i,\alpha}^{-})\omega_{i,\alpha}^{-}\geq 0,\end{aligned}$$

which indicates that $[u\circledast\omega]^{\alpha}\boxminus_{gH(I)}[v\circledast\omega]^{\alpha}$ exists, so we obtain

$$\begin{aligned}[(u\widetilde{-}_{gH(I)}v)\circledast\omega]^{\alpha} &= \times_{i=1}^{n}[(u_{i,\alpha}^{-}-v_{i,\alpha}^{-})\omega_{i,\alpha}^{-},(u_{i,\alpha}^{+}-v_{i,\alpha}^{+})\omega_{i,\alpha}^{+}]\\ &= [u\circledast\omega]^{\alpha}\boxminus_{gH(I)}[v\circledast\omega]^{\alpha}.\end{aligned}$$

(ix). Since $I_{u_i,\omega_i}^{\alpha,(II)}, I_{v_i,\omega_i}^{\alpha,(II)} \leq 0$, we have

$$[u\circledast\omega]^{\alpha} = \times_{i=1}^{n}[u_{i,\alpha}^{+}\omega_{i,\alpha}^{+},u_{i,\alpha}^{-}\omega_{i,\alpha}^{-}], \quad [v\circledast\omega]^{\alpha} = \times_{i=1}^{n}[v_{i,\alpha}^{+}\omega_{i,\alpha}^{+},v_{i,\alpha}^{-}\omega_{i,\alpha}^{-}].$$

Moreover, it follows from $I_{u_i\widetilde{-}_{gH(I)}v_i,\omega_i}^{\alpha,(II)} \geq 0$ that

$$\begin{aligned} I^{\alpha,(II)}_{u_i \tilde{-}_{gH(I)} v_i, \omega_i} &= \begin{vmatrix} u^+_{i,\alpha} - v^+_{i,\alpha} & u^-_{i,\alpha} - v^-_{i,\alpha} \\ \omega^-_{i,\alpha} & \omega^+_{i,\alpha} \end{vmatrix} \\ &= (u^+_{i,\alpha} - v^+_{i,\alpha})\omega^+_{i,\alpha} - (u^-_{i,\alpha} - v^-_{i,\alpha})\omega^-_{i,\alpha} \geq 0, \end{aligned}$$

which indicates that $[u \circledast \omega]^\alpha \boxminus_{gH(II)} [v \circledast \omega]^\alpha$ exists, so we obtain

$$\begin{aligned} [(u \tilde{-}_{gH(I)} v) \circledast \omega]^\alpha &= \times^n_{i=1}[(u^-_{i,\alpha} - v^-_{i,\alpha})\omega^-_{i,\alpha}, (u^+_{i,\alpha} - v^+_{i,\alpha})\omega^+_{i,\alpha}] \\ &= [u \circledast \omega]^\alpha \boxminus_{gH(II)} [v \circledast \omega]^\alpha. \end{aligned}$$

(x). Since $I^{\alpha,(II)}_{u_i,\omega_i}, I^{\alpha,(II)}_{v_i,\omega_i} \geq 0$, we have

$$[u \circledast \omega]^\alpha = \times^n_{i=1}[u^-_{i,\alpha}\omega^-_{i,\alpha}, u^+_{i,\alpha}\omega^+_{i,\alpha}], \quad [v \circledast \omega]^\alpha = \times^n_{i=1}[v^-_{i,\alpha}\omega^-_{i,\alpha}, v^+_{i,\alpha}\omega^+_{i,\alpha}].$$

Moreover, it follows from $I^{\alpha,(II)}_{u_i \tilde{-}_{gH(I)} v_i, \omega_i} \leq 0$ that

$$\begin{aligned} I^{\alpha,(II)}_{u_i \tilde{-}_{gH(I)} v_i, \omega_i} &= \begin{vmatrix} u^+_{i,\alpha} - v^+_{i,\alpha} & u^-_{i,\alpha} - v^-_{i,\alpha} \\ \omega^-_{i,\alpha} & \omega^+_{i,\alpha} \end{vmatrix} \\ &= (u^+_{i,\alpha} - v^+_{i,\alpha})\omega^+_{i,\alpha} - (u^-_{i,\alpha} - v^-_{i,\alpha})\omega^-_{i,\alpha} \leq 0, \end{aligned}$$

which indicates that $[u \circledast \omega]^\alpha \boxminus_{gH(II)} [v \circledast \omega]^\alpha$ exists, so we obtain

$$\begin{aligned} [(u \tilde{-}_{gH(I)} v) \circledast \omega]^\alpha &= \times^n_{i=1}[(u^+_{i,\alpha} - v^+_{i,\alpha})\omega^+_{i,\alpha}, (u^-_{i,\alpha} - v^-_{i,\alpha})\omega^-_{i,\alpha}] \\ &= [u \circledast \omega]^\alpha \boxminus_{gH(II)} [v \circledast \omega]^\alpha. \end{aligned}$$

(xi). Since $I^{\alpha,(II)}_{u_i,\omega_i}, I^{\alpha,(II)}_{v_i,\omega_i} \leq 0$, we have

$$[u \circledast \omega]^\alpha = \times^n_{i=1}[u^+_{i,\alpha}\omega^+_{i,\alpha}, u^-_{i,\alpha}\omega^-_{i,\alpha}], \quad [v \circledast \omega]^\alpha = \times^n_{i=1}[v^+_{i,\alpha}\omega^+_{i,\alpha}, v^-_{i,\alpha}\omega^-_{i,\alpha}].$$

Moreover, it follows from $I^{\alpha,(II)}_{u_i \tilde{-}_{gH(I)} v_i, \omega_i} \leq 0$ that

$$\begin{aligned} I^{\alpha,(II)}_{u_i \tilde{-}_{gH(I)} v_i, \omega_i} &= \begin{vmatrix} u^+_{i,\alpha} - v^+_{i,\alpha} & u^-_{i,\alpha} - v^-_{i,\alpha} \\ \omega^-_{i,\alpha} & \omega^+_{i,\alpha} \end{vmatrix} \\ &= (u^+_{i,\alpha} - v^+_{i,\alpha})\omega^+_{i,\alpha} - (u^-_{i,\alpha} - v^-_{i,\alpha})\omega^-_{i,\alpha} \leq 0, \end{aligned}$$

which indicates that $[u \circledast \omega]^\alpha \boxminus_{gH(I)} [v \circledast \omega]^\alpha$ exists, so we obtain

$$\begin{aligned} [(u \tilde{-}_{gH(I)} v) \circledast \omega]^\alpha &= \times^n_{i=1}[(u^+_{i,\alpha} - v^+_{i,\alpha})\omega^+_{i,\alpha}, (u^-_{i,\alpha} - v^-_{i,\alpha})\omega^-_{i,\alpha}] \\ &= [u \circledast \omega]^\alpha \boxminus_{gH(I)} [v \circledast \omega]^\alpha. \end{aligned}$$

This completes the proof. □

Theorem 2.7 *Let* $u, v, \omega \in [\mathbb{R}^n_{\mathcal{F}}]$, *we have the following related properties with respect to the multiplications* $u * v$ *and* $u \circledast v$. *For* $(i) - (viii)$ *we assume that* $u\tilde{-}_{gH_{(II)}}v$ *exists.*

(i) *If* $I^{\alpha,(II)}_{u_i\tilde{-}_{gH_{(II)}}v_i,\omega_i} \geq 0$, $I^{\alpha,(I)}_{u_i,\omega_i}, I^{\alpha,(I)}_{v_i,\omega_i} \geq 0$, *then* $(u\tilde{-}_{gH_{(II)}}v) \circledast \omega = u * \omega\tilde{-}_{gH_{(I)}}v * \omega$.

(ii) *If* $I^{\alpha,(II)}_{u_i\tilde{-}_{gH_{(II)}}v_i,\omega_i} \geq 0$, $I^{\alpha,(I)}_{u_i,\omega_i}, I^{\alpha,(I)}_{v_i,\omega_i} \leq 0$, *then* $(u\tilde{-}_{gH_{(II)}}v) \circledast \omega = u * \omega\tilde{-}_{gH_{(I)}}v * \omega$.

(iii) *If* $I^{\alpha,(II)}_{u_i\tilde{-}_{gH_{(II)}}v_i,\omega_i} \leq 0$, $I^{\alpha,(I)}_{u_i,\omega_i}, I^{\alpha,(I)}_{v_i,\omega_i} \geq 0$, *then* $(u\tilde{-}_{gH_{(II)}}v) \circledast \omega = u * \omega\tilde{-}_{gH_{(II)}}v * \omega$.

(iv) *If* $I^{\alpha,(II)}_{u_i\tilde{-}_{gH_{(II)}}v_i,\omega_i} \leq 0$, $I^{\alpha,(I)}_{u_i,\omega_i}, I^{\alpha,(I)}_{v_i,\omega_i} \leq 0$, *then* $(u\tilde{-}_{gH_{(II)}}v) \circledast \omega = u * \omega\tilde{-}_{gH_{(I)}}v * \omega$.

(v) *If* $I^{\alpha,(I)}_{u_i\tilde{-}_{gH_{(II)}}v_i,\omega_i} \geq 0$, $I^{\alpha,(II)}_{u_i,\omega_i}, I^{\alpha,(II)}_{v_i,\omega_i} \geq 0$, *then* $(u\tilde{-}_{gH_{(II)}}v) * \omega = u \circledast \omega\tilde{-}_{gH_{(I)}}v \circledast \omega$.

(vi) *If* $I^{\alpha,(I)}_{u_i\tilde{-}_{gH_{(II)}}v_i,\omega_i} \geq 0$, $I^{\alpha,(II)}_{u_i,\omega_i}, I^{\alpha,(II)}_{v_i,\omega_i} \leq 0$, *then* $(u\tilde{-}_{gH_{(II)}}v) * \omega = u \circledast \omega\tilde{-}_{gH_{(II)}}v \circledast \omega$.

(vii) *If* $I^{\alpha,(I)}_{u_i\tilde{-}_{gH_{(II)}}v_i,\omega_i} \leq 0$, $I^{\alpha,(II)}_{u_i,\omega_i}, I^{\alpha,(II)}_{v_i,\omega_i} \geq 0$, *then* $(u\tilde{-}_{gH_{(II)}}v) * \omega = u \circledast \omega\tilde{-}_{gH_{(I)}}v \circledast \omega$.

$(viii)$ *If* $I^{\alpha,(I)}_{u_i\tilde{-}_{gH_{(II)}}v_i,\omega_i} \leq 0$, $I^{\alpha,(II)}_{u_i,\omega_i}, I^{\alpha,(II)}_{v_i,\omega_i} \leq 0$, *then* $(u\tilde{-}_{gH_{(II)}}v) * \omega = u \circledast \omega\tilde{-}_{gH_{(I)}}v \circledast \omega$.

Proof (i). Since $I^{\alpha,(I)}_{u_i,\omega_i}, I^{\alpha,(I)}_{v_i,\omega_i} \geq 0$, we have

$$[u * \omega]^\alpha = \times_{i=1}^n [u^+_{i,\alpha}\omega^-_{i,\alpha}, u^-_{i,\alpha}\omega^+_{i,\alpha}], \quad [v * \omega]^\alpha = \times_{i=1}^n [v^+_{i,\alpha}\omega^-_{i,\alpha}, v^-_{i,\alpha}\omega^+_{i,\alpha}].$$

Moreover, it follows from $I^{\alpha,(II)}_{u_i\tilde{-}_{gH_{(II)}}v_i,\omega_i} \geq 0$ that

$$\begin{aligned} I^{\alpha,(II)}_{u_i\tilde{-}_{gH_{(II)}}v_i,\omega_i} &= \begin{vmatrix} u^-_{i,\alpha} - v^-_{i,\alpha} & u^+_{i,\alpha} - v^+_{i,\alpha} \\ \omega^-_{i,\alpha} & \omega^+_{i,\alpha} \end{vmatrix} \\ &= (u^-_{i,\alpha} - v^-_{i,\alpha})\omega^+_{i,\alpha} - (u^+_{i,\alpha} - v^+_{i,\alpha})\omega^-_{i,\alpha} \geq 0, \end{aligned}$$

which indicates that $[u * \omega]^\alpha \boxminus_{gH_{(I)}} [v * \omega]^\alpha$ exists, so we obtain

$$\begin{aligned} [(u\tilde{-}_{gH_{(II)}}v) \circledast \omega]^\alpha &= \times_{i=1}^n [(u^+_{i,\alpha} - v^+_{i,\alpha})\omega^-_{i,\alpha}, (u^-_{i,\alpha} - v^-_{i,\alpha})\omega^+_{i,\alpha}] \\ &= [u * \omega]^\alpha \boxminus_{gH_{(I)}} [v * \omega]^\alpha. \end{aligned}$$

(ii). Since $I^{\alpha,(I)}_{u_i,\omega_i}, I^{\alpha,(I)}_{v_i,\omega_i} \leq 0$, we have

$$[u * \omega]^\alpha = \times_{i=1}^n [u^-_{i,\alpha}\omega^+_{i,\alpha}, u^+_{i,\alpha}\omega^-_{i,\alpha}], \quad [v * \omega]^\alpha = \times_{i=1}^n [v^-_{i,\alpha}\omega^+_{i,\alpha}, v^+_{i,\alpha}\omega^-_{i,\alpha}].$$

Moreover, it follows from $I^{\alpha,(II)}_{u_i\tilde{-}_{gH_{(II)}}v_i,\omega_i} \geq 0$ that

$$\begin{aligned} I^{\alpha,(II)}_{u_i \tilde{-}_{gH(II)} v_i, \omega_i} &= \begin{vmatrix} u^-_{i,\alpha} - v^-_{i,\alpha} & u^+_{i,\alpha} - v^+_{i,\alpha} \\ \omega^-_{i,\alpha} & \omega^+_{i,\alpha} \end{vmatrix} \\ &= (u^-_{i,\alpha} - v^-_{i,\alpha})\omega^+_{i,\alpha} - (u^+_{i,\alpha} - v^+_{i,\alpha})\omega^-_{i,\alpha} \geq 0, \end{aligned}$$

which indicates that $[u * \omega]^\alpha \boxminus_{gH_{(II)}} [v * \omega]^\alpha$ exists, so we obtain

$$\begin{aligned} [(u \tilde{-}_{gH_{(II)}} v) \circledast \omega]^\alpha &= \times^n_{i=1}[(u^+_{i,\alpha} - v^+_{i,\alpha})\omega^+_{i,\alpha}, (u^-_{i,\alpha} - v^-_{i,\alpha})\omega^-_{i,\alpha}] \\ &= [u * \omega]^\alpha \boxminus_{gH_{(II)}} [v * \omega]^\alpha. \end{aligned}$$

(iii). Since $I^{\alpha,(I)}_{u_i,\omega_i}, I^{\alpha,(I)}_{v_i,\omega_i} \geq 0$, we have

$$[u * \omega]^\alpha = \times^n_{i=1}[u^+_{i,\alpha}\omega^-_{i,\alpha}, u^-_{i,\alpha}\omega^+_{i,\alpha}], \quad [v * \omega]^\alpha = \times^n_{i=1}[v^+_{i,\alpha}\omega^-_{i,\alpha}, v^-_{i,\alpha}\omega^+_{i,\alpha}].$$

Moreover, it follows from $I^{\alpha,(II)}_{u_i \tilde{-}_{gH(II)} v_i, \omega_i} \leq 0$ that

$$\begin{aligned} I^{\alpha,(II)}_{u_i \tilde{-}_{gH(II)} v_i, \omega_i} &= \begin{vmatrix} u^-_{i,\alpha} - v^-_{i,\alpha} & u^+_{i,\alpha} - v^+_{i,\alpha} \\ \omega^-_{i,\alpha} & \omega^+_{i,\alpha} \end{vmatrix} \\ &= (u^-_{i,\alpha} - v^-_{i,\alpha})\omega^+_{i,\alpha} - (u^+_{i,\alpha} - v^+_{i,\alpha})\omega^-_{i,\alpha} \leq 0, \end{aligned}$$

which indicates that $[u * \omega]^\alpha \boxminus_{gH_{(II)}} [v * \omega]^\alpha$ exists, so we obtain

$$\begin{aligned} [(u \tilde{-}_{gH_{(II)}} v) \circledast \omega]^\alpha &= \times^n_{i=1}[(u^-_{i,\alpha} - v^-_{i,\alpha})\omega^+_{i,\alpha}, (u^+_{i,\alpha} - v^+_{i,\alpha})\omega^-_{i,\alpha}] \\ &= [u * \omega]^\alpha \boxminus_{gH_{(II)}} [v * \omega]^\alpha. \end{aligned}$$

(iv). Since $I^{\alpha,(I)}_{u_i,\omega_i}, I^{\alpha,(I)}_{v_i,\omega_i} \leq 0$, we have

$$[u * \omega]^\alpha = \times^n_{i=1}[u^-_{i,\alpha}\omega^+_{i,\alpha}, u^+_{i,\alpha}\omega^-_{i,\alpha}], \quad [v * \omega]^\alpha = \times^n_{i=1}[v^-_{i,\alpha}\omega^+_{i,\alpha}, v^+_{i,\alpha}\omega^-_{i,\alpha}].$$

Moreover, it follows from $I^{\alpha,(II)}_{u_i \tilde{-}_{gH(II)} v_i, \omega_i} \leq 0$ that

$$\begin{aligned} I^{\alpha,(II)}_{u_i \tilde{-}_{gH(II)} v_i, \omega_i} &= \begin{vmatrix} u^-_{i,\alpha} - v^-_{i,\alpha} & u^+_{i,\alpha} - v^+_{i,\alpha} \\ \omega^-_{i,\alpha} & \omega^+_{i,\alpha} \end{vmatrix} \\ &= (u^-_{i,\alpha} - v^-_{i,\alpha})\omega^+_{i,\alpha} - (u^+_{i,\alpha} - v^+_{i,\alpha})\omega^-_{i,\alpha} \leq 0, \end{aligned}$$

which indicates that $[u * \omega]^\alpha \boxminus_{gH_{(I)}} [v * \omega]^\alpha$ exists, so we obtain

$$\begin{aligned} [(u \tilde{-}_{gH_{(II)}} v) \circledast \omega]^\alpha &= \times^n_{i=1}[(u^-_{i,\alpha} - v^-_{i,\alpha})\omega^+_{i,\alpha}, (u^+_{i,\alpha} - v^+_{i,\alpha})\omega^-_{i,\alpha}] \\ &= [u * \omega]^\alpha \boxminus_{gH_{(I)}} [v * \omega]^\alpha. \end{aligned}$$

(v). Since $I^{\alpha,(II)}_{u_i,\omega_i}, I^{\alpha,(II)}_{v_i,\omega_i} \geq 0$, we have

$$[u \circledast \omega]^\alpha = \times_{i=1}^n [u_{i,\alpha}^- \omega_{i,\alpha}^-, u_{i,\alpha}^+ \omega_{i,\alpha}^+], \quad [v \circledast \omega]^\alpha = \times_{i=1}^n [v_{i,\alpha}^- \omega_{i,\alpha}^-, v_{i,\alpha}^+ \omega_{i,\alpha}^+].$$

Moreover, it follows from $I^{\alpha,(I)}_{u_i \tilde{-}_{gH(II)} v_i, \omega_i} \geq 0$ that

$$\begin{aligned} I^{\alpha,(I)}_{u_i \tilde{-}_{gH(II)} v_i, \omega_i} &= \begin{vmatrix} u_{i,\alpha}^+ - v_{i,\alpha}^+ & u_{i,\alpha}^- - v_{i,\alpha}^- \\ \omega_{i,\alpha}^- & \omega_{i,\alpha}^+ \end{vmatrix} \\ &= (u_{i,\alpha}^+ - v_{i,\alpha}^+)\omega_{i,\alpha}^+ - (u_{i,\alpha}^- - v_{i,\alpha}^-)\omega_{i,\alpha}^- \geq 0, \end{aligned}$$

which indicates that $[u \circledast \omega]^\alpha \boxminus_{gH_{(I)}} [v \circledast \omega]^\alpha$ exists, so we obtain

$$\begin{aligned} [(u \tilde{-}_{gH_{(II)}} v) * \omega]^\alpha &= \times_{i=1}^n [(u_{i,\alpha}^- - v_{i,\alpha}^-)\omega_{i,\alpha}^-, (u_{i,\alpha}^+ - v_{i,\alpha}^+)\omega_{i,\alpha}^+] \\ &= [u \circledast \omega]^\alpha \boxminus_{gH_{(I)}} [v \circledast \omega]^\alpha. \end{aligned}$$

(vi). Since $I^{\alpha,(II)}_{u_i, \omega_i}, I^{\alpha,(II)}_{v_i, \omega_i} \leq 0$, we have

$$[u \circledast \omega]^\alpha = \times_{i=1}^n [u_{i,\alpha}^+ \omega_{i,\alpha}^+, u_{i,\alpha}^- \omega_{i,\alpha}^-], \quad [v \circledast \omega]^\alpha = \times_{i=1}^n [v_{i,\alpha}^+ \omega_{i,\alpha}^+, v_{i,\alpha}^- \omega_{i,\alpha}^-].$$

Moreover, it follows from $I^{\alpha,(I)}_{u_i \tilde{-}_{gH(II)} v_i, \omega_i} \geq 0$ that

$$\begin{aligned} I^{\alpha,(I)}_{u_i \tilde{-}_{gH(II)} v_i, \omega_i} &= \begin{vmatrix} u_{i,\alpha}^+ - v_{i,\alpha}^+ & u_{i,\alpha}^- - v_{i,\alpha}^- \\ \omega_{i,\alpha}^- & \omega_{i,\alpha}^+ \end{vmatrix} \\ &= (u_{i,\alpha}^+ - v_{i,\alpha}^+)\omega_{i,\alpha}^+ - (u_{i,\alpha}^- - v_{i,\alpha}^-)\omega_{i,\alpha}^- \geq 0, \end{aligned}$$

which indicates that $[u \circledast \omega]^\alpha \boxminus_{gH_{(II)}} [v \circledast \omega]^\alpha$ exists, so we obtain

$$\begin{aligned} [(u \tilde{-}_{gH_{(II)}} v) * \omega]^\alpha &= \times_{i=1}^n [(u_{i,\alpha}^- - v_{i,\alpha}^-)\omega_{i,\alpha}^-, (u_{i,\alpha}^+ - v_{i,\alpha}^+)\omega_{i,\alpha}^+] \\ &= [u \circledast \omega]^\alpha \boxminus_{gH_{(II)}} [v \circledast \omega]^\alpha. \end{aligned}$$

(vii). Since $I^{\alpha,(II)}_{u_i, \omega_i}, I^{\alpha,(II)}_{v_i, \omega_i} \geq 0$, we have

$$[u \circledast \omega]^\alpha = \times_{i=1}^n [u_{i,\alpha}^- \omega_{i,\alpha}^-, u_{i,\alpha}^+ \omega_{i,\alpha}^+], \quad [v \circledast \omega]^\alpha = \times_{i=1}^n [v_{i,\alpha}^- \omega_{i,\alpha}^-, v_{i,\alpha}^+ \omega_{i,\alpha}^+].$$

Moreover, it follows from $I^{\alpha,(I)}_{u_i \tilde{-}_{gH(II)} v_i, \omega_i} \leq 0$ that

$$\begin{aligned} I^{\alpha,(I)}_{u_i \tilde{-}_{gH(II)} v_i, \omega_i} &= \begin{vmatrix} u_{i,\alpha}^+ - v_{i,\alpha}^+ & u_{i,\alpha}^- - v_{i,\alpha}^- \\ \omega_{i,\alpha}^- & \omega_{i,\alpha}^+ \end{vmatrix} \\ &= (u_{i,\alpha}^+ - v_{i,\alpha}^+)\omega_{i,\alpha}^+ - (u_{i,\alpha}^- - v_{i,\alpha}^-)\omega_{i,\alpha}^- \leq 0, \end{aligned}$$

which indicates that $[u \circledast \omega]^\alpha \boxminus_{gH_{(II)}} [v \circledast \omega]^\alpha$ exists, so we obtain

$$\begin{aligned}[(u\tilde{-}_{gH(II)}v)*\omega]^\alpha &= \times_{i=1}^n[(u_{i,\alpha}^+ - v_{i,\alpha}^+)\omega_{i,\alpha}^+, (u_{i,\alpha}^- - v_{i,\alpha}^-)\omega_{i,\alpha}^-] \\ &= [u \circledast \omega]^\alpha \boxminus_{gH(II)} [v \circledast \omega]^\alpha.\end{aligned}$$

$(viii)$. Since $I_{u_i,\omega_i}^{\alpha,(II)}, I_{v_i,\omega_i}^{\alpha,(II)} \leq 0$, we have

$$[u \circledast \omega]^\alpha = \times_{i=1}^n[u_{i,\alpha}^+\omega_{i,\alpha}^+, u_{i,\alpha}^-\omega_{i,\alpha}^-], \quad [v \circledast \omega]^\alpha = \times_{i=1}^n[v_{i,\alpha}^+\omega_{i,\alpha}^+, v_{i,\alpha}^-\omega_{i,\alpha}^-].$$

Moreover, it follows from $I_{u_i\tilde{-}_{gH(II)}v_i,\omega_i}^{\alpha,(I)} \leq 0$ that

$$\begin{aligned}I_{u_i\tilde{-}_{gH(II)}v_i,\omega_i}^{\alpha,(I)} &= \begin{vmatrix} u_{i,\alpha}^+ - v_{i,\alpha}^+ & u_{i,\alpha}^- - v_{i,\alpha}^- \\ \omega_{i,\alpha}^- & \omega_{i,\alpha}^+ \end{vmatrix} \\ &= (u_{i,\alpha}^+ - v_{i,\alpha}^+)\omega_{i,\alpha}^+ - (u_{i,\alpha}^- - v_{i,\alpha}^-)\omega_{i,\alpha}^- \leq 0,\end{aligned}$$

which indicates that $[u \circledast \omega]^\alpha \boxminus_{gH(I)} [v \circledast \omega]^\alpha$ exists, so we obtain

$$\begin{aligned}[(u\tilde{-}_{gH(II)}v)*\omega]^\alpha &= \times_{i=1}^n[(u_{i,\alpha}^+ - v_{i,\alpha}^+)\omega_{i,\alpha}^+, (u_{i,\alpha}^- - v_{i,\alpha}^-)\omega_{i,\alpha}^-] \\ &= [u \circledast \omega]^\alpha \boxminus_{gH(I)} [v \circledast \omega]^\alpha.\end{aligned}$$

This completes the proof. □

Theorem 2.8 *If $u, v \in [\mathbb{R}_{\mathscr{F}}^n]$, then $\|u * v\|_{\mathscr{F}} \leq \|u\|_{\mathscr{F}} \cdot \|v\|_{\mathscr{F}}$ and $\|u \circledast v\|_{\mathscr{F}} \leq \|u\|_{\mathscr{F}} \cdot \|v\|_{\mathscr{F}}$.*

Proof From Theorem 2.5, we have

$$\begin{aligned}&\|u * v\|_{\mathscr{F}} = \|j(u * v)\|_{\times_{i=1}^n(\bar{C}\times\bar{C})} \\ &= \|(u_i^- v_i^+, u_i^+ v_i^-)\|_{\times_{i=1}^n(\bar{C}\times\bar{C})} \\ &= \sup_{\alpha\in[0,1]} \max\left\{\left(\sum_{i=1}^n (u_i^- v_i^+)^2\right)^{\frac{1}{2}}, \left(\sum_{i=1}^n (u_i^+ v_i^-)^2\right)^{\frac{1}{2}}\right\} \\ &\leq \sup_{\alpha\in[0,1]} \max\left\{\sum_{i=1}^n |u_i^- v_i^+|, \sum_{i=1}^n |u_i^+ v_i^-|\right\} \\ &\leq \sup_{\alpha\in[0,1]} \max\left\{\left(\sum_{i=1}^n |u_i^-|^2\right)^{\frac{1}{2}} \cdot \left(\sum_{i=1}^n |v_i^+|^2\right)^{\frac{1}{2}}, \left(\sum_{i=1}^n |u_i^+|^2\right)^{\frac{1}{2}} \cdot \left(\sum_{i=1}^n |v_i^-|^2\right)^{\frac{1}{2}}\right\} \\ &\leq \sup_{\alpha\in[0,1]} \max\left\{\left(\sum_{i=1}^n |u_i^-|^2\right)^{\frac{1}{2}}, \left(\sum_{i=1}^n |u_i^+|^2\right)^{\frac{1}{2}}\right\} \\ &\quad \times \sup_{\alpha\in[0,1]} \max\left\{\left(\sum_{i=1}^n |v_i^-|^2\right)^{\frac{1}{2}}, \left(\sum_{i=1}^n |v_i^+|^2\right)^{\frac{1}{2}}\right\}\end{aligned}$$

$$= \| \times_{i=1}^{n} (u_i^-, u_i^+)\|_{\times_{i=1}^{n}(\bar{C}\times\bar{C})} \cdot \| \times_{i=1}^{n} (v_i^+, v_i^-)\|_{\times_{i=1}^{n}(\bar{C}\times\bar{C})}$$
$$= \|j(u)\|_{\times_{i=1}^{n}(\bar{C}\times\bar{C})} \cdot \|j(v)\|_{\times_{i=1}^{n}(\bar{C}\times\bar{C})}$$
$$= \|u\|_{\mathscr{F}} \cdot \|u\|_{\mathscr{F}}.$$

Similar to the above discussion, we also have

$$\|u \circledast v\|_{\mathscr{F}} = \|j(u \circledast v)\|_{\times_{i=1}^{n}(\bar{C}\times\bar{C})}$$
$$= \|(u_i^+ v_i^+, u_i^- v_i^-)\|_{\times_{i=1}^{n}(\bar{C}\times\bar{C})}$$
$$= \sup_{\alpha\in[0,1]} \max\left\{\left(\sum_{i=1}^{n}(u_i^+ v_i^+)^2\right)^{\frac{1}{2}}, \left(\sum_{i=1}^{n}(u_i^- v_i^-)^2\right)^{\frac{1}{2}}\right\}$$
$$\le \sup_{\alpha\in[0,1]} \max\left\{\sum_{i=1}^{n}|u_i^+ v_i^+|, \sum_{i=1}^{n}|u_i^- v_i^-|\right\}$$
$$\le \sup_{\alpha\in[0,1]} \max\left\{\left(\sum_{i=1}^{n}|u_i^+|^2\right)^{\frac{1}{2}} \cdot \left(\sum_{i=1}^{n}|v_i^+|^2\right)^{\frac{1}{2}}, \left(\sum_{i=1}^{n}|u_i^-|^2\right)^{\frac{1}{2}} \cdot \left(\sum_{i=1}^{n}|v_i^-|^2\right)^{\frac{1}{2}}\right\}$$
$$\le \sup_{\alpha\in[0,1]} \max\left\{\left(\sum_{i=1}^{n}|u_i^-|^2\right)^{\frac{1}{2}}, \left(\sum_{i=1}^{n}|u_i^+|^2\right)^{\frac{1}{2}}\right\}$$
$$\times \sup_{\alpha\in[0,1]} \max\left\{\left(\sum_{i=1}^{n}|v_i^-|^2\right)^{\frac{1}{2}}, \left(\sum_{i=1}^{n}|v_i^+|^2\right)^{\frac{1}{2}}\right\}$$
$$= \| \times_{i=1}^{n} (u_i^-, u_i^+)\|_{\times_{i=1}^{n}(\bar{C}\times\bar{C})} \cdot \| \times_{i=1}^{n} (v_i^+, v_i^-)\|_{\times_{i=1}^{n}(\bar{C}\times\bar{C})}$$
$$= \|j(u)\|_{\times_{i=1}^{n}(\bar{C}\times\bar{C})} \cdot \|j(v)\|_{\times_{i=1}^{n}(\bar{C}\times\bar{C})}$$
$$= \|u\|_{\mathscr{F}} \cdot \|v\|_{\mathscr{F}}.$$

This completes the proof. □

From Theorem 2.5 and the definition embedding j, we can prove the following properties easily.

Theorem 2.9 *For $u, v, w \in [\mathbb{R}^n_{\mathscr{F}}]$, if the gH-difference among them exist, then the following properties hold:*

(*i*) $D_\infty(u\tilde{\pm}_{gH}w, v\tilde{\pm}_{gH}w) = D_\infty(u, v)$;
(*ii*) $D_\infty(u\tilde{\pm}_{gH}w, v\tilde{\pm}_{gH}e) \le D_\infty(u, v) + D_\infty(w, e)$;
(*iii*) $D_\infty(\mu \cdot u, \mu \cdot v) = |\mu| D_\infty(u, v)$ *for* $\mu \in \mathbb{R}$;
(*iv*) $D_\infty(u * w, v * w) \le \|w\|_{\mathscr{F}} D_\infty(u, v)$ *if* $(u\tilde{-}_{gH}v) * \omega = u * \omega\tilde{-}_{gH}v * \omega$;
$D_\infty(u \circledast w, v \circledast w) \le \|w\|_{\mathscr{F}} D_\infty(u, v)$ *if* $(u\tilde{-}_{gH}v) \circledast \omega = u \circledast \omega\tilde{-}_{gH}v \circledast \omega$;

(*v*) $D_\infty(\mu \cdot u, \nu \cdot u) = |\mu - \nu|\|u\|_{\mathscr{F}}$ *for* $\mu, \nu \geq 0$ *or* $\mu, \nu \leq 0$.

Proof (i). For $u, v, w \in [\mathbb{R}^n_{\mathscr{F}}]$, we have

$$\begin{aligned}
D_\infty(u\tilde{\pm}_{gH}w, v\tilde{\pm}_{gH}w) &= \|j(u\tilde{\pm}_{gH}w) - j(v\tilde{\pm}_{gH}w)\|_{\times_{i=1}^n(\bar{C}\times\bar{C})} \\
&= \|j(u) \pm j(w) - j(v) \mp j(w)\|_{\times_{i=1}^n(\bar{C}\times\bar{C})} \\
&= \|j(u) - j(v)\|_{\times_{i=1}^n(\bar{C}\times\bar{C})} = D_\infty(u, v).
\end{aligned}$$

(ii). For $u, v, w, e \in [\mathbb{R}^n_{\mathscr{F}}]$, we obtain

$$\begin{aligned}
D_\infty(u\tilde{\pm}_{gH}w, v\tilde{\pm}_{gH}e) &= \|j(u\tilde{\pm}_{gH}w) - j(v\tilde{\pm}_{gH}e)\|_{\times_{i=1}^n(\bar{C}\times\bar{C})} \\
&= \|j(u) \pm j(w) - j(v) \mp j(e)\|_{\times_{i=1}^n(\bar{C}\times\bar{C})} \\
&= \|(j(u) - j(v)) \pm (j(w) - j(e))\|_{\times_{i=1}^n(\bar{C}\times\bar{C})} \\
&\leq \|j(u) - j(v)\| + \|j(w) - j(e)\|_{\times_{i=1}^n(\bar{C}\times\bar{C})} \\
&= D_\infty(u, v) + D_\infty(w, e).
\end{aligned}$$

(iii). For $u, v \in [\mathbb{R}^n_{\mathscr{F}}]$ and $\mu \in \mathbb{R}$, we have

$$\begin{aligned}
D_\infty(\mu \cdot u, \mu \cdot v) &= \big\|\mu\big(j(u) - j(v)\big)\big\|_{\times_{i=1}^n(\bar{C}\times\bar{C})} \\
&= |\mu|\|j_{\alpha,p}(u) - j_{\alpha,p}(v)\|_{\times_{i=1}^n(\bar{C}\times\bar{C})} = |\mu| D_\infty(u, v).
\end{aligned}$$

(iv). By Theorem 2.8, we have

$$\begin{aligned}
D_\infty(u * w, v * w) &= \|j(u * w) - j(v * w)\|_{\times_{i=1}^n(\bar{C}\times\bar{C})} \\
&= \big\|j\big((u\tilde{-}_{gH}v) * w\big)\big\|_{\times_{i=1}^n(\bar{C}\times\bar{C})} \\
&\leq \|u\tilde{-}_{gH}v\|_{\mathscr{F}}\|w\|_{\mathscr{F}} = \|w\|_{\mathscr{F}} D_\infty(u, v).
\end{aligned}$$

Similarly, we also have

$$\begin{aligned}
D_\infty(u \circledast w, v \circledast w) &= \|j(u \circledast w) - j(v \circledast w)\|_{\times_{i=1}^n(\bar{C}\times\bar{C})} \\
&= \big\|j\big((u\tilde{-}_{gH}v) \circledast w\big)\big\|_{\times_{i=1}^n(\bar{C}\times\bar{C})} \\
&\leq \|u\tilde{-}_{gH}v\|_{\mathscr{F}}\|w\|_{\mathscr{F}} = \|w\|_{\mathscr{F}} D_\infty(u, v).
\end{aligned}$$

(v).

$$\begin{aligned}
D_\infty(\mu \cdot u, \nu \cdot u) &= \|\mu j(u) - \nu j(u)\|_{\times_{i=1}^n(\bar{C}\times\bar{C})} \\
&= \|(\mu - \nu)(j(u) - j(\tilde{0}))\|_{\times_{i=1}^n(\bar{C}\times\bar{C})} = |\mu - \nu|\|u\|_{\mathscr{F}}.
\end{aligned}$$

This completes the proof. □

3 Calculus of Fuzzy Vector-Valued Functions on Time Scales

In this chapter, we will establish some basic results of calculus of fuzzy vector-valued functions on time scales.

For convenience, we introduce the following notations.

Let $f, g : \mathbb{T} \to [\mathbb{R}^n_{\mathcal{F}}]$, where $f = (f_1, f_2, \dots, f_n)$, $g = (g_1, g_2, \dots, g_n)$ with the box α-level sets $(0 \le \alpha < 1)$ as follows

$$\begin{aligned}[f(t)]^\alpha &= [f^-_{1,\alpha}(t), f^+_{1,\alpha}(t)] \times [f^-_{2,\alpha}(t), f^+_{2,\alpha}(t)] \times \cdots \times [f^-_{n,\alpha}(t), f^+_{n,\alpha}(t)] \\ &= \times^n_{i=1}[f^-_{i,\alpha}(t), f^+_{i,\alpha}(t)]\end{aligned}$$

and

$$\begin{aligned}[g(t)]^\alpha &= [g^-_{1,\alpha}(t), g^+_{1,\alpha}(t)] \times [g^-_{2,\alpha}(t), g^+_{2,\alpha}(t)] \times \cdots \times [g^-_{n,\alpha}(t), g^+_{n,\alpha}(t)] \\ &= \times^n_{i=1}[g^-_{i,\alpha}(t), g^+_{i,\alpha}(t)].\end{aligned}$$

Next, we will introduce some knowledge of time scales which will be used in later chapters.

A time scale $\mathbb{T}$ is a closed subset of $\mathbb{R}$. It follows that the jump operators $\sigma, \varrho : \mathbb{T} \to \mathbb{T}$ defined by $\sigma(t) = \inf\{s \in \mathbb{T} : s > t\}$ and $\rho(t) = \sup\{s \in \mathbb{T} : s < t\}$ (supplemented by $\inf \phi := \sup \mathbb{T}$ and $\sup \phi := \inf \mathbb{T}$) are well defined. The point $t \in \mathbb{T}$ is left-dense, left-scattered, right-dense, right-scattered if $\rho(t) = t$, $\rho(t) < t$, $\sigma(t) = t$, $\sigma(t) > t$, respectively. If $\mathbb{T}$ has a right scattered minimum m, define $\mathbb{T}_k := \mathbb{T} \setminus m$; if $\mathbb{T}$ has a left scattered maximum M, define $\mathbb{T}^k = \mathbb{T} \setminus M$. The notations $[a, b]_{\mathbb{T}}$, $[a, b)_{\mathbb{T}}$ and so on, we will denote time scale intervals

$$[a, b]_{\mathbb{T}} = \{t \in \mathbb{T} : a \le t \le b\},$$

C. Wang and R. P. Agarwal, *Dynamic Equations and Almost Periodic Fuzzy Functions on Time Scales*, Synthesis Lectures on Mathematics & Statistics,
https://doi.org/10.1007/978-3-031-11236-2_3

where $a, b \in \mathbb{T}$ with $a < \varrho(b)$. The graininess function is defined by $\mu : \mathbb{T} \to [0, \infty) : \mu(t) := \sigma(t) - t$, for all $t \in \mathbb{T}$. For more details, we refer the reader to [4, 20, 41].

3.1 gH-Δ-Derivative of Fuzzy Vector-Valued Functions on Time Scales

In this subsection, we will introduce the definition of gH-Δ-derivative of fuzzy vector-valued functions on time scales.

Definition 3.1 For $f : \mathbb{T} \to [\mathbb{R}^n_{\mathcal{F}}]$ and $t \in \mathbb{T}^\kappa$, we define the gH-Δ-derivative of $f(t)$, $f^\Delta(t) = (f_1^\Delta, f_2^\Delta, \ldots, f_n^\Delta)$, to be the fuzzy vector (if it exists) with the property that for a given $\varepsilon > 0$, there exists a neighborhood U of t (i.e., $U = (t - \delta, t + \delta)_{\mathbb{T}}$ for some $\delta > 0$) such that

$$D_\infty^{(i)}\big(f_i(\sigma(t)) \tilde{-}_{gH} f_i(s), f_i^\Delta(t)(\sigma(t) - s)\big) < \varepsilon|\sigma(t) - s|, \ i = 1, 2, \ldots, n$$

for all $s \in U$. That is, the limit

$$f_i^\Delta(t) = \lim_{s \to t} \frac{f_i\big(\sigma(t)\big) \tilde{-}_{gH} f_i(s)}{\sigma(t) - s}$$

exists for each $i = 1, 2, \ldots, n$.

The following definition is obviously equivalent to Definition 3.1.

Definition 3.2 For $f : \mathbb{T} \to \mathbb{R}^n_{\mathcal{F}}$ and $t \in \mathbb{T}^\kappa$, we define the gH-Δ-derivative of $f(t)$, $f^\Delta(t) = (f_1^\Delta, f_2^\Delta, \ldots, f_n^\Delta)$, to be the fuzzy vector (if it exists) with the property that for a given $\varepsilon > 0$, there exists a $\delta > 0$ such that $|h| < \delta$ implies

$$D_\infty^{(i)}\big(f_i(\sigma(t)) \tilde{-}_{gH} f_i(t + h), f_i^\Delta(t)(\mu(t) - h)\big) \leq \varepsilon|\mu(t) - h|,$$

i.e.

$$\lim_{h \to 0} \frac{f_i\big(\sigma(t)\big) \tilde{-}_{gH} f_i(t + h)}{\mu(t) - h} = f_i^\Delta(t)$$

exists for each $i = 1, 2, \ldots, n$.

A sufficient and necessary condition for gH-Δ-differentiability of functions is given by the following theorem.

Theorem 3.1 *Let $f : \mathbb{T} \to \mathbb{R}^n_{\mathscr{F}}$ be a function and $[f(t)]^\alpha = \times_{i=1}^n [f_{i,\alpha}^-(t), f_{i,\alpha}^+(t)]$, $\alpha \in [0,1]$. The function $f(t)$ is gH-Δ-differentiable if $f_{i,\alpha}^-(t)$ and $f_{i,\alpha}^+(t)$ are Δ-differentiable real-valued functions for each $i = 1, 2, \ldots, n$. Furthermore*

$$[f^\Delta(t)]^\alpha = \times_{i=1}^n \big[\min\{(f_{i,\alpha}^-)^\Delta(t), (f_{i,\alpha}^+)^\Delta(t)\}, \max\{(f_{i,\alpha}^-)^\Delta(t), (f_{i,\alpha}^+)^\Delta(t)\}\big].$$

Proof In fact, for sufficiently small $|h| > 0$, we have

$$\begin{aligned}
&\left[\frac{f(\sigma(t))\tilde{-}_{gH} f(t+h)}{\mu(t)-h}\right]^\alpha \\
&= \times_{i=1}^n \left[\frac{f_i(\sigma(t))\tilde{-}_{gH} f_i(t+h)}{\mu(t)-h}\right]^\alpha \\
&= \times_{i=1}^n \left[\frac{1}{\mu(t)-h}\cdot [f_i(\sigma(t))\tilde{-}_{gH} f_i(t+h)]^\alpha\right] \\
&= \times_{i=1}^n \left[\frac{1}{\mu(t)-h}\cdot \big([f_i(\sigma(t))]^\alpha \boxminus_{gH} [f_i(t+h)]^\alpha\big)\right] \\
&= \times_{i=1}^n \left[\frac{1}{\mu(t)-h}\cdot \big([f_{i,\alpha}^-(\sigma(t)), f_{i,\alpha}^+(\sigma(t))] \boxminus_{gH} [f_{i,\alpha}^-(t+h), f_{i,\alpha}^+(t+h)]\big)\right] \\
&= \times_{i=1}^n \left[\frac{1}{\mu(t)-h}\cdot \big[\min\{f_{i,\alpha}^-(\sigma(t)) - f_{i,\alpha}^-(t+h), f_{i,\alpha}^+(\sigma(t)) - f_{i,\alpha}^+(t+h)\},\right. \\
&\quad \left.\max\{f_{i,\alpha}^-(\sigma(t)) - f_{i,\alpha}^-(t+h), f_{i,\alpha}^+(\sigma(t)) - f_{i,\alpha}^+(t+h)\}\big]\right] \\
&= \times_{i=1}^n \left[\min\left\{\frac{f_{i,\alpha}^-(\sigma(t)) - f_{i,\alpha}^-(t+h)}{\mu(t)-h}, \frac{f_{i,\alpha}^+(\sigma(t)) - f_{i,\alpha}^+(t+h)}{\mu(t)-h}\right\},\right. \\
&\quad \left.\max\left\{\frac{f_{i,\alpha}^-(\sigma(t)) - f_{i,\alpha}^-(t+h)}{\mu(t)-h}, \frac{f_{i,\alpha}^+(\sigma(t)) - f_{i,\alpha}^+(t+h)}{\mu(t)-h}\right\}\right].
\end{aligned}$$

Let $h \to 0$, by Definition 3.2 we have

$$[f^\Delta(t)]^\alpha = \times_{i=1}^n \big[\min\{(f_{i,\alpha}^-)^\Delta, (f_{i,\alpha}^+)^\Delta\}, \max\{(f_{i,\alpha}^-)^\Delta, (f_{i,\alpha}^+)^\Delta\}\big].$$

This completes the proof. □

Remark 3.1 Note that the condition of Theorem 3.1 is sufficient but not necessary, that is, if $f(t)$ is gH-Δ-differentiable, then $f_{i,\alpha}^-(t)$ and $f_{i,\alpha}^+(t)$ may not be Δ-differentiable real-valued functions for each $i = 1, 2, \ldots, n$ and $\alpha \in [0, 1]$. For example, consider the function $f : [-1, 1] \to \mathbb{R}_{\mathscr{F}}$, where $[f(t)]^\alpha = [f_\alpha^-(t), f_\alpha^+(t)] = [-|t|, |t|]$ for all $\alpha \in [0, 1]$. Given $t = 0$, then $t = 0$ is the right and left dense point. It is obvious that f_α^- and f_α^+ are not Δ-differentiable at $t = 0$. However, f is gH-Δ-differentiable at $t = 0$. In fact, since

$$\begin{aligned}\left[\lim_{s\to 0^+}\frac{f(0)\tilde{-}_{gH}f(s)}{-s}\right]^\alpha &= \lim_{s\to 0^+}\frac{-1}{s}\big([f(0)]^\alpha \boxminus_{gH} [f(s)]^\alpha\big)\\ &= \lim_{s\to 0^+}\frac{-1}{s}\big([0,0]\boxminus_{gH}[-|s|,|s|]\big)\\ &= \lim_{s\to 0^+}\left[\frac{-|s|}{s},\frac{|s|}{s}\right]\\ &= \lim_{s\to 0^+}[-1,1]=[-1,1] \text{ for all } \alpha\in[0,1],\end{aligned}$$

$$\begin{aligned}\left[\lim_{s\to 0^-}\frac{f(0)\tilde{-}_{gH}f(s)}{-s}\right]^\alpha &= \lim_{s\to 0^-}\frac{-1}{s}\big([f(0)]^\alpha \boxminus_{gH} [f(s)]^\alpha\big)\\ &= \lim_{s\to 0^-}\frac{-1}{s}\big([0,0]\boxminus_{gH}[-|s|,|s|]\big)\\ &= \lim_{s\to 0^-}\left[\frac{|s|}{s},\frac{-|s|}{s}\right]\\ &= \lim_{s\to 0^-}[-1,1]=[-1,1] \text{ for all } \alpha\in[0,1],\end{aligned}$$

it follows that f is gH-Δ-differentiable at $t=0$ and its α-levels are given by $[f^\Delta(0)]^\alpha = [-1,1]$ for all $\alpha\in[0,1]$.

Now, we introduce the definition of continuous fuzzy vector-valued functions on time scales.

Definition 3.3 A function $f:\mathbb{T}\to\mathbb{R}^n_{\mathcal{F}}$ is said a fuzzy continuous function, if for arbitrary fixed $t_0\in\mathbb{T}$ and $\varepsilon>0$, there exists a $\delta>0$ such that $|t-t_0|<\delta$ implies $D_\infty\big(f(t),f(t_0)\big)<\varepsilon$.

Throughout the paper, the set of all fuzzy continuous vector-valued functions is denoted by $C(\mathbb{T},\mathbb{R}^n_{\mathcal{F}})$.

From Definition 3.3, the following lemma is clear.

Theorem 3.2 *Let the fuzzy continuous function $f:[a,b]_{\mathbb{T}}\to\mathbb{R}^n_{\mathcal{F}}$. Then f is bounded, i.e., there exists a positive number M such that $\|f(t)\|_{\mathcal{F}}\le M$ for all $t\in[a,b]_{\mathbb{T}}$.*

Theorem 3.3 *Assume $f:\mathbb{T}\to\mathbb{R}^n_{\mathcal{F}}$ is a function and let $t\in\mathbb{T}^k$ then we have the following:*

(i) If f is continuous at t and t is right-scattered, then f is gH-Δ-differentiable at t with

$$f^\Delta(t)=\frac{f(\sigma(t))\tilde{-}_{gH}f(t)}{\mu(t)}.$$

(ii) If t is right-dense, then f is gH-Δ-differentiable at t if and only if the limits

$$\lim_{h\to 0}\frac{f(t+h)\tilde{-}_{gH}f(t)}{h} \tag{3.1}$$

exist and satisfy the following

$$\lim_{h\to 0}\frac{f(t+h)\tilde{-}_{gH}f(t)}{h}=f^{\Delta}(t). \tag{3.2}$$

Proof (i). If f is continuous at t and t is right-scattered, then we obtain

$$\lim_{h\to 0}\frac{f_i\big(\sigma(t)\big)\tilde{-}_{gH}f_i(t+h)}{\mu(t)-h}=\frac{f_i\big(\sigma(t)\big)\tilde{-}_{gH}f_i(t)}{\mu(t)},$$

then for any $\varepsilon>0$, there exists $\delta>0$ such that $|h|<\delta$ implies

$$D_{\infty}^{(i)}\left(\frac{f_i\big(\sigma(t)\big)\tilde{-}_{gH}f_i(t+h)}{\mu(t)-h},\frac{f_i\big(\sigma(t)\big)\tilde{-}_{gH}f_i(t)}{\mu(t)}\right)<\varepsilon,$$

which yields that

$$\frac{1}{|\mu(t)-h|}D_{\infty}^{(i)}\left(f_i\big(\sigma(t)\big)\tilde{-}_{gH}f_i(t+h),\frac{f_i\big(\sigma(t)\big)\tilde{-}_{gH}f_i(t)}{\mu(t)}(\mu(t)-h)\right)<\varepsilon.$$

Hence we obtain that

$$f_i^{\Delta}(t)=\frac{f_i\big(\sigma(t)\big)\tilde{-}_{gH}f_i(t)}{\mu(t)},\quad i=1,2,\ldots,n.$$

(ii). If t is right-dense, then $\sigma(t)=t$, i.e., $\mu(t)=0$. If f is gH-Δ-differentiable at t, then for a given $\varepsilon>0$, there exists a $\delta>0$ such that $|h|<\delta$ implies

$$D_{\infty}^{(i)}\big(f_i(t)\tilde{-}_{gH}f_i(t+h),f_i^{\Delta}(t)\cdot(-h)\big)\le\varepsilon|h|,$$

i.e.

$$\lim_{h\to 0}\frac{f_i(t)\tilde{-}_{gH}f_i(t+h)}{-h}=\lim_{h\to 0}\frac{f_i(t+h)\tilde{-}_{gH}f_i(t)}{h}=f_i^{\Delta}(t).$$

exists for $i=1,2,\ldots,n$.

Conversely, if (3.1) and (3.2) hold, then for a given $\varepsilon>0$, there exists a $\delta>0$ such that $|h|<\delta$ implies

$$\begin{aligned}&D_{\infty}^{(i)}\big(f_i(\sigma(t))\tilde{-}_{gH}f_i(t+h),f_i^{\Delta}(t)(\mu(t)-h)\big)\\&=D_{\infty}^{(i)}\left(f_i(t)\tilde{-}_{gH}f_i(t+h),\frac{f_i(t+h)\tilde{-}_{gH}f_i(t)}{h}(-h)\right)\\&=D_{\infty}^{(i)}\big(f_i(t)\tilde{-}_{gH}f_i(t+h),f_i(t)\tilde{-}_{gH}f_i(t+h)\big)=0.\end{aligned}$$

Hence, we can easily obtain the desired results. This completes the proof. □

Theorem 3.4 *Assume $f : \mathbb{T} \to \mathbb{R}^n_{\mathscr{F}}$ is gH-Δ-differentiable at $t \in \mathbb{T}^k$, then f is continuous at t.*

Proof Assume that f is gH-Δ-differentiable at t, let $\varepsilon \in (0, 1)$ and define $\varepsilon^* = \varepsilon\big[\| f_i^{\Delta}(t)\|_{\mathscr{F}_0} + 1 + 2\mu(t)\big]^{-1}$. Obviously, $\varepsilon^* \in (0, 1)$. By Definition 3.1, for each $i = 1, 2, \ldots, n$, we have

$$D_\infty^{(i)}\big(f_i(\sigma(t))\tilde{-}_{gH} f_i(t+h),\, f_i^{\Delta}(t)(\mu(t)-h)\big) \le \varepsilon^*|\mu(t)-h|.$$

Therefore, for $t + h \in U \cap (t - \varepsilon^*, t + \varepsilon^*)$ with $h \in [0, \varepsilon^*)$, since

$$\begin{aligned} D_\infty^{(i)}\big((\mu(t)-h) f_i^{\Delta}(t), \tilde{0}\big) &= |\mu(t)-h| D_\infty^{(i)}\big(f_i^{\Delta}(t), \tilde{0}\big) \\ &= |\mu(t)-h| \| f_i^{\Delta}(t)\|_{\mathscr{F}_0} \\ &= D_\infty^{(i)}\big(\mu(t) f_i^{\Delta}(t), h f_i^{\Delta}(t)\big), \end{aligned}$$

we obtain

$$\begin{aligned} & D_\infty^{(i)}\big(f_i(t+h), f_i(t)\big) \\ &= D_\infty^{(i)}\big(f_i(\sigma(t))\tilde{-}_{gH} f_i(t+h), f_i(\sigma(t))\tilde{-}_{gH} f_i(t)\big) \\ &\le D_\infty^{(i)}\big(f_i(\sigma(t))\tilde{-}_{gH} f_i(t+h), \tilde{0}\big) + D_\infty^{(i)}\big(f_i(\sigma(t))\tilde{-}_{gH} f_i(t), \tilde{0}\big) \\ &\le D_\infty^{(i)}\big(f_i(\sigma(t))\tilde{-}_{gH} f_i(t+h), f_i^{\Delta}(t)(\mu(t)-h)\big) + D_\infty^{(i)}\big(f_i^{\Delta}(t)(\mu(t)-h), \tilde{0}\big) \\ &\quad + D_\infty^{(i)}\big(f_i(\sigma(t))\tilde{-}_{gH} f_i(t), \tilde{0}\big) \\ &= D_\infty^{(i)}\big(f_i(\sigma(t))\tilde{-}_{gH} f_i(t+h), f_i^{\Delta}(t)(\mu(t)-h)\big) + D_\infty^{(i)}\big(\mu(t) f_i^{\Delta}(t), h f_i^{\Delta}(t)\big) \\ &\quad + D_\infty^{(i)}\big(f_i(\sigma(t))\tilde{-}_{gH} f_i(t), \tilde{0}\big) \\ &\le D_\infty^{(i)}\big(f_i(\sigma(t))\tilde{-}_{gH} f_i(t+h), f^{\Delta}(t)(\mu(t)-h)\big) \\ &\quad + D_\infty^{(i)}\big(f_i(\sigma(t))\tilde{-}_{gH} f_i(t), f_i^{\Delta}(t)\mu(t)\big) + h D_\infty^{(i)}\big(f_i^{\Delta}(t), \tilde{0}\big) \\ &\le \varepsilon^*|\mu(t)-h| + \varepsilon^*\mu(t) + h\| f_i^{\Delta}(t)\|_{\mathscr{F}_0} \\ &= \varepsilon^*\big[1 + \| f_i^{\Delta}(t)\|_{\mathscr{F}_0} + 2\mu(t)\big] = \varepsilon. \end{aligned} \tag{3.3}$$

Similarly, for $h \in (-\varepsilon^*, 0)$, since

$$\begin{aligned} D_\infty^{(i)}\big((\mu(t)-h) f_i^{\Delta}(t), \tilde{0}\big) &= |\mu(t)-h| D_\infty^{(i)}\big(f_i^{\Delta}(t), \tilde{0}\big) \\ &= |\mu(t)-h| \| f_i^{\Delta}(t)\|_{\mathscr{F}_0} \\ &= D_\infty^{(i)}\big(-\mu(t) f_i^{\Delta}(t), h f_i^{\Delta}(t)\big) \\ &= D_\infty^{(i)}\big(\mu(t) f_i^{\Delta}(t), -h f_i^{\Delta}(t)\big). \end{aligned}$$

Similar to the proof of (3.3), we can also obtain $D_\infty^{(i)}\big(f_i(t+h), f_i(t)\big) < \varepsilon$. This completes the proof. □

By Theorem 3.1, for the definition of gH-Δ-differentiability, we distinguished two cases, corresponding to (I) and (II) of (2.6).

Definition 3.4 Let $f : \mathbb{T} \to \mathbb{R}^n_{\mathcal{F}}$ be a function and $[f(t)]^\alpha = \times_{i=1}^n [f_{i,\alpha}^-(t), f_{i,\alpha}^+(t)]$, $\alpha \in [0, 1]$. Let $f_{i,\alpha}^-(t)$ and $f_{i,\alpha}^+(t)$ be Δ-differentiable real-valued functions at $t_0 \in (a, b)_\mathbb{T}$ for each $i = 1, 2, \ldots, n$ and $\alpha \in [0, 1]$. We say that f is (I)-gH-Δ-differentiable at $t_0 \in (a, b)_\mathbb{T}$ if $f^{\Delta_I}(t) = \big(f_1^{\Delta_I}(t), f_2^{\Delta_I}(t), \ldots, f_n^{\Delta_I}(t)\big)$ with α-level set

$$[f^{\Delta_I}(t)]^\alpha = \times_{i=1}^n [(f_{i,\alpha}^-)^\Delta(t), (f_{i,\alpha}^+)^\Delta(t)], \tag{3.4}$$

and f is (II)-gH-Δ-differentiable at $t_0 \in (a, b)_\mathbb{T}$ if

$$f^{\Delta_{II}}(t) = \big(f_1^{\Delta_{II}}(t), f_2^{\Delta_{II}}(t), \ldots, f_n^{\Delta_{II}}(t)\big)$$

with α-level set

$$[f^{\Delta_{II}}(t)]^\alpha = \times_{i=1}^n [(f_{i,\alpha}^+)^\Delta(t), (f_{i,\alpha}^-)^\Delta(t)]. \tag{3.5}$$

Similar to the literature [74], we will introduce and study the switch between the two cases (I) and (II) in Definition 3.4.

Definition 3.5 We say a point $t_0 \in (a, b)_\mathbb{T}$ is a switching point for the gH-Δ-differentiability of f, if in any neighborhood U of t_0 there exists points $t_1 < t_0 < t_2$ such that

(i) (type-I switch) at t_1 (3.4) holds while (3.5) does not hold and at t_2 (3.5) holds while (3.4) does not hold, or

(ii) (type-II switch) at t_1 (3.5) holds while (3.4) does not hold and at t_2 (3.4) holds while (3.5) does not hold.

Theorem 3.5 *Let $f : \mathbb{T} \to \mathbb{R}^n_{\mathcal{F}}$ be a function with $[f(t)]^\alpha = \times_{i=1}^n [f_{i,\alpha}^-(t), f_{i,\alpha}^+(t)]$, $\alpha \in [0, 1]$, and $f_{i,\alpha}^-(t)$ and $f_{i,\alpha}^+(t)$ are Δ-differentiable real-valued functions for each $i = 1, 2, \ldots, n$ and $\alpha \in [0, 1]$. Let the length function be $len\big(f(t)\big) = \sum_{i=1}^n f_i^+(t) - f_i^-(t)$, if a point $t_0 \in (a, b)_\mathbb{T}$ is a switching point, then in any neighborhood U of t_0 (i.e. $U = (t_0 - \delta, t_0 + \delta)_\mathbb{T}$) there exists points $t_1, t_2 \in U$ with $t_1 < t_0 < t_2$ such that $len\big(f^\Delta(t_1)\big)len\big(f^\Delta(t_2)\big) < 0$.*

Proof Without loss of generality, we assume that t_0 belongs to the switching point of type I in Definition 3.5 (the proof of the other case type II is similar).

(i). If t_0 is a switching point and right scattered, the α-level set of $f(t)$ for $t = t_1$ is that

$$[f^{\Delta}(t_1)]^{\alpha} = \times_{i=1}^{n}[(f_i^-)^{\Delta}(t_1), (f_i^+)^{\Delta}(t_1)]_{\alpha},$$

which implies that $(f_i^+)^{\Delta}(t_1) - (f_i^-)^{\Delta}(t_1) = \big(f_i^+ - f_i^-\big)^{\Delta}(t_1) > 0$ for $i = 1, 2, \ldots, n$. Also the α-level set of $f(t)$ for $t = t_2$ is that

$$[f^{\Delta}(t_2)]^{\alpha} = \times_{i=1}^{n}[(f_i^+)^{\Delta}(t_2), (f_i^-)^{\Delta}(t_2)]_{\alpha},$$

which implies that $(f_i^+)^{\Delta}(t_2) - (f_i^-)^{\Delta}(t_2) = \big(f_i^+ - f_i^-\big)^{\Delta}(t_2) < 0$ for $i = 1, 2, \ldots, n$. Hence we have

$$\begin{aligned} &len\big(f^{\Delta}(t_1)\big)len\big(f^{\Delta}(t_2)\big) \\ &= \Bigg(\sum_{i=1}^{n}(f_i^+)^{\Delta}(t_1) - (f_i^-)^{\Delta}(t_1)\Bigg)\Bigg(\sum_{i=1}^{n}(f_i^+)^{\Delta}(t_2) - (f_i^-)^{\Delta}(t_2)\Bigg) < 0. \end{aligned}$$

(ii). By Definition 3.5, similar to the proof of (i), we can also easily obtain the desired result. This completes the proof. □

Theorem 3.6 *Let $f : \mathbb{T} \to \mathbb{R}^n_{\mathscr{F}}$ be a function with $[f(t)]^{\alpha} = \times_{i=1}^{n}[f_i^-(t), f_i^+(t)]_{\alpha}$, $\alpha \in [0, 1]$, and $f_i^-(t)$ and $f_i^+(t)$ are Δ-differentiable real-valued functions for each $i = 1, 2, \ldots, n$. Let the length function be $len\big(f(t)\big) = \sum_{i=1}^{n} f_i^+(t) - f_i^-(t)$, if a point $t_0 \in (a, b)_{\mathbb{T}}$ is a switching point, then there exist a point $c \in (t_0 - \delta, t_0 + \delta)_{\mathbb{T}}$ such that*

$$\begin{aligned} &len\big(f^{\Delta}(c)\big) = 0 \text{ if } c \text{ is right dense} \\ &\text{or } len\big(f^{\Delta}(c)\big)len\big(f^{\Delta}(\sigma(c))\big) < 0 \text{ if } c \text{ is right scattered.} \end{aligned}$$

Proof According to Theorem 3.5, there exists points $t_1, t_2 \in U$ such that

$$len\big(f^{\Delta}(t_1)\big)len\big(f^{\Delta}(t_2)\big) < 0.$$

Hence, by the Intermediate Value Theorem (Theorem 1.115 from [20]), we can obtain the desired result immediately. This completes the proof. □

Theorem 3.7 *Let $f : (a, b)_{\mathbb{T}} \to \mathbb{R}^n_{\mathscr{F}}$ be a fuzzy vector-valued function with the α-level set $[f(t)]^{\alpha} = \times_{i=1}^{n}[f_i^-(t), f_i^+(t)]_{\alpha}$ and $f_i^-(t)$ and $f_i^+(t)$ are Δ-differentiable real-valued functions for each $i = 1, 2, \ldots, n$. Then the set of switching point is finite.*

Proof We argue by contradiction. Suppose that the set of switching point is infinite. Then this set has a cluster point $t_0 \in [a, b]_{\mathbb{T}}$. Let $t_n \to t_0$ be a convergent sequence of switching points. For $\delta > 0$, according to the definition of a switching point there exist $t_{n_1}, t_{n_2} \in (t_0 - \delta, t_0 + \delta)_{\mathbb{T}}$ such that t_{n_1} and t_{n_2} are in opposite differentiability case (I)-gH-Δ-differentiability or (II)-gH-Δ-differentiability exclusively. Then the gH-differences $f(t_0 + h)\tilde{-}_{gH} f(t_0)$ cannot exist for any h with $0 \leq h < \delta$, and this is a contradiction because $f_i^-(t)$ and $f_i^+(t)$ are Δ-differentiable real-valued functions for each $i = 1, 2, \ldots, n$. This completes the proof. □

Remark 3.2 Note that f is gH-Δ differentiable at t_0, the set of switching point may be infinite since that gH-Δ differentiable at t_0 does not imply $f_i^-(t)$ and $f_i^+(t)$ are Δ-differentiable real-valued functions for each $i = 1, 2, \ldots, n$ (see Remark 3.1). For example, given $f : [-1, 1] \to \mathbb{R}_{\mathscr{F}}$ be the function whose α-levels are given by $[f(t)]^\alpha = g(t) \cdot [-1, 1]$ for all $\alpha \in [0, 1]$, where

$$g(t) = \begin{cases} 1 + t^2 \sin \frac{1}{t} & \text{if } t \neq 0, \\ 1 & \text{if } t = 0. \end{cases}$$

In fact, we have

$$\begin{aligned}
&\left[\lim_{s \to 0^+} \frac{f(0)\tilde{-}_{gH} f(s)}{-s}\right]^\alpha \\
&= \lim_{s \to 0^+} \frac{-1}{s}\left([f(0)]^\alpha \boxminus_{gH} [f(s)]^\alpha\right) \\
&= \lim_{s \to 0^+} \frac{-1}{s}\left([-1, 1] \boxminus_{gH} \left[-1 - s^2 \sin\frac{1}{s}, 1 + s^2 \sin\frac{1}{s}\right]\right) \\
&= \lim_{s \to 0^+} \frac{-1}{s}\left[\min\left\{s^2 \sin\frac{1}{s}, -s^2 \sin\frac{1}{s}\right\}, \max\left\{s^2 \sin\frac{1}{s}, -s^2 \sin\frac{1}{s}\right\}\right] \\
&= \lim_{s \to 0^+} (-1)\left[\min\left\{s \sin\frac{1}{s}, -s \sin\frac{1}{s}\right\}, \max\left\{s \sin\frac{1}{s}, -s \sin\frac{1}{s}\right\}\right] \\
&= \lim_{s \to 0^+} \left[(-1)\max\left\{s \sin\frac{1}{s}, -s \sin\frac{1}{s}\right\}, (-1)\min\left\{s \sin\frac{1}{s}, -s \sin\frac{1}{s}\right\}\right] \\
&= \lim_{s \to 0^+} \left[\min\left\{-s \sin\frac{1}{s}, s \sin\frac{1}{s}\right\}, \max\left\{-s \sin\frac{1}{s}, s \sin\frac{1}{s}\right\}\right] = [0, 0]
\end{aligned}$$

and

$$\begin{aligned}
&\left[\lim_{s\to 0^-}\frac{f(0)\widetilde{-}_{gH}f(s)}{-s}\right]^{\alpha}\\
&=\lim_{s\to 0^-}\frac{-1}{s}([f(0)]^{\alpha}\boxminus_{gH}[f(s)]^{\alpha})\\
&=\lim_{s\to 0^-}\left([-1,1]\boxminus_{gH}\left[-1-s^2\sin\frac{1}{s},1+s^2\sin\frac{1}{s}\right]\right)\\
&=\lim_{s\to 0^-}\frac{-1}{s}\left[\min\left\{s^2\sin\frac{1}{s},-s^2\sin\frac{1}{s}\right\},\max\left\{s^2\sin\frac{1}{s},-s^2\sin\frac{1}{s}\right\}\right]\\
&=\lim_{s\to 0^-}\left[\min\left\{-s\sin\frac{1}{s},s\sin\frac{1}{s}\right\},\max\left\{-s\sin\frac{1}{s},s\sin\frac{1}{s}\right\}\right]=[0,0],
\end{aligned}$$

so f is gH-Δ-differentiable at $t=0$ and $[f^{\Delta}(0)]^{\alpha}=[0,0]$ for all $\alpha\in[0,1]$.

Further, for any $t\in(-1,1)\backslash\{0\}$, it follows that

$$\begin{aligned}
&\left[\lim_{s\to t}\frac{f(t)\widetilde{-}_{gH}f(s)}{t-s}\right]^{\alpha}\\
&=\lim_{s\to t}\frac{1}{t-s}([f(t)]^{\alpha}\boxminus_{gH}[f(s)]^{\alpha})\\
&=\lim_{s\to t}\frac{1}{t-s}\left(\left[-1-t^2\sin\frac{1}{t},1+t^2\sin\frac{1}{t}\right]\boxminus_{gH}\left[-1-s^2\sin\frac{1}{s},1+s^2\sin\frac{1}{s}\right]\right)\\
&=\lim_{s\to t}\frac{1}{t-s}\left[\min\left\{-t^2\sin\frac{1}{t}+s^2\sin\frac{1}{s},t^2\sin\frac{1}{t}-s^2\sin\frac{1}{s}\right\},\right.\\
&\qquad\left.\max\left\{-t^2\sin\frac{1}{t}+s^2\sin\frac{1}{s},t^2\sin\frac{1}{t}-s^2\sin\frac{1}{s}\right\}\right]\\
&=\left[\min\left\{\cos\frac{1}{t}-2t\sin\frac{1}{t},-\cos\frac{1}{t}+2t\sin\frac{1}{t}\right\},\right.\\
&\qquad\left.\max\left\{\cos\frac{1}{t}-2t\sin\frac{1}{t},-\cos\frac{1}{t}+2t\sin\frac{1}{t}\right\}\right].
\end{aligned}$$

Therefore, f is gH-Δ-differentiable for $t\in(-1,1)\backslash\{0\}$. However, there are infinite numbers of switching points. In fact, note that $f_{\alpha}^{-}(t)=-g(t)$ and $f_{\alpha}^{+}(t)=g(t)$ are not Δ-differentiable at $t=0$ which leads to the infinite numbers of switching points in the neighborhood of zero.

Theorem 3.8 *If $f,g:\mathbb{T}\to\mathbb{R}^n_{\mathscr{F}}$ is gH-Δ-differentiable at $t\in\mathbb{T}^k$, then*

(i) $f(\sigma(t))=f(t)\widetilde{+}\mu(t)\cdot f^{\Delta}(t)$ *or* $f(t)=f(\sigma(t))\widetilde{+}(-1)\mu(t)\cdot f^{\Delta}(t)$, *i.e.,*

$$f\big(\sigma(t)\big)\widetilde{-}_{gH}f(t)=\mu(t)\cdot f^{\Delta}(t).$$

(ii) Let f,g be (I)-gH-Δ-differentiable at $t\in(a,b)_{\mathbb{T}}$ or (II)-gH-Δ-differentiable at $t\in(a,b)_{\mathbb{T}}$, then $f\widetilde{+}g:\mathbb{T}\to\mathbb{R}^n_{\mathscr{F}}$ is gH-Δ-differentiable at t and

$$(f\widetilde{+}g)^{\Delta} = f^{\Delta}(t)\widetilde{+}g^{\Delta}(t). \tag{3.6}$$

(iii) For any nonnegative constant $\lambda \in \mathbb{R}$, $\lambda \cdot f : \mathbb{T} \to \mathbb{R}^n_{\mathcal{F}}$ is gH-Δ-differentiable at t with

$$(\lambda \cdot f)^{\Delta}(t) = \lambda \cdot f^{\Delta}(t).$$

Proof (i). If t is the right dense point, then $\sigma(t) = t$, we have $f\big(\sigma(t)\big)\widetilde{-}_{gH} f(t) = \mu(t) \cdot f^{\Delta}(t)$. If t is the right scattered point, then $\sigma(t) > t$, and it follows from Theorem 3.3 (i) that $f\big(\sigma(t)\big)\widetilde{-}_{gH} f(t) = \mu(t) \cdot f^{\Delta}(t)$.

(ii). Case I. If f, g are (I)-gH-Δ-differentiable at $t \in (a,b)_{\mathbb{T}}$, then

$$[f^{\Delta}(t)]^{\alpha} = \times_{i=1}^{n}[(f_i^-)^{\Delta}, (f_i^+)^{\Delta}]_{\alpha}, \;\; [g^{\Delta}(t)]^{\alpha} = \times_{i=1}^{n}[(g_i^-)^{\Delta}, (g_i^+)^{\Delta}]_{\alpha},$$

then one obtains

$$[f^{\Delta}(t)]^{\alpha} + [g^{\Delta}(t)]^{\alpha} = \times_{i=1}^{n}[(f_i^-)^{\Delta} + (g_i^-)^{\Delta}, (f_i^+)^{\Delta} + (g_i^+)^{\Delta}]_{\alpha}.$$

Moreover, by Theorem 3.1, we have

$$\begin{aligned}
&[(f+g)^{\Delta}]^{\alpha}\\
&= \times_{i=1}^{n}\big[\min\{(f_i^- + g_1^-)^{\Delta}, (f_i^+ + g_1^+)^{\Delta}\}, \max\{(f_i^- + g_1^-)^{\Delta}, (f_i^+ + g_1^+)^{\Delta}\}\big]_{\alpha}\\
&= \times_{i=1}^{n}\big[\min\{(f_i^-)^{\Delta} + (g_1^-)^{\Delta}, (f_i^+)^{\Delta} + (g_1^+)^{\Delta}\},\\
&\quad \max\{(f_i^-)^{\Delta} + (g_1^-)^{\Delta}, (f_i^+)^{\Delta} + (g_1^+)^{\Delta}\}\big]_{\alpha}\\
&= = \times_{i=1}^{n}[(f_i^-)^{\Delta} + (g_i^-)^{\Delta}, (f_i^+)^{\Delta} + (g_i^+)^{\Delta}]_{\alpha}.
\end{aligned}$$

Hence, we have $[f^{\Delta}(t)]^{\alpha} + [g^{\Delta}(t)]^{\alpha} = [(f+g)^{\Delta}]^{\alpha}$ for any $0 \le \alpha \le 1$.

Case II. If f, g are (II)-gH-Δ-differentiable at $t \in (a,b)_{\mathbb{T}}$, the proof of Case II is similar to Case I.

(iii). For any nonnegative constant $\lambda \in \mathbb{R}$ and any $\alpha \in [0,1]$, we have

$$\big[(\lambda \cdot f)^{\Delta}(t)\big]^{\alpha} = \times_{i=1}^{n}\big[(\lambda \cdot f_i)^{\Delta}(t)\big]^{\alpha}.$$

Case I. If f is (I)-gH-Δ-differentiable at $t \in (a,b)_{\mathbb{T}}$, then

$$\begin{aligned}
\big[(\lambda \cdot f)^{\Delta}(t)\big]^{\alpha} &= \times_{i=1}^{n}\big[\lambda(f_i^-)^{\Delta}, \lambda(f_i^+)^{\Delta}\big]_{\alpha}\\
&= \lambda \cdot \times_{i=1}^{n}[(f_i^-)^{\Delta}, (f_i^+)^{\Delta}]_{\alpha} = \lambda \cdot [f^{\Delta}(t)]^{\alpha}.
\end{aligned}$$

Case II. If f is (II)-gH-Δ-differentiable at $t \in (a,b)_{\mathbb{T}}$, then

$$\begin{aligned}
\big[(\lambda \cdot f)^{\Delta}(t)\big]^{\alpha} &= \times_{i=1}^{n}\big[\lambda(f_i^+)^{\Delta}, \lambda(f_i^-)^{\Delta}\big]_{\alpha}\\
&= \lambda \cdot \times_{i=1}^{n}[(f_i^+)^{\Delta}, (f_i^-)^{\Delta}]_{\alpha} = \lambda \cdot [f^{\Delta}(t)]^{\alpha}.
\end{aligned}$$

This completes the proof. □

Remark 3.3 In Theorem 3.8 (ii), the condition that f, g are (I)-gH-Δ-differentiable at $t \in (a,b)_{\mathbb{T}}$ or (II)-gH-Δ-differentiable at $t \in (a,b)_{\mathbb{T}}$ simultaneously is essential, and without this condition, (3.6) will be invalid. For example, let $f, g : [0,5]_{\mathbb{T}} \to \mathbb{R}_{\mathscr{F}}$ with

$$[f(t)]^{\alpha} = \big[e_{-1}(t,0) - t + 2, e_{-1}(t,0) + t + 5\big]_{\alpha},$$

$$[g(t)]^{\alpha} = \big[-e_{-1}(t,0) + t - 10, -e_{-1}(t,0) - t + 10]_{\alpha},$$

where $f(t)$ is (I)-gH-Δ differentiable and $g(t)$ is (II)-gH-Δ differentiable at any $t_0 \in (0,5)_{\mathbb{T}}$. We have $[(f+g)(t)]^{\alpha} = [-8, 15]$, so $[(f+g)^{\Delta}(t)]^{\alpha} = \{0\}$. However, we also obtain

$$\begin{aligned}[f^{\Delta}(t) + g^{\Delta}(t)]^{\alpha} &= \big[-e_{-1}(t,0) - 1, -e_{-1}(t,0) + 1\big]_{\alpha} \\ &\quad + \big[e_{-1}(t,0) - 1, e_{-1}(t,0) + 1\big]_{\alpha} \\ &= [-2, 2],\end{aligned}$$

which implies that $f^{\Delta} + g^{\Delta} \not\subseteq (f+g)^{\Delta}$. Hence, this condition is necessary. Note that if we let $\mathbb{T} = \mathbb{R}$ and $\mathbb{Z}$, it turns into the examples of the continuous situation and discrete situation for fuzzy vector-valued functions.

Now, we will provide some arithmetic properties of the gH-Δ-derivatives of the product of two fuzzy vector-valued functions on time scales. For convenience, we adopt the notation $f\big(\sigma(t)\big) = f^{\sigma}(t)$ in some statement.

Theorem 3.9 *Let f, g be (I)-gH-Δ-differentiable, then*

(i) *if* $I^{\alpha,(I)}_{f_i,g_i} < 0$, $I^{\alpha,(I)}_{f_i^{\sigma},g_i^{\Delta_I}} < 0$, $I^{\alpha,(I)}_{f_i^{\Delta_I},g_i} < 0$ *and* $f * g$ *is (I)-gH-Δ-differentiable, then*

$$(f * g)^{\Delta_I} = f^{\sigma} * g^{\Delta_I} \widetilde{+} f^{\Delta_I} * g.$$

(ii) *if* $I^{\alpha,(I)}_{f_i,g_i} < 0$, $I^{\alpha,(I)}_{f_i^{\sigma},g_i^{\Delta_I}} > 0$, $I^{\alpha,(I)}_{f_i^{\Delta_I},g_i} > 0$ *and* $f * g$ *is (II)-gH-Δ-differentiable, then*

$$(f * g)^{\Delta_{II}} = f^{\sigma} * g^{\Delta_I} \widetilde{+} f^{\Delta_I} * g.$$

(iii) *if* $I^{\alpha,(II)}_{f_i,g_i} < 0$, $I^{\alpha,(II)}_{f_i^{\sigma},g_i^{\Delta_I}} < 0$, $I^{\alpha,(II)}_{f_i^{\Delta_I},g_i} < 0$ *and* $f \circledast g$ *is (I)-gH-Δ-differentiable, then*

$$(f \circledast g)^{\Delta_I} = f^{\sigma} \circledast g^{\Delta_I} \widetilde{+} g \circledast f^{\Delta_I}.$$

(iv) *if* $I^{\alpha,(II)}_{f_i,g_i} < 0$, $I^{\alpha,(II)}_{f_i^{\sigma},g_i^{\Delta_I}} > 0$, $I^{\alpha,(II)}_{f_i^{\Delta_I},g_i} > 0$ *and* $f \circledast g$ *is (II)-gH-Δ-differentiable, then*

$$(f \circledast g)^{\Delta_{II}} = f^{\sigma} \circledast g^{\Delta_I} \widetilde{+} f^{\Delta_I} \circledast g.$$

(v) *if* $I^{\alpha,(I)}_{f_i,g_i} > 0$, $I^{\alpha,(I)}_{f_i^\sigma,g_i^{\Delta_I}} > 0$, $I^{\alpha,(I)}_{f_i^{\Delta_I},g_i} > 0$ *and* $f * g$ *is* (I)-gH-Δ-*differentiable, then*

$$(f * g)^{\Delta_I} = f^\sigma * g^{\Delta_I} \widetilde{\mp} f^{\Delta_I} * g.$$

(vi) *if* $I^{\alpha,(I)}_{f_i,g_i} > 0$, $I^{\alpha,(I)}_{f_i^\sigma,g_i^{\Delta_I}} < 0$, $I^{\alpha,(I)}_{f_i^{\Delta_I},g_i} < 0$ *and* $f * g$ *is* (II)-gH-Δ-*differentiable, then*

$$(f * g)^{\Delta_{II}} = f^\sigma * g^{\Delta_I} \widetilde{\mp} f^{\Delta_I} * g.$$

(vii) *if* $I^{\alpha,(II)}_{f_i,g_i} > 0$, $I^{\alpha,(II)}_{f_i^\sigma,g_i^{\Delta_I}} > 0$, $I^{\alpha,(II)}_{f_i^{\Delta_I},g_i} > 0$ *and* $f \circledast g$ *is* (I)-gH-Δ-*differentiable, then*

$$(f \circledast g)^{\Delta_I} = f^\sigma \circledast g^{\Delta_I} \widetilde{\mp} g \circledast f^{\Delta_I}.$$

(viii) *if* $I^{\alpha,(II)}_{f_i,g_i} > 0$, $I^{\alpha,(II)}_{f_i^\sigma,g_i^{\Delta_I}} < 0$, $I^{\alpha,(II)}_{f_i^{\Delta_I},g_i} < 0$ *and* $f \circledast g$ *is* (II)-gH-Δ-*differentiable, then*

$$(f \circledast g)^{\Delta_{II}} = f^\sigma \circledast g^{\Delta_I} \widetilde{\mp} f^{\Delta_I} \circledast g.$$

Proof Since f, g be (I)-gH-Δ-differentiable, then

$$[f^{\Delta_I}]^\alpha = \times_{i=1}^n \big[(f_i^-)^\Delta, (f_i^+)^\Delta\big]_\alpha, \quad [g^{\Delta_I}]^\alpha = \times_{i=1}^n \big[(g_i^-)^\Delta, (g_i^+)^\Delta\big]_\alpha.$$

(i). Since $I^{\alpha,(I)}_{f_i,g_i} < 0$, we have $[f * g]^\alpha = \times_{i=1}^n [f_i^- g_i^+, f_i^+ g_i^-]_\alpha$. Moreover, because $f * g$ is (I)-gH-Δ-differentiable, we obtain

$$\begin{aligned}\big[(f * g)^{\Delta_I}\big]^\alpha &= \times_{i=1}^n \big[(f_i^- g_i^+)^\Delta, (f_i^+ g_i^-)^\Delta\big]_\alpha \\ &= \times_{i=1}^n \big[(f_i^-)^\sigma (g_i^+)^\Delta + g_i^+ (f_i^-)^\Delta, (f_i^+)^\sigma (g_i^-)^\Delta + g_i^- (f_i^+)^\Delta\big]_\alpha.\end{aligned}$$

In addition, from $I^{\alpha,(I)}_{f_i^\sigma,g_i^{\Delta_I}} < 0$, $I^{\alpha,(I)}_{f_i^{\Delta_I},g_i} < 0$, we have

$$\big[f^\sigma * g^{\Delta_I}\big]^\alpha = \times_{i=1}^n [(f_i^-)^\sigma (g_i^+)^\Delta, (f_i^+)^\sigma (g_i^-)^\Delta]_\alpha,$$

$$\big[f^{\Delta_I} * g\big]^\alpha = \times_{i=1}^n \big[g_i^+ (f_i^-)^\Delta, g_i^- (f_i^+)^\Delta\big]_\alpha.$$

Hence, we obtain $\big[f^\sigma * g^{\Delta_I}\big]^\alpha + \big[f^{\Delta_I} * g\big]^\alpha = \big[(f * g)^{\Delta_I}\big]^\alpha$.

(ii). Since $f * g$ is (II)-gH-Δ-differentiable, through calculation, we have

$$\begin{aligned}\big[(f * g)^{\Delta_{II}}\big]^\alpha &= \times_{i=1}^n \big[(f_i^+ g_i^-)^\Delta, (f_i^- g_i^+)^\Delta\big]_\alpha \\ &= \times_{i=1}^n \big[(f_i^+)^\sigma (g_i^-)^\Delta + g_i^- (f_i^+)^\Delta, (f_i^-)^\sigma (g_i^+)^\Delta + g_i^+ (f_i^-)^\Delta\big]_\alpha.\end{aligned}$$

In addition, from $I^{\alpha,(I)}_{f_i^\sigma,g_i^{\Delta_I}} > 0$, $I^{\alpha,(I)}_{f_i^{\Delta_I},g_i} > 0$, we have

$$\big[f^\sigma * g^{\Delta_I}\big]^\alpha = \times_{i=1}^n [(f_i^+)^\sigma (g_i^-)^\Delta, (f_i^-)^\sigma (g_i^+)^\Delta]_\alpha,$$

$$[f^{\Delta_I} * g]^\alpha = \times_{i=1}^n [g_i^-(f_i^+)^\Delta, g_i^+(f_i^-)^\Delta]_\alpha,$$

and thus, we obtain $[f^\sigma * g^{\Delta_I}]^\alpha + [f^{\Delta_I} * g]^\alpha = [(f * g)^{\Delta_{II}}]^\alpha$.

(iii). Since $I_{f_i,g_i}^{\alpha,(II)} < 0$, we have $[f \circledast g]^\alpha = \times_{i=1}^n [f_i^+ g_i^+, f_i^- g_i^-]_\alpha$. Moreover, because $f \circledast g$ is (I)-gH-Δ-differentiable, we obtain

$$\begin{aligned}[(f \circledast g)^{\Delta_I}]^\alpha &= \times_{i=1}^n [(f_i^+ g_i^+)^\Delta, (f_i^- g_i^-)^\Delta]_\alpha \\ &= \times_{i=1}^n [(f_i^+)^\sigma (g_i^+)^\Delta + g_i^+ (f_i^+)^\Delta, (f_i^-)^\sigma (g_i^-)^\Delta + g_i^- (f_i^-)^\Delta]_\alpha.\end{aligned}$$

In addition, from $I_{f_i^\sigma, g_i^{\Delta_I}}^{\alpha,(II)} < 0$, $I_{f_i^{\Delta_I}, g_i}^{\alpha,(II)} < 0$, we have

$$[f^\sigma \circledast g^{\Delta_I}]^\alpha = \times_{i=1}^n [(f_i^+)^\sigma (g_i^+)^\Delta, (f_i^-)^\sigma (g_i^-)^\Delta]_\alpha,$$

$$[f^{\Delta_I} \circledast g]^\alpha = \times_{i=1}^n [g_i^+(f_i^+)^\Delta, g_i^-(f_i^-)^\Delta]_\alpha,$$

and thus, it follows that $[f^\sigma \circledast g^{\Delta_I}]^\alpha + [f^{\Delta_I} \circledast g]^\alpha = [(f \circledast g)^{\Delta_I}]^\alpha$.

(iv). Since $f \circledast g$ is (II)-gH-Δ-differentiable, through calculation, we have

$$\begin{aligned}[(f \circledast g)^{\Delta_{II}}]^\alpha &= \times_{i=1}^n [(f_i^- g_i^-)^\Delta, (f_i^+ g_i^+)^\Delta]_\alpha \\ &= \times_{i=1}^n [(f_i^-)^\sigma (g_i^-)^\Delta + g_i^- (f_i^-)^\Delta, (f_i^+)^\sigma (g_i^+)^\Delta + g_i^+ (f_i^+)^\Delta]_\alpha.\end{aligned}$$

In addition, from $I_{f_i^\sigma, g_i^{\Delta_I}}^{\alpha,(II)} > 0$, $I_{f_i^{\Delta_I}, g_i}^{\alpha,(II)} > 0$, we have

$$[f^\sigma \circledast g^{\Delta_I}]^\alpha = \times_{i=1}^n [(f_i^-)^\sigma (g_i^-)^\Delta, (f_i^+)^\sigma (g_i^+)^\Delta]_\alpha,$$

$$[f^{\Delta_I} \circledast g]^\alpha = \times_{i=1}^n [g_i^-(f_i^-)^\Delta, g_i^+(f_i^+)^\Delta]_\alpha,$$

and thus, we have $[f^\sigma \circledast g^{\Delta_I}]^\alpha + [f^{\Delta_I} \circledast g]^\alpha = [(f \circledast g)^{\Delta_{II}}]^\alpha$.

(v). Since $I_{f_i,g_i}^{\alpha,(I)} > 0$, we have $[f * g]^\alpha = \times_{i=1}^n [f_i^+ g_i^-, f_i^- g_i^+]_\alpha$. Moreover, because $f * g$ is (I)-gH-Δ-differentiable, we obtain

$$\begin{aligned}[(f * g)^{\Delta_I}]^\alpha &= \times_{i=1}^n [(f_i^+ g_i^-)^\Delta, (f_i^- g_i^+)^\Delta]_\alpha \\ &= \times_{i=1}^n [(f_i^+)^\sigma (g_i^-)^\Delta + g_i^- (f_i^+)^\Delta, (f_i^-)^\sigma (g_i^+)^\Delta + g_i^+ (f_i^-)^\Delta]_\alpha.\end{aligned}$$

In addition, from $I_{f_i^\sigma, g_i^{\Delta_I}}^{\alpha,(I)} > 0$, $I_{f_i^{\Delta_I}, g_i}^{\alpha,(I)} > 0$, we have

$$[f^\sigma * g^{\Delta_I}]^\alpha = \times_{i=1}^n [(f_i^+)^\sigma (g_i^-)^\Delta, (f_i^-)^\sigma (g_i^+)^\Delta]_\alpha,$$

$$[f^{\Delta_I} * g]^\alpha = \times_{i=1}^n [g_i^-(f_i^+)^\Delta, g_i^+(f_i^-)^\Delta]_\alpha,$$

and thus, we can obtain $[f^\sigma * g^{\Delta_I}]^\alpha + [f^{\Delta_I} * g]^\alpha = [(f * g)^{\Delta_I}]^\alpha$.

(vi). Since $I_{f_i,g_i}^{\alpha,(I)} > 0$, we have $[f * g]^\alpha = \times_{i=1}^n [f_i^+ g_i^-, f_i^- g_i^+]_\alpha$. Moreover, because $f * g$ is (II)-gH-Δ-differentiable, we obtain

$$\begin{aligned}\left[(f * g)^{\Delta_I}\right]^\alpha &= \times_{i=1}^n\left[(f_i^- g_i^+)^\Delta, (f_i^+ g_i^-)^\Delta\right]_\alpha \\ &= \times_{i=1}^n\left[(f_i^-)^\sigma (g_i^+)^\Delta + g_i^+ (f_i^-)^\Delta, (f_i^+)^\sigma (g_i^-)^\Delta + g_i^- (f_i^+)^\Delta\right]_\alpha.\end{aligned}$$

In addition, from $I^{\alpha,(I)}_{f_i^\sigma, g_i^{\Delta_I}} < 0$, $I^{\alpha,(I)}_{f_i^{\Delta_I}, g_i} < 0$, we have

$$\left[f^\sigma * g^{\Delta_I}\right]^\alpha = \times_{i=1}^n\left[(f_i^-)^\sigma (g_i^+)^\Delta, (f_i^+)^\sigma (g_i^-)^\Delta\right]_\alpha,$$

$$\left[f^{\Delta_I} * g\right]^\alpha = \times_{i=1}^n\left[g_i^+ (f_i^-)^\Delta, g_i^- (f_i^+)^\Delta\right]_\alpha,$$

and thus, we can obtain $\left[f^\sigma * g^{\Delta_I}\right]^\alpha + \left[f^{\Delta_I} * g\right]^\alpha = \left[(f * g)^{\Delta_{II}}\right]^\alpha$.

(vii). Since $I^{\alpha,(II)}_{f_i, g_i} > 0$, we have $[f \circledast g]^\alpha = \times_{i=1}^n [f_i^- g_i^-, f_i^+ g_i^+]_\alpha$. Moreover, because $f \circledast g$ is (I)-gH-Δ-differentiable, we obtain

$$\begin{aligned}\left[(f \circledast g)^{\Delta_I}\right]^\alpha &= \times_{i=1}^n\left[(f_i^- g_i^-)^\Delta, (f_i^+ g_i^+)^\Delta\right]_\alpha \\ &= \times_{i=1}^n\left[(f_i^-)^\sigma (g_i^-)^\Delta + g_i^- (f_i^-)^\Delta, (f_i^+)^\sigma (g_i^+)^\Delta + g_i^+ (f_i^+)^\Delta\right]_\alpha.\end{aligned}$$

In addition, from $I^{\alpha,(II)}_{f_i^\sigma, g_i^{\Delta_I}} > 0$, $I^{\alpha,(II)}_{f_i^{\Delta_I}, g_i} > 0$, we have

$$\left[f^\sigma \circledast g^{\Delta_I}\right]^\alpha = \times_{i=1}^n\left[(f_i^-)^\sigma (g_i^-)^\Delta, (f_i^+)^\sigma (g_i^+)^\Delta\right]_\alpha,$$

$$\left[f^{\Delta_I} \circledast g\right]^\alpha = \times_{i=1}^n\left[g_i^- (f_i^-)^\Delta, g_i^+ (f_i^+)^\Delta\right]_\alpha,$$

and thus, we can obtain $\left[f^\sigma \circledast g^{\Delta_I}\right]^\alpha + \left[f^{\Delta_I} \circledast g\right]^\alpha = \left[(f \circledast g)^{\Delta_I}\right]^\alpha$.

$(viii)$. Since $I^{\alpha,(II)}_{f_i, g_i} > 0$, we have $[f \circledast g]^\alpha = \times_{i=1}^n [f_i^- g_i^-, f_i^+ g_i^+]_\alpha$. Moreover, because $f \circledast g$ is (II)-gH-Δ-differentiable, we obtain

$$\begin{aligned}\left[(f \circledast g)^{\Delta_{II}}\right]^\alpha &= \times_{i=1}^n\left[(f_i^+ g_i^+)^\Delta, (f_i^- g_i^-)^\Delta\right]_\alpha \\ &= \times_{i=1}^n\left[(f_i^+)^\sigma (g_i^+)^\Delta + g_i^+ (f_i^+)^\Delta, (f_i^-)^\sigma (g_i^-)^\Delta + g_i^- (f_i^-)^\Delta\right]_\alpha.\end{aligned}$$

In addition, from $I^{\alpha,(II)}_{f_i^\sigma, g_i^{\Delta_I}} < 0$, $I^{\alpha,(II)}_{f_i^{\Delta_I}, g_i} < 0$, we have

$$\left[f^\sigma \circledast g^{\Delta_I}\right]^\alpha = \times_{i=1}^n\left[(f_i^+)^\sigma (g_i^+)^\Delta, (f_i^-)^\sigma (g_i^-)^\Delta\right]_\alpha,$$

$$\left[f^{\Delta_I} \circledast g\right]^\alpha = \times_{i=1}^n\left[g_i^+ (f_i^+)^\Delta, g_i^- (f_i^-)^\Delta\right]_\alpha,$$

and thus, we can obtain $\left[f^\sigma \circledast g^{\Delta_I}\right]^\alpha + \left[f^{\Delta_I} \circledast g\right]^\alpha = \left[(f \circledast g)^{\Delta_{II}}\right]^\alpha$. This completes the proof. □

Theorem 3.10 *Let f, g be (II)-gH-Δ-differentiable, then*

(i) *if* $I^{\alpha,(II)}_{f_i,g_i} < 0$, $I^{\alpha,(I)}_{f_i^{\sigma},g_i^{\Delta_{II}}} > 0$, $I^{\alpha,(I)}_{f_i^{\Delta_{II}},g_i} < 0$ *and $f \circledast g$ is (I)-gH-Δ-differentiable, then*

$$(f \circledast g)^{\Delta_I} = f^{\sigma} * g^{\Delta_{II}} \widetilde{\mp} f^{\Delta_{II}} * g.$$

(ii) *if* $I^{\alpha,(II)}_{f_i,g_i} < 0$, $I^{\alpha,(I)}_{f_i^{\sigma},g_i^{\Delta_{II}}} < 0$, $I^{\alpha,(I)}_{f_i^{\Delta_{II}},g_i} > 0$ *and $f \circledast g$ is (II)-gH-Δ-differentiable, then*

$$(f \circledast g)^{\Delta_{II}} = f^{\sigma} * g^{\Delta_I} \widetilde{\mp} f^{\Delta_I} * g.$$

(iii) *if* $I^{\alpha,(II)}_{f_i,g_i} > 0$, $I^{\alpha,(I)}_{f_i^{\sigma},g_i^{\Delta_{II}}} < 0$, $I^{\alpha,(I)}_{f_i^{\Delta_{II}},g_i} > 0$ *and $f \circledast g$ is (I)-gH-Δ-differentiable, then*

$$(f \circledast g)^{\Delta_I} = f^{\sigma} * g^{\Delta_{II}} \widetilde{\mp} g * f^{\Delta_{II}}.$$

(iv) *if* $I^{\alpha,(II)}_{f_i,g_i} > 0$, $I^{\alpha,(I)}_{f_i^{\sigma},g_i^{\Delta_{II}}} > 0$, $I^{\alpha,(I)}_{f_i^{\Delta_{II}},g_i} < 0$ *and $f \circledast g$ is (II)-gH-Δ-differentiable, then*

$$(f \circledast g)^{\Delta_{II}} = f^{\sigma} * g^{\Delta_{II}} \widetilde{\mp} f^{\Delta_{II}} * g.$$

(v) *if* $I^{\alpha,(I)}_{f_i,g_i} < 0$, $I^{\alpha,(II)}_{f_i^{\sigma},g_i^{\Delta_{II}}} > 0$, $I^{\alpha,(II)}_{f_i^{\Delta_{II}},g_i} < 0$ *and $f * g$ is (I)-gH-Δ-differentiable, then*

$$(f * g)^{\Delta_I} = f^{\sigma} \circledast g^{\Delta_{II}} \widetilde{\mp} f^{\Delta_{II}} \circledast g.$$

(vi) *if* $I^{\alpha,(I)}_{f_i,g_i} < 0$, $I^{\alpha,(II)}_{f_i^{\sigma},g_i^{\Delta_{II}}} < 0$, $I^{\alpha,(II)}_{f_i^{\Delta_{II}},g_i} > 0$ *and $f * g$ is (II)-gH-Δ-differentiable, then*

$$(f * g)^{\Delta_{II}} = f^{\sigma} \circledast g^{\Delta_{II}} \widetilde{\mp} f^{\Delta_{II}} \circledast g.$$

(vii) *if* $I^{\alpha,(I)}_{f_i,g_i} > 0$, $I^{\alpha,(II)}_{f_i^{\sigma},g_i^{\Delta_{II}}} < 0$, $I^{\alpha,(II)}_{f_i^{\Delta_{II}},g_i} > 0$ *and $f * g$ is (I)-gH-Δ-differentiable, then*

$$(f * g)^{\Delta_I} = f^{\sigma} \circledast g^{\Delta_{II}} \widetilde{\mp} g \circledast f^{\Delta_{II}}.$$

(viii) *if* $I^{\alpha,(I)}_{f_i,g_i} > 0$, $I^{\alpha,(II)}_{f_i^{\sigma},g_i^{\Delta_{II}}} > 0$, $I^{\alpha,(II)}_{f_i^{\Delta_{II}},g_i} < 0$ *and $f * g$ is (II)-gH-Δ-differentiable, then*

$$(f * g)^{\Delta_{II}} = f^{\sigma} \circledast g^{\Delta_{II}} \widetilde{\mp} f^{\Delta_{II}} \circledast g.$$

Proof Since f, g be (II)-gH-Δ-differentiable, then

$$[f^{\Delta_{II}}]^{\alpha} = \times_{i=1}^{n}\big[(f_i^{+})^{\Delta}, (f_i^{-})^{\Delta}\big]_{\alpha}, \quad [g^{\Delta_{II}}]^{\alpha} = \times_{i=1}^{n}\big[(g_i^{+})^{\Delta}, (g_i^{-})^{\Delta}\big]_{\alpha}.$$

(i). Since $I^{\alpha,(II)}_{f_i,g_i} < 0$, we have $[f \circledast g]^{\alpha} = \times_{i=1}^{n}[f_i^{+}g_i^{+}, f_i^{-}g_i^{-}]_{\alpha}$. Moreover, because $f \circledast g$ is (I)-gH-Δ-differentiable, we obtain

$$\begin{aligned}\big[(f \circledast g)^{\Delta_I}\big]^{\alpha} &= \times_{i=1}^{n}\big[(f_i^{+}g_i^{+})^{\Delta}, (f_i^{-}g_i^{-})^{\Delta}\big]_{\alpha} \\ &= \times_{i=1}^{n}\big[(f_i^{+})^{\sigma}(g_i^{+})^{\Delta} + g_i^{+}(f_i^{+})^{\Delta}, (f_i^{-})^{\sigma}(g_i^{-})^{\Delta} + g_i^{-}(f_i^{-})^{\Delta}\big]_{\alpha}.\end{aligned}$$

In addition, from $I^{\alpha,(I)}_{f_i^\sigma, g_i^{\Delta_{II}}} > 0$, $I^{\alpha,(I)}_{f_i^{\Delta_{II}}, g_i} < 0$, we have

$$[f^\sigma * g^{\Delta_{II}}]^\alpha = \times_{i=1}^n[(f_i^+)^\sigma (g_i^+)^\Delta, (f_i^-)^\sigma (g_i^-)^\Delta]_\alpha,$$

$$\left[f^{\Delta_{II}} * g\right]^\alpha = \times_{i=1}^n\left[g_i^+(f_i^+)^\Delta, g_i^-(f_i^-)^\Delta\right]_\alpha,$$

and thus, we can obtain $\left[f^\sigma * g^{\Delta_{II}}\right]^\alpha + \left[f^{\Delta_{II}} * g\right]^\alpha = \left[(f \circledast g)^{\Delta_I}\right]^\alpha$.

(ii). Since $I^{\alpha,(II)}_{f_i,g_i} < 0$, we have $[f \circledast g]^\alpha = \times_{i=1}^n[f_i^+ g_i^+, f_i^- g_i^-]_\alpha$. Moreover, because $f \circledast g$ is (II)-gH-Δ-differentiable, we obtain

$$\begin{aligned}\left[(f \circledast g)^{\Delta_{II}}\right]^\alpha &= \times_{i=1}^n\left[(f_i^- g_i^-)^\Delta, (f_i^+ g_i^+)^\Delta\right]_\alpha \\ &= \times_{i=1}^n\left[(f_i^-)^\sigma (g_i^-)^\Delta + g_i^-(f_i^-)^\Delta, (f_i^+)^\sigma (g_i^+)^\Delta + g_i^+(f_i^+)^\Delta\right]_\alpha.\end{aligned}$$

In addition, from $I^{\alpha,(I)}_{f_i^\sigma, g_i^{\Delta_{II}}} < 0$, $I^{\alpha,(I)}_{f_i^{\Delta_{II}}, g_i} > 0$, we have

$$[f^\sigma * g^{\Delta_{II}}]^\alpha = \times_{i=1}^n[(f_i^-)^\sigma (g_i^-)^\Delta, (f_i^+)^\sigma (g_i^+)^\Delta]_\alpha,$$

$$\left[f^{\Delta_{II}} * g\right]^\alpha = \times_{i=1}^n\left[g_i^-(f_i^-)^\Delta, g_i^+(f_i^+)^\Delta\right]_\alpha,$$

and thus, we can obtain $\left[f^\sigma * g^{\Delta_{II}}\right]^\alpha + \left[f^{\Delta_{II}} * g\right]^\alpha = \left[(f \circledast g)^{\Delta_{II}}\right]^\alpha$.

(iii). Since $I^{\alpha,(II)}_{f_i,g_i} > 0$, we have $[f \circledast g]^\alpha = \times_{i=1}^n[f_i^- g_i^-, f_i^+ g_i^+]_\alpha$. Moreover, because $f \circledast g$ is (I)-gH-Δ-differentiable, we obtain

$$\begin{aligned}\left[(f \circledast g)^{\Delta_I}\right]^\alpha &= \times_{i=1}^n\left[(f_i^- g_i^-)^\Delta, (f_i^+ g_i^+)^\Delta\right]_\alpha \\ &= \times_{i=1}^n\left[(f_i^-)^\sigma (g_i^-)^\Delta + g_i^-(f_i^-)^\Delta, (f_i^+)^\sigma (g_i^+)^\Delta + g_i^+(f_i^+)^\Delta\right]_\alpha.\end{aligned}$$

In addition, from $I^{\alpha,(I)}_{f_i^\sigma, g_i^{\Delta_{II}}} < 0$, $I^{\alpha,(I)}_{f_i^{\Delta_{II}}, g_i} > 0$, we have

$$[f^\sigma * g^{\Delta_{II}}]^\alpha = \times_{i=1}^n[(f_i^-)^\sigma (g_i^-)^\Delta, (f_i^+)^\sigma (g_i^+)^\Delta]_\alpha,$$

$$\left[f^{\Delta_{II}} * g\right]^\alpha = \times_{i=1}^n\left[g_i^-(f_i^-)^\Delta, g_i^+(f_i^+)^\Delta\right]_\alpha,$$

and thus, we can obtain $\left[f^\sigma * g^{\Delta_{II}}\right]^\alpha + \left[f^{\Delta_{II}} * g\right]^\alpha = \left[(f \circledast g)^{\Delta_I}\right]^\alpha$.

(iv). Since $I^{\alpha,(II)}_{f_i,g_i} > 0$, we have $[f \circledast g]^\alpha = \times_{i=1}^n[f_i^- g_i^-, f_i^+ g_i^+]_\alpha$. Moreover, because $f \circledast g$ is (II)-gH-Δ-differentiable, we obtain

$$\begin{aligned}\left[(f \circledast g)^{\Delta_{II}}\right]^\alpha &= \times_{i=1}^n\left[(f_i^+ g_i^+)^\Delta, (f_i^- g_i^-)^\Delta\right]_\alpha \\ &= \times_{i=1}^n\left[(f_i^+)^\sigma (g_i^+)^\Delta + g_i^+(f_i^+)^\Delta, (f_i^-)^\sigma (g_i^-)^\Delta + g_i^-(f_i^-)^\Delta\right]_\alpha.\end{aligned}$$

In addition, from $I^{\alpha,(I)}_{f_i^\sigma, g_i^{\Delta_{II}}} > 0$, $I^{\alpha,(I)}_{f_i^{\Delta_{II}}, g_i} < 0$, we have

$$[f^\sigma * g^{\Delta_{II}}]^\alpha = \times_{i=1}^n[(f_i^+)^\sigma (g_i^+)^\Delta, (f_i^-)^\sigma (g_i^-)^\Delta]_\alpha,$$

$$\left[f^{\Delta_{II}} * g\right]^{\alpha} = \times_{i=1}^{n}\left[g_i^{+}(f_i^{+})^{\Delta}, g_i^{-}(f_i^{-})^{\Delta}\right]_{\alpha},$$

and thus, we can obtain $\left[f^{\sigma} * g^{\Delta_{II}}\right]^{\alpha} + \left[f^{\Delta_{II}} * g\right]^{\alpha} = \left[(f \circledast g)^{\Delta_{II}}\right]^{\alpha}$.

(v). Since $I_{f_i,g_i}^{\alpha,(I)} < 0$, we have $[f * g]^{\alpha} = \times_{i=1}^{n}[f_i^{-}g_i^{+}, f_i^{+}g_i^{-}]_{\alpha}$. Moreover, because $f * g$ is (I)-gH-Δ-differentiable, we obtain

$$\begin{aligned}\left[(f * g)^{\Delta_I}\right]^{\alpha} &= \times_{i=1}^{n}\left[(f_i^{-}g_i^{+})^{\Delta}, (f_i^{+}g_i^{-})^{\Delta}\right]_{\alpha} \\ &= \times_{i=1}^{n}\left[(f_i^{-})^{\sigma}(g_i^{+})^{\Delta} + g_i^{+}(f_i^{-})^{\Delta}, (f_i^{+})^{\sigma}(g_i^{-})^{\Delta} + g_i^{-}(f_i^{+})^{\Delta}\right]_{\alpha}.\end{aligned}$$

In addition, from $I_{f_i^{\sigma},g_i^{\Delta_{II}}}^{\alpha,(II)} > 0$, $I_{f_i^{\Delta_{II}},g_i}^{\alpha,(II)} < 0$, we have

$$\left[f^{\sigma} \circledast g^{\Delta_{II}}\right]^{\alpha} = \times_{i=1}^{n}\left[(f_i^{-})^{\sigma}(g_i^{+})^{\Delta}, (f_i^{+})^{\sigma}(g_i^{-})^{\Delta}\right]_{\alpha},$$

$$\left[f^{\Delta_{II}} \circledast g\right]^{\alpha} = \times_{i=1}^{n}\left[g_i^{+}(f_i^{-})^{\Delta}, g_i^{-}(f_i^{+})^{\Delta}\right]_{\alpha},$$

and thus, we can obtain $\left[f^{\sigma} \circledast g^{\Delta_{II}}\right]^{\alpha} + \left[f^{\Delta_{II}} \circledast g\right]^{\alpha} = \left[(f * g)^{\Delta_I}\right]^{\alpha}$.

(vi). Since $I_{f_i,g_i}^{\alpha,(I)} < 0$, we have $[f * g]^{\alpha} = \times_{i=1}^{n}[f_i^{-}g_i^{+}, f_i^{+}g_i^{-}]_{\alpha}$. Moreover, because $f * g$ is (II)-gH-Δ-differentiable, we obtain

$$\begin{aligned}\left[(f * g)^{\Delta_I}\right]^{\alpha} &= \times_{i=1}^{n}\left[(f_i^{+}g_i^{-})^{\Delta}, (f_i^{-}g_i^{+})^{\Delta}\right]_{\alpha} \\ &= \times_{i=1}^{n}\left[(f_i^{+})^{\sigma}(g_i^{-})^{\Delta} + g_i^{-}(f_i^{+})^{\Delta}, (f_i^{-})^{\sigma}(g_i^{+})^{\Delta} + g_i^{+}(f_i^{-})^{\Delta}\right]_{\alpha}.\end{aligned}$$

In addition, from $I_{f_i^{\sigma},g_i^{\Delta_{II}}}^{\alpha,(II)} < 0$, $I_{f_i^{\Delta_{II}},g_i}^{\alpha,(II)} > 0$, we have

$$\left[f^{\sigma} \circledast g^{\Delta_{II}}\right]^{\alpha} = \times_{i=1}^{n}\left[(f_i^{+})^{\sigma}(g_i^{-})^{\Delta}, (f_i^{-})^{\sigma}(g_i^{+})^{\Delta}\right]_{\alpha},$$

$$\left[f^{\Delta_{II}} \circledast g\right]^{\alpha} = \times_{i=1}^{n}\left[g_i^{-}(f_i^{+})^{\Delta}, g_i^{+}(f_i^{-})^{\Delta}\right]_{\alpha},$$

and thus, we can obtain $\left[f^{\sigma} \circledast g^{\Delta_{II}}\right]^{\alpha} + \left[f^{\Delta_{II}} \circledast g\right]^{\alpha} = \left[(f * g)^{\Delta_{II}}\right]^{\alpha}$.

(vii). Since $I_{f_i,g_i}^{\alpha,(I)} > 0$, we have $[f * g]^{\alpha} = \times_{i=1}^{n}[f_i^{+}g_i^{-}, f_i^{-}g_i^{+}]_{\alpha}$. Moreover, because $f * g$ is (I)-gH-Δ-differentiable, we obtain

$$\begin{aligned}\left[(f * g)^{\Delta_I}\right]^{\alpha} &= \times_{i=1}^{n}\left[(f_i^{+}g_i^{-})^{\Delta}, (f_i^{-}g_i^{+})^{\Delta}\right]_{\alpha} \\ &= \times_{i=1}^{n}\left[(f_i^{+})^{\sigma}(g_i^{-})^{\Delta} + g_i^{-}(f_i^{+})^{\Delta}, (f_i^{-})^{\sigma}(g_i^{+})^{\Delta} + g_i^{+}(f_i^{-})^{\Delta}\right]_{\alpha}.\end{aligned}$$

In addition, from $I_{f_i^{\sigma},g_i^{\Delta_{II}}}^{\alpha,(II)} < 0$, $I_{f_i^{\Delta_{II}},g_i}^{\alpha,(II)} > 0$, we have

$$\left[f^{\sigma} \circledast g^{\Delta_{II}}\right]^{\alpha} = \times_{i=1}^{n}\left[(f_i^{+})^{\sigma}(g_i^{-})^{\Delta}, (f_i^{-})^{\sigma}(g_i^{+})^{\Delta}\right]_{\alpha},$$

$$\left[f^{\Delta_{II}} \circledast g\right]^{\alpha} = \times_{i=1}^{n}\left[g_i^{-}(f_i^{+})^{\Delta}, g_i^{+}(f_i^{-})^{\Delta}\right]_{\alpha},$$

and thus, we can obtain $\left[f^{\sigma} \circledast g^{\Delta_{II}}\right]^{\alpha} + \left[f^{\Delta_{II}} \circledast g\right]^{\alpha} = \left[(f * g)^{\Delta_I}\right]^{\alpha}$.

$(viii)$. Since $I^{\alpha,(I)}_{f_i,g_i} > 0$, we have $[f * g]^\alpha = \times_{i=1}^n [f_i^+ g_i^-, f_i^- g_i^+]_\alpha$. Moreover, because $f * g$ is (II)-gH-Δ-differentiable, we obtain

$$\begin{aligned}\left[(f * g)^{\Delta_{II}}\right]^\alpha &= \times_{i=1}^n \left[(f_i^- g_i^+)^\Delta, (f_i^+ g_i^-)^\Delta\right]_\alpha \\ &= \times_{i=1}^n \left[(f_i^-)^\sigma (g_i^+)^\Delta + g_i^+ (f_i^-)^\Delta, (f_i^+)^\sigma (g_i^-)^\Delta + g_i^- (f_i^+)^\Delta\right]_\alpha.\end{aligned}$$

In addition, from $I^{\alpha,(II)}_{f_i^\sigma, g_i^{\Delta_{II}}} > 0$, $I^{\alpha,(II)}_{f_i^{\Delta_{II}}, g_i} < 0$, we have

$$\left[f^\sigma \circledast g^{\Delta_{II}}\right]^\alpha = \times_{i=1}^n \left[(f_i^-)^\sigma (g_i^+)^\Delta, (f_i^+)^\sigma (g_i^-)^\Delta\right]_\alpha,$$

$$\left[f^{\Delta_{II}} \circledast g\right]^\alpha = \times_{i=1}^n \left[g_i^+ (f_i^-)^\Delta, g_i^- (f_i^+)^\Delta\right]_\alpha,$$

and thus, we can obtain $\left[f^\sigma \circledast g^{\Delta_{II}}\right]^\alpha + \left[f^{\Delta_{II}} \circledast g\right]^\alpha = \left[(f * g)^{\Delta_{II}}\right]^\alpha$. This completes the proof. □

Example 3.1.1 We provide an example which satisfies the condition of Theorem 3.11 (i).

Let $\mathbb{T} = [\frac{1}{2}, 1]_\mathbb{T}$, let $f, g : \mathbb{T} \to \mathbb{R}_\mathscr{F}$ be given such that

$$[f(t)]^\alpha = [-2t, t]_\alpha, \quad [g(t)]^\alpha = \left[\frac{1}{2}t, t\right]_\alpha,$$

and then

$$[f(t) * g(t)]^\alpha = \begin{vmatrix} -2t & t \\ \frac{1}{2}t & t \end{vmatrix} = -2t^2 - \frac{1}{2}t^2 < 0, \text{ i.e. } [f(t) * g(t)]^\alpha = \left[-2t^2, \frac{1}{2}t^2\right].$$

Thus we have

$$\left[(f(t) * g(t))^\Delta\right]^\alpha = \left[(-2t^2)^\Delta, \left(\frac{1}{2}t^2\right)^\Delta\right]_\alpha = \left[-2\big(t + \sigma(t)\big), \frac{1}{2}\big(t + \sigma(t)\big)\right]_\alpha.$$

In addition, we also have

$$[f^\sigma * g^\Delta]^\alpha = \begin{vmatrix} -2\sigma(t) & \sigma(t) \\ \frac{1}{2} & 1 \end{vmatrix} = -2\sigma(t) - \frac{1}{2}\sigma(t) < 0,$$

i.e.

$$[f^\sigma * g^\Delta]^\alpha = \left[-2\sigma(t), \frac{1}{2}\sigma(t)\right]_\alpha,$$

$$[f^\Delta * g]^\alpha = \begin{vmatrix} -2 & 1 \\ \frac{1}{2}t & t \end{vmatrix} = -2t - \frac{1}{2}t < 0, \text{ i.e. } [f^\Delta * g]^\alpha = \left[-2t, \frac{1}{2}t\right]_\alpha,$$

so we have

$$\begin{aligned}[f^\sigma * g^\Delta]^\alpha + [f^\Delta * g]^\alpha &= [f^\sigma * g^\Delta \tilde{\mp} f^\Delta * g]^\alpha \\ &= \Big[-2\big(t+\sigma(t)\big), \frac{1}{2}\big(t+\sigma(t)\big)\Big]_\alpha \\ &= \big[\big(f(t) * g(t)\big)^\Delta\big]^\alpha.\end{aligned}$$

Remark 3.4 For the traditional multiplication of fuzzy functions, i.e., let $f, g : \mathbb{T} \to \mathbb{R}_{\mathcal{F}}$ with the α-level sets $[f]^\alpha = [f^-, f^+]_\alpha$ and $[g]^\alpha = [g^-, g^+]_\alpha$, the traditional multiplication of fuzzy functions f, g is defined by their α-level set:

$$[fg]^\alpha = [\min\{f^+g^+, f^-g^+, f^+g^-, f^-g^-\}, \max\{f^+g^+, f^-g^+, f^+g^-, f^-g^-\}]_\alpha,$$

and we cannot obtain $(fg)^\Delta = f^\sigma g^\Delta + f^\Delta g$. For example, let $\mathbb{T} = \mathbb{R}$ and $f, g : [\frac{1}{2}, 1] \to \mathbb{R}_{\mathcal{F}}$ be given such that

$$[f(t)]^\alpha = [f^-(t), f^+(t)]_\alpha = [-2, 1],\ [g(t)]^\alpha = [g^-(t), g^+(t)]_\alpha = [t^2, t]_\alpha,$$

then $[f'(t)]^\alpha = [0, 0], [g'(t)]^\alpha = [1, 2t]_\alpha, [(fg)']^\alpha = [-2, 1]$, where $t \in (\frac{1}{2}, 1), \alpha \in [0, 1]$. Consequently,

$$[f(t)g'(t)]^\alpha = [-2, 1][1, 2t]_\alpha = [-4t, 2t]_\alpha,$$

$$[g(t)f'(t)]^\alpha = [t^2, t]_\alpha[0, 0] = [0, 0],$$

where $t \in (\frac{1}{2}, 1)$ and $\alpha \in [0, 1]$. Thus, it follows that $[f(t)g'(t)]^\alpha + [g(t)f'(t)]^\alpha \not\subseteq [(fg)'(t)]^\alpha$, i.e., $(fg)' \neq fg' + f'g$.

Theorem 3.11 *Let $f, g : \mathbb{T} \to [\mathbb{R}^n_{\mathcal{F}}]$ be (I)-gH-Δ-differentiable. Then*

(*i*) *if $I^{\alpha,(I)}_{f_i,g_i} \leq 0$, $I^{\alpha,(I)}_{f_i,g_i^{\Delta_I}} \leq 0$, $I^{\alpha,(I)}_{f_i^{\Delta_I},g_i^\sigma} \leq 0$ and $f * g$ is (I)-gH-Δ-differentiable, then*

$$(f * g)^{\Delta_I} = f * g^{\Delta_I} \tilde{\mp} f^{\Delta_I} * g^\sigma.$$

(*ii*) *if $I^{\alpha,(I)}_{f_i,g_i} \leq 0$, $I^{\alpha,(I)}_{f_i,g_i^{\Delta_I}} \geq 0$, $I^{\alpha,(I)}_{f_i^{\Delta_I},g_i^\sigma} \geq 0$ and $f * g$ is (II)-gH-Δ-differentiable, then*

$$(f * g)^{\Delta_{II}} = f * g^{\Delta_I} \tilde{\mp} f^{\Delta_I} * g^\sigma.$$

(*iii*) *if $I^{\alpha,(II)}_{f_i,g_i} \leq 0$, $I^{\alpha,(II)}_{f_i,g_i^{\Delta_I}} \leq 0$, $I^{\alpha,(II)}_{f_i^{\Delta_I},g_i^\sigma} \leq 0$ and $f \circledast g$ is (I)-gH-Δ-differentiable, then*

$$(f \circledast g)^{\Delta_I} = f \circledast g^{\Delta_I} \tilde{\mp} g^\sigma \circledast f^{\Delta_I}.$$

(*iv*) *if $I^{\alpha,(II)}_{f_i,g_i} \leq 0$, $I^{\alpha,(II)}_{f_i,g_i^{\Delta_I}} \geq 0$, $I^{\alpha,(II)}_{f_i^{\Delta_I},g_i^\sigma} \geq 0$ and $f \circledast g$ is (II)-gH-Δ-differentiable, then*

$$(f \circledast g)^{\Delta_{II}} = f \circledast g^{\Delta_I} \tilde{\mp} f^{\Delta_I} \circledast g^\sigma.$$

(v) *if* $I^{\alpha,(I)}_{f_i,g_i} \geq 0,\ I^{\alpha,(I)}_{f_i,g_i^{\Delta_I}} \geq 0,\ I^{\alpha,(I)}_{f_i^{\Delta_I},g_i^{\sigma}} \geq 0$ *and* $f * g$ *is* (I)-gH-Δ-*differentiable, then*

$$(f * g)^{\Delta_I} = f * g^{\Delta_I} \widetilde{\mp} f^{\Delta_I} * g^{\sigma}.$$

(vi) *if* $I^{\alpha,(I)}_{f_i,g_i} \geq 0,\ I^{\alpha,(I)}_{f_i,g_i^{\Delta_I}} \leq 0,\ I^{\alpha,(I)}_{f_i^{\Delta_I},g_i^{\sigma}} \leq 0$ *and* $f * g$ *is* (II)-gH-Δ-*differentiable, then*

$$(f * g)^{\Delta_{II}} = f * g^{\Delta_I} \widetilde{\mp} f^{\Delta_I} * g^{\sigma}.$$

(vii) *if* $I^{\alpha,(II)}_{f_i,g_i} \geq 0,\ I^{\alpha,(II)}_{f_i,g_i^{\Delta_I}} \geq 0,\ I^{\alpha,(II)}_{f_i^{\Delta_I},g_i^{\sigma}} \geq 0$ *and* $f \circledast g$ *is* (I)-gH-Δ-*differentiable, then*

$$(f \circledast g)^{\Delta_I} = f \circledast g^{\Delta_I} \widetilde{\mp} g^{\sigma} \circledast f^{\Delta_I}.$$

(viii) *if* $I^{\alpha,(II)}_{f_i,g_i} \geq 0,\ I^{\alpha,(II)}_{f_i,g_i^{\Delta_I}} \leq 0,\ I^{\alpha,(II)}_{f_i^{\Delta_I},g_i^{\sigma}} \leq 0$ *and* $f \circledast g$ *is* (II)-gH-Δ-*differentiable, then*

$$(f \circledast g)^{\Delta_{II}} = f \circledast g^{\Delta_I} \widetilde{\mp} f^{\Delta_I} \circledast g^{\sigma}.$$

Proof (i). Since f, g be (I)-gH-Δ-differentiable, then

$$[f^{\Delta_I}]^{\alpha} = \times_{i=1}^{n}\big[(f_{i,\alpha}^{-})^{\Delta}, (f_{i,\alpha}^{+})^{\Delta}\big],\quad [g^{\Delta_I}]^{\alpha} = \times_{i=1}^{n}\big[(g_{i,\alpha}^{-})^{\Delta}, (g_{i,\alpha}^{+})^{\Delta}\big].$$

From $I^{\alpha,(I)}_{f_i,g_i} \leq 0$, we have $[f * g]^{\alpha} = \times_{i=1}^{n}[f_{i,\alpha}^{-} g_{i,\alpha}^{+}, f_{i,\alpha}^{+} g_{i,\alpha}^{-}]$. Moreover, because $f * g$ is (I)-gH-Δ-differentiable, we obtain

$$\begin{aligned}\big[(f * g)^{\Delta_I}\big]^{\alpha} &= \times_{i=1}^{n}\big[(f_{i,\alpha}^{-} g_{i,\alpha}^{+})^{\Delta}, (f_{i,\alpha}^{+} g_{i,\alpha}^{-})^{\Delta}\big] \\ &= \times_{i=1}^{n}\big[f_{i,\alpha}^{-}(g_{i,\alpha}^{+})^{\Delta} + (g_{i,\alpha}^{+})^{\sigma}(f_{i,\alpha}^{-})^{\Delta}, f_{i,\alpha}^{+}(g_{i,\alpha}^{-})^{\Delta} + (g_{i,\alpha}^{-})^{\sigma}(f_{i,\alpha}^{+})^{\Delta}\big].\end{aligned}$$

In addition, from $I^{\alpha,(I)}_{f_i,g_i^{\Delta_I}} \leq 0,\ I^{\alpha,(I)}_{f_i^{\Delta_I},g_i^{\sigma}} \leq 0$, we have

$$\big[f * g^{\Delta_I}\big]^{\alpha} = \times_{i=1}^{n}\big[f_{i,\alpha}^{-}(g_{i,\alpha}^{+})^{\Delta}, f_{i,\alpha}^{+}(g_{i,\alpha}^{-})^{\Delta}\big],$$

$$\big[f^{\Delta_I} * g^{\sigma}\big]^{\alpha} = \times_{i=1}^{n}\big[(g_{i,\alpha}^{+})^{\sigma}(f_{i,\alpha}^{-})^{\Delta}, (g_{i,\alpha}^{-})^{\sigma}(f_{i,\alpha}^{+})^{\Delta}\big],$$

and thus, we obtain $\big[f * g^{\Delta_I}\big]^{\alpha} + \big[f^{\Delta_I} * g^{\sigma}\big]^{\alpha} = \big[(f * g)^{\Delta_I}\big]^{\alpha}$.

By applying a similar discussion as in (i) to $(ii) - (viii)$, one will obtain the results $(ii) - (viii)$ immediately, so we will not repeat these similar proof steps here. □

Theorem 3.12 *Let* $f, g : \mathbb{T} \to [\mathbb{R}^n_{\mathscr{F}}]$ *be* (II)-gH-Δ-*differentiable. Then*

(i) *if* $I^{\alpha,(II)}_{f_i,g_i} \leq 0,\ I^{\alpha,(I)}_{f_i^{\sigma},g_i^{\Delta_{II}}} \geq 0,\ I^{\alpha,(I)}_{f_i^{\Delta_{II}},g_i} \leq 0$ *and* $f \circledast g$ *is* (I)-gH-Δ-*differentiable, then*

$$(f \circledast g)^{\Delta_I} = f * g^{\Delta_{II}} \widetilde{\mp} f^{\Delta_{II}} * g^{\sigma}.$$

(ii) *if* $I^{\alpha,(II)}_{f_i,g_i} \leq 0$, $I^{\alpha,(I)}_{f_i,g_i^{\Delta_{II}}} \leq 0$, $I^{\alpha,(I)}_{f_i^{\Delta_{II}},g_i^{\sigma}} \geq 0$ *and* $f \circledast g$ *is* (II)-gH-Δ-*differentiable, then*

$$(f \circledast g)^{\Delta_{II}} = f * g^{\Delta_I} \widetilde{+} f^{\Delta_I} * g^{\sigma}.$$

(iii) *if* $I^{\alpha,(II)}_{f_i,g_i} \geq 0$, $I^{\alpha,(I)}_{f_i,g_i^{\Delta_{II}}} \leq 0$, $I^{\alpha,(I)}_{f_i^{\Delta_{II}},g_i^{\sigma}} \geq 0$ *and* $f \circledast g$ *is* (I)-gH-Δ-*differentiable, then*

$$(f \circledast g)^{\Delta_I} = f * g^{\Delta_{II}} \widetilde{+} g^{\sigma} * f^{\Delta_{II}}.$$

(iv) *if* $I^{\alpha,(II)}_{f_i,g_i} \geq 0$, $I^{\alpha,(I)}_{f_i,g_i^{\Delta_{II}}} \geq 0$, $I^{\alpha,(I)}_{f_i^{\Delta_{II}},g_i^{\sigma}} \leq 0$ *and* $f \circledast g$ *is* (II)-gH-Δ-*differentiable, then*

$$(f \circledast g)^{\Delta_{II}} = f * g^{\Delta_{II}} \widetilde{+} f^{\Delta_{II}} * g^{\sigma}.$$

(v) *if* $I^{\alpha,(I)}_{f_i,g_i} \leq 0$, $I^{\alpha,(II)}_{f_i,g_i^{\Delta_{II}}} \geq 0$, $I^{\alpha,(II)}_{f_i^{\Delta_{II}},g_i^{\sigma}} \leq 0$ *and* $f * g$ *is* (I)-gH-Δ-*differentiable, then*

$$(f * g)^{\Delta_I} = f \circledast g^{\Delta_{II}} \widetilde{+} f^{\Delta_{II}} \circledast g^{\sigma}.$$

(vi) *if* $I^{\alpha,(I)}_{f_i,g_i} \leq 0$, $I^{\alpha,(II)}_{f_i,g_i^{\Delta_{II}}} \leq 0$, $I^{\alpha,(II)}_{f_i^{\Delta_{II}},g_i^{\sigma}} \geq 0$ *and* $f * g$ *is* (II)-gH-Δ-*differentiable, then*

$$(f * g)^{\Delta_{II}} = f \circledast g^{\Delta_{II}} \widetilde{+} f^{\Delta_{II}} \circledast g^{\sigma}.$$

(vii) *if* $I^{\alpha,(I)}_{f_i,g_i} \geq 0$, $I^{\alpha,(II)}_{f_i,g_i^{\Delta_{II}}} \leq 0$, $I^{\alpha,(II)}_{f_i^{\Delta_{II}},g_i^{\sigma}} \geq 0$ *and* $f * g$ *is* (I)-gH-Δ-*differentiable, then*

$$(f * g)^{\Delta_I} = f \circledast g^{\Delta_{II}} \widetilde{+} g^{\sigma} \circledast f^{\Delta_{II}}.$$

(viii) *if* $I^{\alpha,(I)}_{f_i,g_i} \geq 0$, $I^{\alpha,(II)}_{f_i,g_i^{\Delta_{II}}} \geq 0$, $I^{\alpha,(II)}_{f_i^{\Delta_{II}},g_i^{\sigma}} \leq 0$ *and* $f * g$ *is* (II)-gH-Δ-*differentiable, then*

$$(f * g)^{\Delta_{II}} = f \circledast g^{\Delta_{II}} \widetilde{+} f^{\Delta_{II}} \circledast g^{\sigma}.$$

Proof (i). Since f, g be (II)-gH-Δ-differentiable, then

$$[f^{\Delta_{II}}]^{\alpha} = \times_{i=1}^{n}\big[(f^{+}_{i,\alpha})^{\Delta}, (f^{-}_{i,\alpha})^{\Delta}\big], \quad [g^{\Delta_{II}}]^{\alpha} = \times_{i=1}^{n}\big[(g^{+}_{i,\alpha})^{\Delta}, (g^{-}_{i,\alpha})^{\Delta}\big].$$

From $I^{\alpha,(II)}_{f_i,g_i} \leq 0$, we have $[f \circledast g]^{\alpha} = \times_{i=1}^{n}[f^{+}_{i,\alpha}g^{+}_{i,\alpha}, f^{-}_{i,\alpha}g^{-}_{i,\alpha}]$. Moreover, because $f \circledast g$ is (I)-gH-Δ-differentiable, we obtain

$$\begin{aligned}\big[(f \circledast g)^{\Delta_I}\big]^{\alpha} &= \times_{i=1}^{n}\big[(f^{+}_{i,\alpha}g^{+}_{i,\alpha})^{\Delta}, (f^{-}_{i,\alpha}g^{-}_{i,\alpha})^{\Delta}\big] \\ &= \times_{i=1}^{n}\big[(f^{+}_{i,\alpha})(g^{+}_{i,\alpha})^{\Delta} + (g^{+}_{i,\alpha})^{\sigma}(f^{+}_{i,\alpha})^{\Delta}, f^{-}_{i,\alpha}(g^{-}_{i,\alpha})^{\Delta} + (g^{-}_{i,\alpha})^{\sigma}(f^{-}_{i,\alpha})^{\Delta}\big].\end{aligned}$$

In addition, from $I^{\alpha,(I)}_{f_i,g_i^{\Delta_{II}}} \geq 0$, $I^{\alpha,(I)}_{f_i^{\Delta_{II}},g_i^{\sigma}} \leq 0$, we have

$$\big[f * g^{\Delta_{II}}\big]^{\alpha} = \times_{i=1}^{n}\big[f^{+}_{i,\alpha}(g^{+}_{i,\alpha})^{\Delta}, f^{-}_{i,\alpha}(g^{-}_{i,\alpha})^{\Delta}\big],$$

$$\big[f^{\Delta_{II}} * g^{\sigma}\big]^{\alpha} = \times_{i=1}^{n}\big[(g^{+}_{i,\alpha})^{\sigma}(f^{+}_{i,\alpha})^{\Delta}, (g^{-}_{i,\alpha})^{\sigma}(f^{-}_{i,\alpha})^{\Delta}\big],$$

and thus, we obtain $\left[f * g^{\Delta_{II}}\right]^{\alpha} + \left[f^{\Delta_{II}} * g^{\sigma}\right]^{\alpha} = \left[(f \circledast g)^{\Delta_{I}}\right]^{\alpha}$.

By applying a similar discussion as in (i) to $(ii) - (viii)$, the results $(ii) - (viii)$ may be checked immediately and we will not repeat these similar proof steps here. □

3.2 Δ-Integral of Fuzzy Vector-Valued Functions on Time Scales

In this section, we examine the relations between gH-Δ-differentiability and the integral of fuzzy vector-valued functions on time scales.

Definition 3.6 The fuzzy Aumann Δ-integral (or Δ-integral for short) of $f : [a, b]_{\mathbb{T}} \to \mathbb{R}^n_{\mathscr{F}}$ is defined level-wise by

$$\begin{aligned}\left[\int_a^b f(t)\Delta t\right]^{\alpha} &= \int_a^b [f(t)]^{\alpha} \Delta t = \times_{i=1}^{n}\left[\int_a^b [f_i(t)]^{\alpha} \Delta t\right] \\ &= \times_{i=1}^{n}\left[\int_a^b f_i^{-}(t)\Delta t, \int_a^b f_i^{+}(t)\Delta t\right], \quad \alpha \in [0, 1].\end{aligned}$$

Theorem 3.13 *Let $f : [a, b]_{\mathbb{T}} \to \mathbb{R}^n_{\mathscr{F}}$ be continuous with $[f(t)]^{\alpha} = \times_{i=1}^{n}[f_i^{-}, f_i^{+}]_{\alpha}$. Then*

(i) *the function $F(t) = \int_a^t f(s)\Delta s$ is gH-Δ-differentiable and $F^{\Delta}(t) = f(t)$;*
(ii) *the function $F(t) = \int_t^b f(s)\Delta s$ is gH-Δ-differentiable and $G^{\Delta}(t) = -f(t)$;*

Proof We have

$$\begin{aligned}[F(t)]^{\alpha} &= \times_{i=1}^{n}[F_i^{-}, F_i^{+}]_{\alpha} = \left[\int_a^t f(s)\Delta s\right]^{\alpha} \\ &= \times_{i=1}^{n}\left[\int_a^t f_i^{-}(s)\Delta s, \int_a^t f_i^{+}(s)\Delta s\right]_{\alpha},\end{aligned}$$

$$\begin{aligned}[G(t)]^{\alpha} &= \times_{i=1}^{n}[G_i^{-}, G_i^{+}]_{\alpha} = \left[\int_t^b f(s)\Delta s\right]^{\alpha} \\ &= \times_{i=1}^{n}\left[\int_t^b f_i^{-}(s)\Delta s, \int_t^b f_i^{+}(s)\Delta s\right]_{\alpha}.\end{aligned}$$

Then

$$\begin{aligned}\left[F^{\Delta}(t)\right]^{\alpha} &= \times_{i=1}^{n}\left[\left(\int_a^t f_i^{-}(s)\Delta s\right)^{\Delta}, \left(\int_a^t f_i^{+}(s)\Delta s\right)^{\Delta}\right]_{\alpha} \\ &= \times_{i=1}^{n}[f_i^{-}, f_i^{+}]_{\alpha} = [f(t)]^{\alpha},\end{aligned}$$

$$[G^{\Delta}(t)]^{\alpha} = \times_{i=1}^{n}\left[\left(\int_t^b f_i^{-}(s)\Delta s\right)^{\Delta}, \left(\int_t^b f_i^{+}(s)\Delta s\right)^{\Delta}\right]_{\alpha}$$
$$= \times_{i=1}^{n}[-f_i^{+}, -f_i^{-}]_{\alpha} = [-f(t)]^{\alpha}.$$

This completes the proof. □

Theorem 3.14 *If $f : [a, b]_{\mathbb{T}} \to \mathbb{R}^n_{\mathcal{F}}$ is Δ-integrable and $c \in [a, b]_{\mathbb{T}}$. Then*

$$\int_a^b f(t)\Delta t = \int_a^c f(t)\Delta t \tilde{+} \int_c^b f(t)\Delta t.$$

Proof Clearly the integrability of f implies that f is integrable over any subinterval of $[a, b]_{\mathbb{T}}$. Now let $\alpha \in [0, 1]$ and f be a measurable selection for $\times_{i=1}^{n}[f_i^{-}(t), f_i^{+}(t)]_{\alpha}$. Since $\int_a^b f_i^{+}(t) = \int_a^c f_i^{+}(t)\Delta t + \int_c^b f_i^{+}(t)\Delta t$ and $\int_a^b f_i^{-}(t) = \int_a^c f_i^{-}(t)\Delta t + \int_c^b f_i^{-}(t)\Delta t$ for each $i = 1, 2, \ldots, n$ then we obtain

$$\left[\int_a^b f(t)\Delta t\right]^{\alpha} = \times_{i=1}^{n}\left[\int_a^b f_i^{-}(t)\Delta t, \int_a^b f_i^{+}(t)\Delta t\right]_{\alpha}$$
$$= \times_{i=1}^{n}\left[\int_a^c f_i^{-}(t)\Delta t + \int_c^b f_i^{-}(t)\Delta t, \int_a^c f_i^{+}(t)\Delta t + \int_c^b f_i^{+}(t)\Delta t\right]_{\alpha}$$
$$= \left(\times_{i=1}^{n}\left[\int_a^c f_i^{-}(t)\Delta t, \int_a^c f_i^{+}(t)\Delta t\right]_{\alpha}\right)$$
$$+\left(\times_{i=1}^{n}\left[\int_c^b f_i^{-}(t)\Delta t, \int_c^b f_i^{+}(t)\Delta t\right]_{\alpha}\right)$$
$$= \left[\int_a^c f(t)\Delta t\right]^{\alpha} + \left[\int_c^b f(t)\Delta t\right]^{\alpha}.$$

This completes the proof. □

Theorem 3.15 *If f is gH-Δ differentiable with no switching point in the interval $[a, b]_{\mathbb{T}}$, then*

$$\int_a^b f^{\Delta}(t)\Delta t = f(b)\tilde{-}_{gH} f(a).$$

Proof If there is no switching point in the interval $[a, b]_{\mathbb{T}}$ then f is (I) or (II) gH-Δ-differentiable as in Definition 3.4. Without loss of generality, we assume that f is (II)-gH-Δ-differentiable (the proof for the (I)-gH-Δ-differentiability case being similar).

We obtain

$$\begin{aligned}\left[\int_a^b f^{\Delta}(t)\Delta t\right]^{\alpha} &= \times_{i=1}^{n}\left[\int_a^b \left[(f_i^{+})^{\Delta}(t), (f_i^{-})^{\Delta}(t)\right]_{\alpha}\Delta t\right]_{\alpha} \\ &= \times_{i=1}^{n}\left[\int_a^b (f_i^{+})^{\Delta}(t)\Delta t, \int_a^b (f_i^{-})^{\Delta}(t)\Delta t\right]_{\alpha} \\ &= \times_{i=1}^{n}\left[f_i^{+}(b) - f_i^{+}(a), f_i^{-}(b) - f_i^{-}(a)\right]_{\alpha} \\ &= [f(b)]^{\alpha} \boxminus_{gH} [f(a)]^{\alpha}.\end{aligned}$$

This completes the proof. □

Theorem 3.16 *Assume that function f is gH-Δ-differentiable with n switching points at c_i, $i = 1, 2 \ldots, n$, $a = c_0 < c_1 < c_2 < \cdots < c_n < c_{n+1} = b$ and exactly at these points. Then*

$$f(b)\tilde{-}_{gH} f(a) = \sum_{i=1}^{n}\left[\int_{c_{i-1}}^{c_i} f^{\Delta}(t)\Delta t \tilde{-}_{gH}(-1)\int_{c_i}^{c_{i+1}} f^{\Delta}(t)\Delta t\right]. \tag{3.1}$$

Also,

$$\int_a^b f^{\Delta}(t)\Delta t = \sum_{i=1}^{n+1}\left(f(c_i)\tilde{-}_{gH} f(c_{i-1})\right), \tag{3.2}$$

where summation denotes standard fuzzy addition in this statement.

Proof For simplicity we consider only one switch point, the case of a finite number of switch-points follows similarly. Suppose that f is (I)-gH-Δ-differentiable on $[a, c]_{\mathbb{T}}$ and (II)–gH-Δ-differentiable on $[c, b]_{\mathbb{T}}$. Then by calculation we obtain

$$\begin{aligned}&\left[\int_a^c f^{\Delta}(t)\Delta t \tilde{-}_{gH}\int_c^b f^{\Delta}(t)\Delta t\right]^{\alpha} \\ &= \left[\left(f(c)\tilde{-}_{gH} f(a)\right)\tilde{-}_{gH}(-1)\left(f(b)\tilde{-}_{gH} f(c)\right)\right]^{\alpha} \\ &= \left[\left(f(c) - f(a)\right)\tilde{-}_{gH}\left(f(c)\tilde{-}_{gH} f(b)\right)\right]^{\alpha} \\ &= \times_{i=1}^{n}\{[f_i^{-}(c) - f_i^{-}(a), f_i^{+}(c) - f_i^{+}(a)]_{\alpha} \\ &\quad \boxminus_{gH}\left[f_i^{-}(c) - f_i^{-}(b), f_i^{+}(c) - f_i^{+}(b)\right]_{\alpha}\} \\ &= \times_{i=1}^{n}\left[\min\{f_i^{-}(b) - f_i^{-}(a), f_i^{+}(b) - f_i^{+}(a)\},\right. \\ &\quad \left.\max\{f_i^{-}(b) - f_i^{-}(a), f_i^{+}(b) - f_i^{+}(a)\}\right]_{\alpha} \\ &= [f(b)\tilde{-}_{gH} f(a)]^{\alpha}.\end{aligned}$$

Also, one can easily check that

$$\int_a^b f^{\Delta}(t)\Delta t = \int_a^c f^{\Delta}(t)\Delta t \tilde{\mp} \int_c^b f^{\Delta}(t)\Delta t = \big(f(c)\tilde{-}_{gH} f(a)\big)\tilde{\mp}\big(f(b)\tilde{-}_{gH} f(c)\big)$$

and $\big(f(c)\tilde{-}_{gH} f(a)\big)\tilde{\mp}\big(f(b)\tilde{-}_{gH} f(c)\big) = f(b)\tilde{-}_{gH} f(a)$ if and only if $f(c) \in \mathbb{R}^n$ is a singleton. This completes the proof. □

Theorem 3.17 *If $f, g : [a, b]_{\mathbb{T}} \to \mathbb{R}^n_{\mathscr{F}}$ are Δ-integrable on $[a, b]_{\mathbb{T}}$, then $\tilde{\alpha} \cdot f + \tilde{\beta} \cdot g$, where $\tilde{\alpha}, \tilde{\beta} \in \mathbb{R}$, is Δ-integrable on $[a, b]_{\mathbb{T}}$ and*

$$\int_a^b \big(\tilde{\alpha} \cdot f(t) + \tilde{\beta} \cdot g(t)\big)\Delta t = \tilde{\alpha} \cdot \int_a^b f(t)\Delta t \tilde{\mp} \tilde{\beta} \cdot \int_a^b g(t)\Delta t.$$

Proof For any $0 \le \alpha \le 1$, we have

$$\begin{aligned}
&\left[\tilde{\alpha} \cdot \int_a^b f(t)\Delta t \tilde{\mp} \tilde{\beta} \cdot \int_a^b g(t)\Delta t\right]^{\alpha} \\
&= \tilde{\alpha} \cdot \left(\times_{i=1}^n \left[\int_a^b f_i^-(t)\Delta t, \int_a^b f_i^+(t)\Delta t\right]_{\alpha}\right) \\
&\quad + \tilde{\beta} \cdot \left(\times_{i=1}^n \left[\int_a^b g_i^-(t)\Delta t, \int_a^b g_i^+(t)\Delta t\right]_{\alpha}\right) \\
&= \left(\times_{i=1}^n \left[\int_a^b \tilde{\alpha} \cdot f_i^-(t)\Delta t, \int_a^b \tilde{\alpha} \cdot f_i^+(t)\Delta t\right]_{\alpha}\right) \\
&\quad + \left(\times_{i=1}^n \left[\int_a^b \tilde{\beta} \cdot g_i^-(t)\Delta t, \int_a^b \tilde{\beta} \cdot g_i^+(t)\Delta t\right]_{\alpha}\right) \\
&= \times_{i=1}^n \left[\int_a^b \big(\tilde{\alpha} \cdot f_i^-(t) + \tilde{\beta} \cdot g_i^-(t)\big)\Delta t, \int_a^b \big(\tilde{\alpha} \cdot f_i^+(t) + \tilde{\beta} \cdot g_i^+(t)\big)\Delta t\right]_{\alpha} \\
&= \left[\int_a^b \big(\tilde{\alpha} \cdot f(t) + \tilde{\beta} \cdot g(t)\big)\Delta t\right]^{\alpha}.
\end{aligned}$$

This completes the proof. □

Theorem 3.18 *Assume $f, g : [a, b]_{\mathbb{T}} \to \mathbb{R}^n_{\mathscr{F}}$ are gH-Δ-differentiable, then $\int_a^b f(t)\Delta t = -\int_b^a f(t)\Delta t$.*

Proof In fact, for any $\alpha \in [0, 1]$, we can obtain

$$
\begin{aligned}
\left[\int_a^b f(t)\Delta t\right]^\alpha &= \times_{i=1}^n \left[\int_a^b f_i^-(t)\Delta t, \int_a^b f_i^+(t)\Delta t\right]_\alpha \\
&= \times_{i=1}^n \left[-\int_b^a f_i^-(t)\Delta t, -\int_b^a f_i^+(t)\Delta t\right]_\alpha \\
&= -\int_b^a \left(\times_{i=1}^n \left[f_i^-(t), f_i^+(t)\right]_\alpha\right)\Delta t \\
&= -\int_b^a [f(t)]^\alpha \Delta t = \left[-\int_b^a f(t)\Delta t\right]^\alpha .
\end{aligned}
$$

This completes the proof. □

Theorem 3.19 *Assume $f, g : [a, b]_{\mathbb{T}} \to \mathbb{R}^n_{\mathscr{F}}$ are (I)-gH-Δ-differentiable and $f * g$ is also (I)-gH-Δ-differentiable. If there is no switching point in $[a, b]_{\mathbb{T}}$ and $I^{\alpha,(I)}_{f_i,g_i} > 0$, $I^{\alpha,(I)}_{f_i^\sigma, g_i^{\Delta_I}} > 0$, $I^{\alpha,(I)}_{f_i^{\Delta_I}, g_i} > 0$ for each $i = 1, 2, \ldots, n$, then*

$$
\int_a^b f(t) * g^{\Delta_I}(t)\Delta t = \big(f(b) * g(b) \tilde{-}_{gH} f(a) * g(a)\big) \tilde{-}_{gH_I} \int_a^b f^{\Delta_I}(t) * g\big(\sigma(t)\big)\Delta t \text{ or} \tag{3.3}
$$

$$
\int_a^b f(t) * g^{\Delta_I}(t)\Delta t = \int_a^b g\big(\sigma(t)\big) * f^{\Delta_I}(t)\Delta t \tilde{-}_{gH_{II}} \big(f(a) * g(a) \tilde{-}_{gH} f(b) * g(b)\big). \tag{3.4}
$$

Proof First, let us prove (3.3). Since

$$
\begin{aligned}
& f(b) * g(b) \tilde{-}_{gH} f(a) * g(a) \\
&= \int_a^b \big(f(t) * g(t)\big)^{\Delta_I} \Delta t \\
&= \int_a^b \big(f(t) * g^{\Delta_I}(t) \tilde{+} f^{\Delta_I}(t) * g\big(\sigma(t)\big)\big)\Delta t \\
&= \int_a^b f(t) * g^{\Delta_I}(t)\Delta t \tilde{+} \int_a^b f^{\Delta_I}(t) * g\big(\sigma(t)\big)\Delta t,
\end{aligned}
$$

which implies that

$$
\int_a^b f(t) * g^{\Delta_I}(t)\Delta t = \big(f(b) * g(b) \tilde{-}_{gH} f(a) * g(a)\big) \tilde{-}_{gH_I} \int_a^b f^{\Delta_I}(t) * g\big(\sigma(t)\big)\Delta t.
$$

Now, we prove (3.4). Because

$$\begin{aligned}
&f(a) * g(a) \widetilde{-}_{gH} f(b) * g(b) \\
&= (-1) \int_a^b \big(f(t) * g(t)\big)^{\Delta_I} \Delta t \\
&= (-1) \int_a^b f(t) * g^{\Delta_I}(t) \Delta t \widetilde{+} (-1) \int_a^b f^{\Delta_I}(t) * g\big(\sigma(t)\big) \Delta t \\
&= \int_a^b g\big(\sigma(t)\big) * f^{\Delta_I}(t) \Delta t \widetilde{+} (-1) \int_a^b f(t) * g^{\Delta_I}(t) \Delta t,
\end{aligned}$$

which indicates that

$$\int_a^b f(t) * g^{\Delta_I}(t) \Delta t = \int_a^b g\big(\sigma(t)\big) * f^{\Delta_I}(t) \Delta t \widetilde{-}_{gH_{II}} \big(f(a) * g(a) \widetilde{-}_{gH} f(b) * g(b)\big).$$

This completes the proof. □

Shift Almost Periodic Fuzzy Vector-Valued Functions

4

Almost periodic functions and their generalized function theory on time scales have been developed in many literatures (see [57, 58, 60, 77–81, 83, 86–88, 90–93, 96–103, 106–114]) and these results have been applied to study various types of dynamic equations on time scales. In this chapter, we will develop a theory of shift almost periodic fuzzy vector-valued functions on time scales.

4.1 Shift Operators on Time Scales

In this section, the concept of shift operators on time scales and their basic properties will be presented which are necessary to study dynamic equations with shift operators on time scales (see [1, 2]).

4.1.1 Shift Operators

Definition 4.1 (*see* [1, 2]) Let $\mathbb{T}^*$ be a non-empty subset of the time scale $\mathbb{T}$ including a fixed number $t_0 \in \mathbb{T}^*$ such that there exist operators $\delta_\pm : [t_0, \infty)_\mathbb{T} \times \mathbb{T}^* \to \mathbb{T}^*$ satisfying the following properties:

P.1 The functions $\delta_\pm$ are strictly increasing with respect to their second arguments, i.e., if

$$(T_0, t), (T_0, u) \in \mathscr{D}_\pm := \{(s, t) \in [t_0, \infty)_\mathbb{T} \times \mathbb{T}^* : \delta_\pm(s, t) \in \mathbb{T}^*\},$$

then

$$T_0 \leqslant t < u \Rightarrow \delta_\pm(T_0, t) < \delta_\pm(T_0, u).$$

P.2 If $(T_1 u), (T_2, u) \in \mathscr{D}_-$ with $T_1 < T_2$, then

C. Wang and R. P. Agarwal, *Dynamic Equations and Almost Periodic Fuzzy Functions on Time Scales*, Synthesis Lectures on Mathematics & Statistics,
https://doi.org/10.1007/978-3-031-11236-2_4

$$\delta_-(T_1, u) > \delta_-(T_2, u),$$

and if $(T_1, u), (T_2, u) \in \mathscr{D}_+$ with $T_1 < T_2$, then

$$\delta_+(T_1, u) < \delta_+(T_2, u).$$

P.3 If $t \in [t_0, \infty)_{\mathbb{T}}$, then $(t, t_0) \in \mathscr{D}_+$ and $\delta_+(t, t_0) = t$. Moreover, if $t \in \mathbb{T}^*$, then $(t_0, t) \in \mathscr{D}_+$ and $\delta_+(t_0, t) = t$ holds.

P.4 If $(s, t) \in \mathscr{D}_\pm$, then $\big(s, \delta_\pm(s, t)\big) \in D$ and $\delta_\mp\big(s, \delta_\pm(s, t)\big) = t$, respectively.

P.5 If $(s, t) \in \mathscr{D}_\pm$ and $\big(u, \delta_\pm(s, t)\big) \in \mathscr{D}_\mp$, then

$$(s, \delta_\mp(u, t)) \in \mathscr{D}_\pm$$

and

$$\delta_\mp(u, \delta_\pm(s, t)) = \delta_\pm\big(s, \delta_\mp(u, t)\big),$$

respectively.

Then the operators δ_- and δ_+ associated with $t_0 \in \mathbb{T}^*$ (called the initial point) are said to be backward and forward shift operators on the set $\mathbb{T}^*$, respectively. The variable $s \in [t_0, \infty)_{\mathbb{T}}$ in $\delta_\pm(s, t)$ is called the shift size. The values $\delta_+(s, t)$ and $\delta_-(s, t)$ in $\mathbb{T}^*$ indicate s units translation of the term $t \in \mathbb{T}^*$ to the right and left, respectively. The sets $\mathscr{D}_\pm$ are the domains of the shift operators $\delta_\pm$, respectively.

Now we shall denote by $\mathbb{T}^*$ the largest subset of the time scale $\mathbb{T}$ such that the shift operators $\delta_\pm : [t_0, \infty)_{\mathbb{T}} \times \mathbb{T}^* \to \mathbb{T}^*$ exist.

Example 4.1.1 Let $\mathbb{T} = \mathbb{R}$ and $t_0 = 1$. The operators

$$\delta_-(s, t) = \begin{cases} t/s & \text{if } t \geqslant 0, \\ st & \text{if } t < 0, \end{cases} \quad \text{for} \quad s \in [1, \infty) \tag{4.1}$$

and

$$\delta_+(s, t) = \begin{cases} st & \text{if } t \geqslant 0, \\ t/s & \text{if } t < 0, \end{cases} \quad \text{for} \quad s \in [1, \infty) \tag{4.2}$$

are backward and forward shift operators (on the set $\mathbb{R}^* = \mathbb{R} - \{0\}$) associated with the initial point $t_0 = 1$. In the table below, we state different time scales with their corresponding shift operators. □

$\mathbb{T}$	t_0	$\mathbb{T}^*$	$\delta_-(s,t)$	$\delta_+(s,t)$
$\mathbb{R}$	0	$\mathbb{R}$	$t-s$	$t+s$
$\mathbb{Z}$	0	$\mathbb{Z}$	$t-s$	$t+s$
$q^{\mathbb{Z}}\cup\{0\}$	1	$q^{\mathbb{Z}}$	$\frac{t}{s}$	st
$\mathbb{N}^{1/2}$	0	$\mathbb{N}^{1/2}$	$\sqrt{t^2-s^2}$	$\sqrt{t^2+s^2}$

The proof of the next lemma is a direct consequence of Definition 4.1.

Lemma 4.1 (see [1, 2]) *Let δ_- and δ_+ be the shift operators associated with the initial point t_0. We have*

(i) $\delta_-(t,t)=t_0$ *for all* $t\in[t_0,\infty)_{\mathbb{T}}$.
(ii) $\delta_-(t_0,t)=t$ *for all* $t\in\mathbb{T}^*$.
(iii) *If* $(s,t)\in\mathscr{D}_+$, *then* $\delta_+(s,t)=u$ *implies* $\delta_-(s,u)=t$. *Conversely, If* $(s,u)\in\mathscr{D}_-$, *then* $\delta_-(s,u)=t$ *implies* $\delta_+(s,t)=u$.
(iv) $\delta_+\big(t,\delta_-(s,t_0)\big)=\delta_-(s,t)$ *for all* $(s,t)\in\mathscr{D}_+$ *with* $t\geqslant t_0$.
(v) $\delta_+\big(u,t\big)=\delta_+(t,u)$ *for all* $(u,t)\in\big([t_0,\infty)_{\mathbb{T}}\times[t_0,\infty)_{\mathbb{T}}\big)\cap\mathscr{D}_+$.
(vi) $\delta_+(s,t)\in[t_0,\infty)_{\mathbb{T}}$ *for all* $(s,t)\in\mathscr{D}_+$ *with* $t\geqslant t_0$.
(vii) $\delta_-(s,t)\in[t_0,\infty)_{\mathbb{T}}$ *for all* $(s,t)\in\big([t_0,\infty)_{\mathbb{T}}\times[s,\infty)_{\mathbb{T}}\big)\cap\mathscr{D}_-$.
(viii) *If* $\delta_+(s,\cdot)$ *is* Δ*-differentiable in its second variable, then* $\delta_+^{\Delta_t}(s,\cdot)>0$.
(ix) $\delta_+\big(\delta_-(u,s),\delta_-(s,v)\big)=\delta_-(u,v)$ *for all* $(s,v)\in\big([t_0,\infty)_{\mathbb{T}}\times[s,\infty)_{\mathbb{T}}\big)\cap\mathscr{D}_-$ *and* $(u,s)\in\big([t_0,\infty)_{\mathbb{T}}\times[u,\infty)_{\mathbb{T}}\big)\cap\mathscr{D}_-$.
(x) *If* $(s,t)\in\mathscr{D}_-$ *and* $\delta_-(s,t)=t_0$, *then* $s=t$.

Proof (i). The result follows from P.3–P.5 and that

$$\delta_-(t,t)=\delta_-\big(t,\delta_+(t,t_0)\big)=t_0\quad\text{for all } t\in[t_0,\infty)_{\mathbb{T}}.$$

(ii). From P.3–P.4 and

$$\delta_-(t_0,t)=\delta_-\big(t_0,\delta_+(t_0,t)\big)=t\quad\text{for all}\quad t\in\mathbb{T}^*.$$

Let $u:=\delta_+(s,t)$. By P.4 we have $(s,u)\in\mathscr{D}_-$ for all $(s,t)\in\mathscr{D}_+$, and hence,

$$\delta_-(s,u)=\delta_-\big(s,\delta_+(s,t)\big)=t.$$

The latter part of (iii) can be done in the similar way.

We have (iv) since P.3 and P.5 yield

$$\delta_+\big(t,\delta_-(s,t_0)\big)=\delta_-\big(s,\delta_+(t,t_0)\big)=\delta_-(s,t).$$

P.3 and P.5 guarantee that

$$t = \delta_+(t, t_0) = \delta_+\big(t, \delta_-(u, u)\big) = \delta_-\big(u, \delta_+(t, u)\big)$$

for all $(u, t) \in \big([t_0, \infty)_{\mathbb{T}} \times [t_0, \infty)_{\mathbb{T}}\big) \cap \mathscr{D}_+$.

Using (iii) we have

$$\delta_+(u, t) = \delta_+\left(u, \delta_-\big(u, \delta_+(t, u)\big)\right) = \delta_+(t, u).$$

This proves (v).

To prove (vi) and (vii), we use P.1–P.2 to get

$$\delta_+(s, t) \geqslant \delta_+(t_0, t) = t \geqslant t_0$$

for all $(s, t) \in \big([t_0, \infty)_{\mathbb{T}} \times [t_0, \infty)_{\mathbb{T}}\big) \cap \mathscr{D}_+$ and

$$\delta_-(s, t) \geqslant \delta_-(s, s) = t_0$$

for all $(s, t) \in ([t_0, \infty)_{\mathbb{T}} \times [s, \infty)_{\mathbb{T}}) \cap \mathscr{D}_-$.

Since $\delta_+(s, t)$ is strictly increasing in its second variable we have (viii) by (see [20]: Corollary 1.16).

(ix) According to P.5 and (v) we have

$$\begin{aligned}
\delta_+\big(\delta_-(u, s), \delta_-(s, v)\big) &= \delta_-\left(s, \delta_+\big(v, \delta_-(u, s)\big)\right) \\
&= \delta_-\big(s, \delta_-(u, \delta_+(v, s))\big) \\
&= \delta_-\big(s, \delta_+(s, \delta_-(u, v))\big) \\
&= \delta_-(u, v)
\end{aligned}$$

for all $(s, v) \in ([t_0, \infty)_{\mathbb{T}} \times [s, \infty)_{\mathbb{T}}) \cap \mathscr{D}_-$ and $(u, s) \in ([t_0, \infty)_{\mathbb{T}} \times [u, \infty)_{\mathbb{T}}) \cap \mathscr{D}_-$.

Suppose $(s, t) \in \mathscr{D}_- = \{(s, t) \in [t_0, \infty)_{\mathbb{T}} \times \mathbb{T}^* : \delta_-(s, t) \in \mathbb{T}^*\}$ and $\delta_-(s, t) = t_0$. Then by P.4 we have

$$t = \delta_+(s, \delta_-(s, t)) = \delta_+(s, t_0) = s.$$

This is (x). The proof is completed. □

Notice that the shift operators $\delta_\pm$ are defined once the initial point $t_0 \in \mathbb{T}$ is fixed. For instance, we choose the initial point $t_0 = 0$ to define shift operators $\delta_\pm(s, t) = t \pm s$ on $\mathbb{T} = \mathbb{R}$. However, if we choose $\lambda \in (0, \infty)$ as the initial point, then the new shift operators associated with λ are defined by $\widetilde{\delta}_\pm(s, t) = t \mp \lambda \pm s$. In terms of $\delta_\pm$ the new shift operators $\widetilde{\delta}_\pm$ can be defined as

$$\widetilde{\delta}_\pm(s, t) = \delta_\pm(\lambda, \delta_\pm(s, t)).$$

Example 4.1.2 Some particular time scales with shift operators associated with different initial points are provided to show the change in the formula of shift operators as the initial point changes. where $\lambda \in \mathbb{Z}_+$, $\mathbb{N}^{1/2} = \{\sqrt{n} : n \in \mathbb{N}\}$, $2^{\mathbb{N}} = \{2^n : n \in \mathbb{N}\}$, and $h\mathbb{Z} = \{hn : n \in \mathbb{Z}\}$. □

	$\mathbb{T} = \mathbb{N}^{1/2}$	$\mathbb{T} = h\mathbb{Z}$	$\mathbb{T} = 2^{\mathbb{N}}$
t_0	$0 \quad \lambda$	$0 \quad h\lambda$	$1 \quad 2^{\lambda}$
$\delta_-(s,t)$	$\sqrt{t^2 - s^2}$	$t - s$	$t/s \; 2^{\lambda} ts^{-1}$
$\delta_+(s,t)$	$\sqrt{t^2 + \lambda^2 - s^2}$	$t + h\lambda - s$	$ts \; 2^{-\lambda} ts$
	$\sqrt{t^2 + s^2}$	$t + s$	
	$\sqrt{t^2 - \lambda^2 + s^2}$	$t - h\lambda + s$	

4.1.2 Periodicity of Time Scales

In this subsection, we will state a classical concept of periodic time scales which was introduced in [47].

Definition 4.2 (*see* [47]) A time scale $\mathbb{T}$ is said to be periodic if there exists a $P > 0$ such that

$$t \pm P \in \mathbb{T} \text{ for all } t \in \mathbb{T}. \tag{4.3}$$

If $\mathbb{T} = \mathbb{R}$, the smallest positive P is called the period of the time scale.

In the following, a new periodicity notion will be presented which does not require the time scale to be closed under the operation $t \pm P$ for a fixed $P > 0$ or to be unbounded.

Definition 4.3 (*see* [1, 2], *Periodicity in shifts*) Let $\mathbb{T}$ be a time scale with the shift operators $\delta_{\pm}$ associated with the initial point $t_0 \in \mathbb{T}^*$. The time scale $\mathbb{T}$ is said to be periodic in shifts $\delta_{\pm}$ if there exists a $p \in (t_0, \infty)_{\mathbb{T}^*}$ such that $(p, t) \in \mathcal{D}_{\mp}$ for all $t \in \mathbb{T}^*$. Furthermore, if

$$P := \inf\{p \in (t_0, \infty)_{\mathbb{T}^*} : (p, t) \in \mathcal{D}_{\mp} \text{ for all } t \in \mathbb{T}^*\} \neq t_0, \tag{4.4}$$

then P is called the period of the time scale $\mathbb{T}$.

A time scale periodic in shifts may be bounded, the following example will be presented to show that a time scale, periodic in shifts, does not have to satisfy (4.3).

Example 4.1.3 The following time scales are not periodic in the sense of Definition 4.2 but periodic with respect to the notion of shift operators given in Definition 4.3 (see [1, 2]).
(1) $\mathbb{T}_1 = \{\pm n^2 : n \in \mathbb{Z}\}$,

$$\delta_{\pm}(P, t) = \begin{cases} (\sqrt{t} \pm \sqrt{P})^2 & \text{if } t > 0, \\ \pm P & \text{if } t = 0, \quad P = 1, \quad t_0 = 0, \\ -(\sqrt{-t} \pm \sqrt{P})^2 & \text{if } t < 0 \end{cases}$$

(2) $\mathbb{T}_2 = \overline{q^{\mathbb{Z}}}$, $\delta_{\pm}(P, t) = P^{\pm 1} t$, $P = q$, $t_0 = 1$,

(3) $\mathbb{T}_3 = \overline{\bigcup_{n\in\mathbb{Z}}[2^{2n}, 2^{2n+1}]}$, $\delta_{\pm}(P,t) = P^{\pm 1}t$, $P = 4$, $t_0 = 1$,

(4) $\mathbb{T}_4 = \left\{\frac{q^n}{1+q^n} : q > 1 \text{ is constant and } n \in \mathbb{Z}\right\} \cup \{0, 1\}$,

$$\delta_{\pm}(P,t) = \frac{q^{\left(\frac{\ln\left(\frac{t}{1-t}\right) \pm \ln\left(\frac{P}{1-P}\right)}{\ln q}\right)}}{1 + q^{\left(\frac{\ln\left(\frac{t}{1-t}\right) \pm \ln\left(\frac{P}{1-P}\right)}{\ln q}\right)}}, \quad P = \frac{q}{1+q}.$$

Note that the time scale $\mathbb{T}_4$ in Example 4.1.3 is bounded above and below and

$$\mathbb{T}_4 = \left\{\frac{q^n}{1+q^n} : q > 1 \text{ is constant and } n \in \mathbb{Z}\right\}.$$

□

Remark 4.1 Let $\mathbb{T}$ be a time scale that is periodic in shifts with the period P. Thorough P.4 of Definition 4.1, it follows that the mapping $\delta_+^P : \mathbb{T}^* \to \mathbb{T}^*$ defined by $\delta_+^P(t) = \delta_+(P,t)$ is surjective. On the other hand, we know by P.1 of Definition 4.1 that shift operators $\delta_{\pm}$ are strictly increasing in their second arguments. That is, the mapping $\delta_+^P(t) := \delta_+(P,t)$ is injective. Hence, δ_+^P is an invertible mapping with the inverse $(\delta_+^P)^{-1} = \delta_-^P$ defined by $\delta_-^P(t) := \delta_-(P,t)$.

For next two results, we will present that if $\mathbb{T}$ is a periodic time scale in shifts $\delta_{\pm}$ with period P, then the operators $\sigma_{\pm}^P : \mathbb{T}^* \to \mathbb{T}^*$ are commutative with the forward jump operator $\sigma : \mathbb{T} \to \mathbb{T}$ given by

$$\sigma(t) := \inf\{s \in \mathbb{T} : s > t\}.$$

That is,

$$\left(\delta_{\pm}^P \circ \sigma\right)(t) = \left(\sigma \circ \delta_{\pm}^P\right)(t) \quad \text{for all } t \in \mathbb{T}^*. \tag{4.5}$$

Lemma 4.2 (see [1, 2]) *The mapping $\delta_+^T : \mathbb{T}^* \to \mathbb{T}^*$ preserves the structure of the points in $\mathbb{T}^*$. That is,*

$$\delta(\widehat{t}) = \widehat{t} \Longrightarrow \sigma\left(\delta_+(P,\widehat{t})\right) = \delta_+(P,\widehat{t}).$$

$$\delta(\widehat{t}) > \widehat{t} \Longrightarrow \sigma\left(\delta_+(P,\widehat{t})\right) > \delta_+(P,\widehat{t}).$$

Proof From the definition, it follows that $\sigma(t) \geqslant t$ for all $t \in \mathbb{T}^*$. Thus, by P.1

$$\delta_+(P,\sigma(t)) \geqslant \delta_+(P,t).$$

Since $\sigma(\delta_+(P,t))$ is the smallest element satisfying

$$\sigma\big(\delta_+(P,t)\big) \geqslant \delta_+(P,t),$$

we get

$$\delta_+(P,\sigma(t)) \geqslant \sigma(\delta_+(P,t)) \quad \text{for all } t \in \mathbb{T}^*. \tag{4.6}$$

If $\sigma(\widehat{t}) = \widehat{t}$, the (4.6) implies

$$\delta_+(P,\widehat{t}) = \delta_+(P,\sigma(\widehat{t})) \geqslant \sigma(\delta_+(P,\widehat{t})).$$

That is,

$$\delta_+(P,\widehat{t}) = \sigma(\delta_+(P,\widehat{t})) \quad \text{if } \sigma(\widehat{t}) = \widehat{t}.$$

If $\sigma(\widehat{t}) > \widehat{t}$, then by definition of σ we have

$$(\widehat{t},\sigma(\widehat{t}))_{\mathbb{T}^*} = \varnothing \tag{4.7}$$

and by P.1

$$\delta_+(P,\sigma(\widehat{t})) > \delta_+(P,\widehat{t}).$$

Suppose to the contrary that $\delta_+(P,\widehat{t})$ is right dense, i.e., $\sigma(\delta_+(P,\widehat{t})) = \delta_+(P,\widehat{t})$. This along with (4.6) implies

$$(\delta_+(P,\widehat{t}),\delta_+(P,\sigma(\widehat{t})))_{\mathbb{T}^*} \neq \varnothing.$$

Pick one element $s \in (\delta_+(P,\widehat{t}),\delta_+(P,\sigma(\widehat{t})))_{\mathbb{T}}$. Since $\delta_+(P,t)$ is strictly increasing in t and invertible there should be an element $t \in (\widehat{t},\sigma(\widehat{t}))_{\mathbb{T}}$ such that $\delta_+(P,t) = s$. This contradicts (4.7). Hence, $\delta_+(P,\widehat{t})$ must be right scattered, i.e., $\sigma(\delta_+(P,\widehat{t})) > \delta_+(P,\widehat{t})$. The proof is completed. □

Corollary 4.1 (see [1, 2]) *The following equalities hold:*

$$\delta_+(P,\sigma(t)) = \sigma(\delta_+(P,t)) \quad \textit{for all } t \in \mathbb{T}^*. \tag{4.8}$$

$$\delta_-(P,\sigma(t)) = \sigma(\delta_-(P,t)) \quad \textit{for all } t \in \mathbb{T}^*. \tag{4.9}$$

Proof The equality (4.8) can be obtained similar to the proof of preceding lemma. By (4.8) we have

$$\delta_+(P,\sigma(s)) = \sigma(\delta_+(P,s)) \quad \text{for all } s \in \mathbb{T}^*.$$

Substituting $s = \delta_-(P,t)$ we obtain

$$\delta_+(P,\sigma(\delta_-(P,t))) = \sigma(\delta_+(P,\delta_-(P,t))) = \sigma(t).$$

This and (iii) of Lemma 4.1 imply

$$\sigma(\delta_-(P,t)) = \delta_-(P,\sigma(t)) \quad \text{for all } t \in \mathbb{T}^*.$$

This completes the proof. □

Note that (4.8) along with (4.9) yields (4.5).

Definition 4.4 (*see* [1, 2], *Periodic function in shifts* $\delta_\pm$) Let $\mathbb{T}$ be a time scale that is periodic in shifts $\delta_\pm$ with the period P. We say that a real valued function f defined on $\mathbb{T}^*$ is periodic in shifts $\delta_\pm$ if there exists a $\mathbb{T} \in [P, \infty)_{\mathbb{T}^*}$ such that

$$(T, t) \in \mathscr{D}_\pm \quad \text{and} \quad f(\delta_\pm^T(t)) = f(t) \quad \text{for all } t \in \mathbb{T}^*, \tag{4.10}$$

where $\delta_\pm^T(t) := \delta_\pm(T, t)$. The smallest number $T \in [P, \infty)_{\mathbb{T}^*}$ such that (4.10) holds is called the period of f.

Example 4.1.4 By Definition 4.3 we know that the real line $\mathbb{R}$ is periodic in shifts $\delta_\pm$ defined by (4.1)–(4.2) associated with the initial point $t_0 = 1$. The function

$$f(t) = \sin\left(\frac{\ln|t|}{\ln(1/2)\pi}\right), \quad t \in \mathbb{R}^* := \mathbb{R} - \{0\}$$

is periodic in shifts $\delta_\pm$ defined by (4.1)–(4.2) with the period $T = 4$ since

$$\begin{aligned} f(\delta_\pm(T, t)) &= \begin{cases} f(t4^{\pm 1}) & \text{if } t \geqslant 0, \\ f(t/4^{\pm 1}) & \text{if } t < 0, \end{cases} = \sin\left(\frac{\ln|t| \pm 2\ln(1/2)}{\ln(1/2)}\pi\right) \\ &= \sin\left(\frac{\ln|t|}{\ln(1/2)}\pi \pm 2\pi\right) \\ &= \sin\left(\frac{\ln|t|}{\ln(1/2)}\pi\right) \\ &= f(t) \end{aligned}$$

for all $t \in \mathbb{R}^*$. □

Example 4.1.5 The time scale $\overline{q^{\mathbb{Z}}} = \{q^n : n \in \mathbb{Z} \text{ and } q > 1\} \cup \{0\}$ is periodic in shifts $\delta_\pm(P, t) = P^{\pm 1}t$ with the period $P = q$. The function f defined by

$$f(t) = (-1)^{\frac{\ln t}{\ln q}}, \quad t \in q^{\mathbb{N}} \tag{4.11}$$

is periodic in shifts $\delta_\pm$ with the period $T = q^2$ since $\delta_+(q^2, t) \in \overline{q^{\mathbb{Z}}}^* = q^{\mathbb{Z}}$ and

$$f\left(\delta_\pm(q^2, t)\right) = (-1)^{\frac{\ln t}{\ln q} \pm 2} = (-1)^{\frac{\ln t}{\ln q}} = f(t)$$

for all $t \in q^{\mathbb{Z}}$. However, f is not periodic in the sense of Definition 4.2 since there is no any positive number T such that $f(t \pm T) = f(t)$ holds. □

Next, we introduce Δ-periodic function in shifts.

Definition 4.5 (*see* [1, 2], *Δ-periodic function in shifts* $\delta_\pm$) Let $\mathbb{T}$ be a time scale that is periodic in shifts $\delta_\pm$ with period P. We say that a real valued function f defined on $\mathbb{T}^*$ is Δ-periodic in shifts on $\mathbb{T}$ if there exists a $T \in [P, \infty)_{\mathbb{T}^*}$ such that

$$(T, t) \in \mathscr{D}_\pm \quad \text{for all } t \in \mathbb{T}^*, \tag{4.12}$$

$$\text{the shifts } \delta_\pm \text{ are } \Delta\text{-differentiable with rd-continuous derivatives,} \tag{4.13}$$

and

$$f(\delta_\pm^T(t))\delta_\pm^{\Delta T}(t) = f(t) \tag{4.14}$$

for all $t \in \mathbb{T}^*$, where $\delta_\pm^T(t) := \delta_\pm(T, t)$. The smallest number $T \in [P, \infty)_{\mathbb{T}^*}$ such that (4.12)–(4.14) hold is called the period of f.

Note that Definitions 4.4 and 4.5 give the classic periodicity definition (i.e. Definition 4.2) on time scales whenever $\delta_\pm^T(t) = t \pm T$ are the shifts satisfying the assumptions of Definitions 4.4 and 4.5.

Example 4.1.6 The real valued function $g(t) = 1/t$ defined on $2^{\mathbb{Z}} = \{2^n : n \in \mathbb{Z}\}$ is Δ-periodic in shifts $\delta_\pm(T, t) = T^{\pm 1}t$ with the period $T = 2$ since

$$f(\delta_\pm(2, t))\delta_\pm^\Delta(2, t) = \frac{1}{2^{\pm 1}t}2^{\pm 1} = \frac{1}{t} = f(t).$$

□

To present the proof of next theorem, the following result is essential.

Theorem 4.1 (Substitution, see [20] Theorem 1.98) *Assume* : $\mathbb{T} \to \mathbb{R}$ *is strictly increasing and* $\widetilde{\mathbb{T}} := v(\mathbb{T})$ *is a time scale. If* $f : \mathbb{T} \to \mathbb{R}$ *is an rd-continuous function and is differentiable with rd-continuous derivative, then for* $a, b \in \mathbb{T}$,

$$\int_a^b g(s)v^\Delta(s)\Delta s = \int_{v(a)}^{v(b)} g(v^{-1}(s))\widetilde{\Delta} s. \tag{4.15}$$

Theorem 4.2 (see [1, 2]) *Let* $\mathbb{T}$ *be a time scale that is periodic in shifts* $\delta_\pm$ *with period* $P \in [t_0, \infty)_{\mathbb{T}^*}$ *and* f *a* Δ*-periodic function in shifts* $\delta_\pm$ *with the period* $T \in [P, \infty)_{\mathbb{T}^*}$. *Suppose that* $f \in C_{rd}(\mathbb{T})$, *then*

$$\int_{t_0}^t f(s)\Delta s = \int_{\delta_\pm^T(t_0)}^{\delta_\pm^T(t)} f(s)\Delta s.$$

Proof Substituting $v(s) = \delta_+^T(s)$ and $g(s) = f(\delta_+^T(s))$ in (4.15) and taking (4.14) into account we have

$$\begin{aligned}\int_{\delta_+^T(t_0)}^{\delta_+^T(t)} f(s)\Delta s &= \int_{v(t_0)}^{v(t)} g(v^{-1}(s))\Delta s \\ &= \int_{t_0}^{t} g(s)v^{\Delta}(s)\Delta s = \int_{t_0}^{t} f(\delta_+^T(s))\delta_+^{\Delta T}(s)\Delta s \\ &= \int_{t_0}^{t} f(s)\Delta s.\end{aligned}$$

The equality

$$\int_{\delta_-^T(t)}^{\delta_-^T(t)} f(s)\Delta s = \int_{t_0}^{t} f(s)\Delta s$$

can be obtained similarly. This completes the proof. □

4.2 Complete-Closed Time Scales Under Non-translational Shifts

To introduce the concept of almost periodic fuzzy vector-valued functions on irregular time scales in the next section, in this part, we will introduce a concept of complete-closed time scales under shift operators.

We assume that $\delta_\pm$ are shift operators fulfilling Definition 3 from [2] and $\tilde{\mathscr{D}}_\pm := \{(s,t) \in [t_0,\infty)_{\mathbb{T}} \times \mathbb{T}^* : \delta_\pm(s,t) \in \mathbb{T}^*\}$, where $\mathbb{T}^*$ is the largest subset of the time scale $\mathbb{T}$, i.e., $\overline{\mathbb{T}^*} = \mathbb{T}$. If $\delta_\pm(s,t)$ is Δ-differentiable to its second argument, then $\delta_\pm^{\Delta}(s,t)$ denotes the Δ-derivative of $\delta_\pm(s,t)$ to its second argument.

Definition 4.6 (*see* [2]) Let $\mathbb{T}$ be a time scale with the shift operators $\delta_\pm$ associated with the initial point $t_0 \in \mathbb{T}^*$. The time scale $\mathbb{T}$ is said to be periodic in shifts $\delta_\pm$ if there exists a $p \in (t_0,\infty)_{\mathbb{T}^*}$ such that $(p,t) \in \tilde{\mathscr{D}}_\mp$ for all $t \in \mathbb{T}^*$. Furthermore, if

$$P := \inf\left\{p \in (t_0,\infty)_{\mathbb{T}^*} : (p,t) \in \tilde{\mathscr{D}}_\mp \text{ for all } t \in \mathbb{T}^*\right\} \neq t_0, \tag{4.16}$$

then P is called the period of the time scale $\mathbb{T}$, where $\tilde{\mathscr{D}}_\pm = \left\{(s,t) \in [t_0,\infty)_{\mathbb{T}} \times \mathbb{T}^* : \delta_\pm(s,t) \in \mathbb{T}^*\right\}$.

Now, we will provide an example to show Definition 4.6 needs a further extension.

Example 4.1.1 Consider the following time scale:

$$\mathbb{T} = \overline{-q^{\mathbb{Z}}} \cup \{1\} = \overline{\left\{-q^n : q > 1,\ n \in \mathbb{Z}\right\}} \cup \{1\}. \tag{4.17}$$

For such a time scale, we obtain that $\mathbb{T}^* = \left\{-q^n : q > 1,\ n \in \mathbb{Z}\right\} \cup \{1\}$. Take the initial point $t_0 = 1$ and the shift operators $\delta_-(s,t) = -st$, $\delta_+(s,t) = -\frac{t}{s}$. Let $\Pi^- = \left\{-q^n : q > 1,\ n \in \mathbb{Z}^+\right\} \subset \mathbb{T}^*$, we obtain $\delta_\pm(s,t) \in \mathbb{T}^*$ for any $s \in \Pi^-$. Hence, according to

Definition 4.6, the time scale $\mathbb{T}$ cannot be regarded as a periodic time scale under shifts $\delta_{\pm}$ since there is **no** number $P \in (1, +\infty)_{\mathbb{T}^*}$ satisfying (4.16). However, this time scale is the opposite number set of the time scale $\overline{q^{\mathbb{Z}}} = \{q^n : q > 1, n \in \mathbb{Z}\} \cup \{0\}$, and this time scale (4.17) also plays an important role in q-difference equations. In fact, from (4.17), it is easy to observe that for any $t \in \mathbb{T}^*$, we have $-(-q)t \in \mathbb{T}^*$ but $-q \notin [1, +\infty)$. □

For convenience, we introduce some notations. Let

$$\mathscr{D}_{\pm} = \{(s, t) \in \mathbb{T}^* \times \mathbb{T}^* : \delta_{\pm}(s, t) \in \mathbb{T}^*\}.$$

For any $s \in \mathbb{T}^*$, denote

$$\mathbb{T}_*^{\delta_s -} := \delta_-(s, \mathbb{T}^*) := \{\delta_-(s, t) : (s, t) \in \mathscr{D}_-, \forall t \in \mathbb{T}^*\}, \tag{4.18}$$

$$\mathbb{T}_*^{\delta_s +} := \delta_+(s, \mathbb{T}^*) := \{\delta_+(s, t) : (s, t) \in \mathscr{D}_+, \forall t \in \mathbb{T}^*\}. \tag{4.19}$$

Definition 4.7 Let $\mathbb{T}$ be a time scale with the shift operators $\delta_{\pm}$ associated with the initial point $t_0 \in \mathbb{T}^*$. The time scale $\mathbb{T}$ is said to be bi-direction shift complete-closed time scales (i.e. S-CCTS for short) in shifts $\delta_{\pm}$ if

$$\Pi := \{p \in \mathbb{T}^* : (p, t) \in \mathscr{D}_{\pm} \text{ for all } t \in \mathbb{T}^*\} = \Pi^{\pm} \notin \{\{t_0\}, \emptyset\}. \tag{4.20}$$

Remark 4.2 Note that from (4.18) and (4.19), it follows that (4.20) can be written into the equivalent form $\Pi = \{p \in \mathbb{T}^* : \mathbb{T}_*^{\delta_p \pm} \subseteq \mathbb{T}^*\} \notin \{\{t_0\}, \emptyset\}$.

Furthermore, from (4.20), we will refine the following the concept of S-CCTS attached with shift direction. For convenience, we will use the notations

$$\Pi^+ := \{p \in \mathbb{T}^* : \mathbb{T}_*^{\delta_p} \subseteq \mathbb{T}^*\}, \ \Pi^- := \{p \in \mathbb{T}^* : \mathbb{T}_*^{\delta_p -} \subseteq \mathbb{T}^*\}.$$

Definition 4.8 Let $\mathbb{T}$ be a S-CCTS, then (i) we say S-CCTS is with positive-direction if $\Pi^+ \notin \{\{t_0\}, \emptyset\}$; (ii) we say S-CCTS is with negative-direction if $\Pi^- \notin \{\{t_0\}, \emptyset\}$; (iii) we say S-CCTS is with bi-direction if $\Pi \notin \{\{t_0\}, \emptyset\}$.

Remark 4.3 From Definition 4.8, it follows that a bi-direction S-CCTS is also with a positive-direction and a negative-direction.

Example 4.1.2 From Definitions 4.7 and 4.8, we provide the following examples of S-CCTS.

(1) Let $\mathbb{T}=\overline{(-q)^{\mathbb{Z}}}\cup\{1\}=\big\{(-q)^n: q>1,\ n\in\mathbb{Z}\big\}\cup\{0,1\}$. Then we obtain that $\Pi^{\pm}=\big\{(-q)^{2n}: q>1,\ n\in\mathbb{Z}^+\big\}$. For any $t\in\mathbb{T}^*$, by taking $t_0=1$, we attach the following shift operators

$$\delta_+(s,t)=\begin{cases} st, & t>0,\\ \frac{t}{s}, & t<0,\end{cases}\qquad \delta_-(s,t)=\begin{cases} \frac{t}{s}, & t>0,\\ st, & t<0.\end{cases}$$

Hence, there exists $q^2\in\Pi^{\pm}$ such that $\delta_{\pm}(q^2,t)\in\mathbb{T}^*$ for all $t\in\mathbb{T}^*$, i.e., $\Pi^{\pm}\notin\big\{\{1\},\emptyset\big\}$. From Definition 4.8, it follows that $\mathbb{T}$ is a S-CCTS with bi-direction.

(2) Consider $\mathbb{T}=\big\{q^n: q>1,\ n\in\mathbb{Z}\big\}\cup\big\{-q^n: q>1,\ n\in\mathbb{Z}\big\}\cup\{0\}$. For any $t\in\mathbb{T}^*$, by taking $t_0=1$, we attach the shift operators

$$\delta_+(s,t)=\begin{cases} st, & t>0,\\ \frac{t}{s}, & t<0,\end{cases}\qquad \delta_-(s,t)=\begin{cases} \frac{t}{s}, & t>0,\\ st, & t<0.\end{cases}$$

Then we obtain that $\Pi^{\pm}=\big\{q^n: q>1,\ n\in\mathbb{Z}^+\big\}$. Hence, there exists $q\in\Pi^{\pm}$ such that $\delta_{\pm}(q,t)\in\mathbb{T}^*$ for all $t\in\mathbb{T}^*$, i.e., $\Pi^{\pm}\notin\big\{\{1\},\emptyset\big\}$. From Definition 4.8, it follows that $\mathbb{T}$ is a S-CCTS with bi-direction.

(3) Consider $\mathbb{N}_{\pm}^{\frac{1}{2}}=\big\{\pm\sqrt{n},\ n\in\mathbb{N}\big\}$, For any $t\in\mathbb{T}^*$, we take $t_0=0$ and attach the shift operators

$$\delta_+(s,t)=\begin{cases} \sqrt{s^2+t^2}, & t>0,\\ -\sqrt{t^2-s^2}, & t<0,\end{cases}\qquad \delta_-(s,t)=\begin{cases} \sqrt{t^2-s^2}, & t>0,\\ -\sqrt{t^2+s^2}, & t<0.\end{cases}$$

Then it follows that $\Pi^{\pm}=\mathbb{N}^{\frac{1}{2}}=\big\{\sqrt{n}: n\in\mathbb{N}\big\}$. Hence, there exists $1\in\Pi^{\pm}$ such that $\delta_{\pm}(1,t)\in\mathbb{T}^*$ for all $t\in\mathbb{T}^*$, i.e., $\Pi^{\pm}\notin\big\{\{0\},\emptyset\big\}$. From Definition 4.8, we obtain $\mathbb{T}$ is a S-CCTS with bi-direction.

(4) Let $\mathbb{T}_1=\big\{q^n: q>1,\ n\in\mathbb{Z}^+\big\}\cup\{1\}$ and $\mathbb{T}_2=\big\{q^n: q>1,\ n\in\mathbb{Z}^-\big\}\cup\{0,1\}$. Now we take $t_0=1$ and

$$\Pi_1^+=\big\{q^n: q>1,\ n\in\mathbb{Z}^+\big\}\subseteq\mathbb{T}_1^*,\ \Pi_2^-=\big\{q^n: q>1,\ n\in\mathbb{Z}^-\big\}\subseteq\mathbb{T}_2^*.$$

Noting that for any $s_1\in\Pi_1^+$, $s_2\in\Pi_2^-$, we have

$$\delta_+(s_1,t_1)=s_1t_1\in\mathbb{T}_1^*\text{ for all }t_1\in\mathbb{T}_1^*,\quad \delta_-(s_1,t_1)=\frac{t_1}{s_1}\notin\mathbb{T}_1^*\text{ for }t_1=q,\ s_1=q^2,$$

and

$$\delta_-(s_2,t_2)=s_2t_2\in\mathbb{T}_2^*\text{ for all }t_2\in\mathbb{T}_2^*,\quad \delta_+(s_2,t_2)=\frac{t_2}{s_2}\notin\mathbb{T}_2^*\text{ for }t_2=\frac{1}{q},\ s_2=\frac{1}{q^2}.$$

Hence, for the shift operator $\delta_+(s,t)=st$, we obatin $\mathbb{T}_1$ is a positive-direction S-CCTS. For the shift operator $\delta_-(s,t)=st$, $\mathbb{T}_2$ is a negative-direction S-CCTS. □

Remark 4.4 We attached the translation direction to time scales in [89] and introduced the concept of translation complete-closed time scales (i.e. T-CCTS). We also introduced the concepts of some special functions arising from differential and difference equations on T-CCTS including almost periodic functions and almost automorphic functions. However, these results will never cover some important and irregular time scales such as $\overline{(-q)^{\mathbb{Z}}}$ and $\overline{q^{\mathbb{Z}}}$, etc. Hence, these results can be applied to fuzzy q-difference equations.

Remark 4.5 Note that if $\mathbb{T}$ is a periodic time scales under translations and $\Pi \subseteq \mathbb{T}^*$, then the shift operators will fulfill $\delta_{\pm}(\tau, t) = t \pm \tau \in \mathbb{T}$ with an initial point $t_0 = 0$. Hence, if $\Pi \subseteq \mathbb{T}^*$, then T-CCTS is included in S-CCTS.

If $\mathbb{T}$ is a bi-direction S-CCTS and t_0 is an initial point, then for any $s \in \Pi$, we define a function $A : \Pi \to \Pi$,

$$A(s) = \begin{cases} \delta_{+}(s, t_0), & s > t_0, \\ \delta_{-}(s, t_0), & s < t_0, \end{cases} \tag{4.21}$$

which will be used later. Note that $A(s) > t_0$ and $A(s) \geq s$.

Remark 4.6 If $\mathbb{T}$ is a T-CCTS with the translation operators $\delta_{\pm}(s, t) = t \pm s$ and $t_0 = 0$. Then (4.21) will become

$$A(s) = \begin{cases} s, & s > 0, \\ -s, & s < 0, \end{cases}$$

that is, $A(s) = |s| \geq s$.

4.3 Shift Almost Periodic Fuzzy Vector-Valued Functions

In this section, we assume that $D \subseteq \mathbb{R}^n_{\mathcal{F}}$ is an open set. In what follows, we will introduce a new type of almost periodic fuzzy vector-valued functions called **shift almost periodic** fuzzy vector-valued functions (or S-almost periodic fuzzy vector-valued functions) on time scales.

Definition 4.9 Let $\mathbb{T}$ be a bi-direction S-CCTS and $f : \mathbb{T} \times D \to \mathbb{R}^n_{\mathcal{F}}$ be continuous on $\mathbb{T} \times D$.

(*i*) A function $f \in C(\mathbb{T} \times D, \mathbb{R}^n_{\mathcal{F}})$ is called **shift almost periodic** fuzzy vector-valued function in $t \in \mathbb{T}$ uniformly for $x \in D$ with shift operators if the ε-shift number set of f

$$E\{\varepsilon, f, S_0\} = \big\{\tau \in \Pi : D_{\infty}\big(f(\delta_{\pm}(\tau, t), x), f(t, x)\big) < \varepsilon, \ \text{for all } t \in \mathbb{T}^* \text{ and } x \in S_0\big\}$$

is a relatively dense set with respect to the pair $(\Pi, \delta_{\pm})$ for all $\varepsilon > 0$ and for each compact subset S_0 of D; that is, for any given $\varepsilon > 0$ and each compact subset S_0 of D, there exists a constant $l(\varepsilon, S_0) > 0$ such that each interval of length $l(\varepsilon, S_0)$ contains a $\tau(\varepsilon, S_0) \in E\{\varepsilon, f, S_0\}$ such that

$$D_\infty\big(f\big(\delta_{\pm}(\tau, t), x\big), f(t, x)\big) < \varepsilon, \quad \text{for all } t \in \mathbb{T}^* \text{ and } x \in S_0.$$

Now τ is called the ε-shift number of f and $l(\varepsilon, S_0)$ is called the inclusion length of $E\{\varepsilon, f, S_0\}$.

(*ii*) A function $f \in C(\mathbb{T} \times D, \mathbb{R}^n_{\mathscr{F}})$ is called shift normal function if for any sequence $F_n : \mathbb{T} \times D \to \mathbb{R}^n_{\mathscr{F}}$ of the form $F_n(t, x) = f\big(\delta_+(h_n, t), x\big)$, $n \in \mathbb{N}$, where $(h_n)_n \subset \Pi$ is a sequence of real numbers, one can extract a subsequence of $(F_n)_n$, converging uniformly on $\mathbb{T} \times D$ (i.e., $\forall (h_n)_n \subset \Pi$, $\exists (h_n)_k$, $\exists F : \mathbb{T} \to \mathbb{R}^n_{\mathscr{F}}$ which may depend on $(h_n)_n$), such that

$$D_\infty\big(F_{n_k}(t, x), F(t, x)\big) \to 0 \text{ as } k \to \infty$$

uniformly with respect to $(t, x) \in \mathbb{T} \times D$.

(*iii*) Let $\delta_{\pm}(s, t)$ be Δ-differentiable to its second argument. A function $f \in C(\mathbb{T} \times D, \mathbb{R}^n_{\mathscr{F}})$ is called shift Δ-almost periodic fuzzy vector-valued function in $t \in \mathbb{T}$ uniformly for $x \in D$ with shift operators if the ε-shift number set of f

$$E\{\varepsilon, f, S_0\} = \big\{\tau \in \Pi : D_\infty\big(f(\delta_{\pm}(\tau, t), x)\delta^{\Delta}_{\pm}(\tau, t), f(t, x)\big) < \varepsilon, \\ \text{for all } t \in \mathbb{T}^* \text{ and } x \in S_0\big\}$$

is a relatively dense set with respect to the pair $(\Pi, \delta_{\pm})$ for all $\varepsilon > 0$ and for each compact subset S_0 of D; that is, for any given $\varepsilon > 0$ and each compact subset S_0 of D, there exists a constant $l(\varepsilon, S_0) > 0$ such that each interval of length $l(\varepsilon, S_0)$ contains a $\tau(\varepsilon, S_0) \in E\{\varepsilon, f, S_0\}$ such that

$$D_\infty\big(f\big(\delta_{\pm}(\tau, t), x\big)\delta^{\Delta}_{\pm}(\tau, t), f(t, x)\big) < \varepsilon, \quad \text{for all } t \in \mathbb{T}^* \text{ and } x \in S_0.$$

Now τ is called the ε-shift number of f and $l(\varepsilon, S_0)$ is called the inclusion length of $E\{\varepsilon, f, S_0\}$.

(*iv*) Let $\delta_{\pm}(s, t)$ be Δ-differentiable to its second argument. A function $f \in C(\mathbb{T} \times D, \mathbb{R}^n_{\mathscr{F}})$ is called shift Δ-normal function if for any sequence $F_n : \mathbb{T} \times D \to \mathbb{R}^n_{\mathscr{F}}$ of the form $F_n(t, x) = f\big(\delta_+(h_n, t), x\big)\delta^{\Delta}_{+}(h_n, t)$, $n \in \mathbb{N}$, where $(h_n)_n \subset \Pi$ is a sequence of real numbers, one can extract a subsequence of $(F_n)_n$, converging uniformly on $\mathbb{T} \times D$ (i.e., $\forall (h_n)_n \subset \Pi$, $\exists (h_n)_k$, $\exists F : \mathbb{T} \to \mathbb{R}^n_{\mathscr{F}}$ which may depend on $(h_n)_n$), such that

$$D_\infty\big(F_{n_k}(t, x), F(t, x)\big) \to 0 \text{ as } k \to \infty$$

uniformly with respect to $(t, x) \in \mathbb{T} \times D$.

For convenience, we denote $AP_S(\mathbb{T})$ the set of all shift almost periodic functions in shifts on $\mathbb{T}$ and we introduce some notations. Let $\alpha = \{\alpha_n\} \subset \Pi$ and $\beta = \{\beta_n\} \subset \Pi$ be two sequences. Then $\beta \subset \alpha$ means that β is a subsequence of α; $\delta_\pm(\alpha, \beta) = \{\delta_\pm(\alpha_n, \beta_n)\}$; $\delta_-(\alpha, t_0) = \{\delta_-(\alpha_n, t_0)\}$, α and β are common subsequences of α' and β', respectively, means that $\alpha_n = \alpha'_{n(k)}$ and $\beta_n = \beta'_{n(k)}$ for some given function $n(k)$.

We introduce the moving-operator T^S, $T^S_\alpha f(t, x) = g(t, x)$ by

$$g(t, x) = \lim_{n\to+\infty} f\big(\delta_+(\alpha_n, t), x\big)$$

and is written only when the limit exists. The mode of convergence, e.g. pointwise, uniform, etc., will be specified at each use of the symbol.

In what follows, we will establish some basic properties of S-almost periodic fuzzy vector-valued functions.

Theorem 4.3 *Let $\mathbb{T}$ be a bi-direction S-CCTS with shifts $\delta_\pm$ and $f \in C(\mathbb{T} \times D, \mathbb{R}^n_{\mathcal{F}})$ be S-almost periodic in t uniformly for $x \in D$, where $\delta_+(\tau, t)$ is continuous in t. Then it is uniformly continuous and bounded on $\mathbb{T}^* \times S_0$.*

Proof For any given $\varepsilon \le 1$ and some compact set $S_0 \subset D$, there exists a constant $l(\varepsilon, S_0)$ such that in any interval of length $l(\varepsilon, S_0)$, there exists $\tau \in E\{\varepsilon, f, S_0\}$ such that

$$D_\infty\big(f\big(\delta_+(\tau, t), x\big), f(t, x)\big) < \varepsilon \le 1, \quad \text{for all } (t, x) \in \mathbb{T}^* \times S_0.$$

Since $f \in C(\mathbb{T} \times D, \mathbb{R}^n_{\mathcal{F}})$, for any $(t, x) \in \big([t_0, \delta_+(l, t_0)]_{\mathbb{T}}\big) \times S_0$, where $t_0 \in \mathbb{T}^*$ is the initial point, there exists an $M > 0$ such that $\|f(t, x)\| < M$. For any given $t \in \mathbb{T}^*$, choose $\tau \in E(\varepsilon, f, S_0) \cap \big[\delta_-(t, t_0), \delta_+\big(l, \delta_-(t, t_0)\big)\big]_{\mathbb{T}}$, then $\delta_+(\tau, t) \in [t_0, \delta_+(l, t_0)]_{\mathbb{T}}$. Hence, for $x \in S_0$, we can obtain

$$\big\|f\big(\delta_+(\tau, t), x\big)\big\|_{\mathcal{F}} < M \quad \text{and} \quad D_\infty\big[f\big(\delta_+(\tau, t), x\big), f(t, x)\big] < 1.$$

Thus for all $(t, x) \in \mathbb{T}^* \times S_0$, we have $\|f(t, x)\|_{\mathcal{F}} < M + 1$.

Moreover, for any $\varepsilon > 0$, let $l_1 = l_1\left(\frac{\varepsilon}{3}, S_0\right)$ be an inclusion length of $E\left(\frac{\varepsilon}{3}, f, S_0\right)$. We can seclect the initial point $t_0 \in \mathbb{T}^*$ such that $f(t, x)$ is uniformly continuous on $\big([t_0, \delta_+(l_1, t_0)]_{\mathbb{T}}\big) \times S_0$. Hence, there exists a positive constant $\delta^* = \delta^*\left(\frac{\varepsilon}{3}, S_0\right)$, for any $t_1, t_2 \in [t_0, \delta_+(l_1, t_0)]_{\mathbb{T}}$ and $|t_1 - t_2| < \delta^*$,

$$D_\infty\big[f(t_1, x), f(t_2, x)\big] < \frac{\varepsilon}{3} \quad \text{for all } x \in S_0.$$

Now, we choose an arbitrary $\upsilon, t \in \mathbb{T}^*$, satisfying $|t - \upsilon| < \delta^*$, and we take

$$\tau \in E\left(\frac{\varepsilon}{3}, f, S_0\right) \cap \left[\delta_-(t, t_0), \delta_+\big(l, \delta_-(t, t_0)\big)\right]_{\mathbb{T}},$$

then $\delta_+(\tau, t), \delta_+(\tau, \upsilon) \in \left[t_0, \delta_+(l, t_0)\right]_{\mathbb{T}}$. Since $\delta_+(\tau, t)$ is continuous in t, then there exists $\delta^{**} > 0$ such that $|t - \upsilon| < \delta^{**}$ implies

$$|\delta_+(\tau, t) - \delta_+(\tau, \upsilon)| < \delta^*,$$

Now, we take $\delta_{**} = \min\left\{\delta^*, \delta^{**}\right\}$, when $|t - \upsilon| < \delta_{**}$ implies

$$D_\infty\big(f\big(\delta_+(\tau, t), x\big), f\big(\delta_+(\tau, \upsilon), x\big)\big) < \frac{\varepsilon}{3} \quad \text{for all } x \in S_0.$$

Therefore, for $(t, x) \in \mathbb{T}^* \times S_0$, we have

$$\begin{aligned} D_\infty(f(t, x), f(\upsilon, x)) \leq\ & D_\infty\big(f(t, x), f\big(\delta_+(\tau, t), x\big)\big) \\ & + D_\infty\big(f\big(\delta_+(\tau, t), x\big), f\big(\delta_+(\tau, \upsilon), x\big)\big) \\ & + D_\infty\big(f\big(\delta_+(\tau, \upsilon), x\big), f(\upsilon, x)\big) < \varepsilon. \end{aligned}$$

The proof is completed. □

In the following, we will establish a shift-convergence theorem of S-almost periodic fuzzy vector-valued functions.

Theorem 4.4 *Let $f \in C(\mathbb{T} \times D, \mathbb{R}^n_{\mathscr{F}})$ be S-almost periodic in t uniformly for $x \in D$ under shifts $\delta_\pm$. Then for any given sequence $\alpha' \subset \Pi$, there exists a subsequence $\beta \subset \alpha'$ and $g \in C(\mathbb{T} \times D, \mathbb{R}^n_{\mathscr{F}})$ such that $T^S_\beta f(t, x) = g(t, x)$ holds uniformly on $\mathbb{T}^* \times S_0$ and $g(t, x)$ is S-almost periodic in t uniformly for $x \in D$ under shifts $\delta_\pm$.*

Proof For any $\varepsilon > 0$ and $S_0 \subset D$, let $l = l\left(\frac{\varepsilon}{4}, S_0\right)$ be an inclusion length of $E\left\{\frac{\varepsilon}{4}, f, S_0\right\}$. For any given subsequence $\alpha' = \{\alpha'_n\} \subset \Pi$, there exists $\gamma'_n \in \Pi$ and $t_0 \leq \gamma'_n \leq l$, $n = 1, 2, \ldots$, such that $\alpha'_n = \delta_+(\tau'_n, \gamma'_n)$, where $\tau'_n \in E\left\{\frac{\varepsilon}{4}, f, S_0\right\}$. In fact, for any interval with length of l, there exists $\tau'_n \in E\left\{\frac{\varepsilon}{4}, f, S_0\right\}$, thus, we can choose a proper interval with length of l such that $t_0 \leq \delta_-\left(\tau'_n, \alpha'_n\right) \leq l$, and from the definition of Π, it is easy to observe that $\gamma'_n = \delta_-\left(\tau'_n, \alpha'_n\right) \in \Pi$. Therefore, there exists a subsequence $\gamma = \{\gamma_n\} \subset \gamma' = \{\gamma'_n\}$ such that $\gamma_n \to s$ as $n \to \infty$, where $t_0 \leq s \leq l$.

Also, it follows from Theorem 4.3 that $f(t, x)$ is uniformly continuous on $\mathbb{T}^* \times S_0$. Hence, there exists $\delta^*(\varepsilon, S_0) > 0$ so that $|t_1 - t_2| < \delta^*$, for $x \in S_0$, implies

$$D_\infty\big(f\big(\delta_+(\tau, t_1), x\big), f\big(\delta_+(\tau, t_2), x\big)\big) < \frac{\varepsilon}{2}.$$

Since γ is a convergent sequence, there exists $N = N(\delta)$ so that $p, m \geq N$ implies $|\gamma_p - \gamma_m| < \delta^*$. Now, one can take $\alpha \subset \alpha'$, $\tau \subset \tau' = \{\tau_n'\}$ such that α, τ common with γ, then for any integers $p, m \geq N$, we obtain

$$\begin{aligned}
&D_\infty\big(f\big(\delta_+(\delta_-(\tau_m,\tau_p),t),x\big), f(t,x)\big)\\
&\leq D_\infty\big(f\big(\delta_-(\tau_m,\delta_+(\tau_p,t)),x\big), f\big(\delta_+(\tau_p,t),x\big)\big) + D_\infty\big(f\big(\delta_+(\tau_p,t),x\big), f(t,x)\big)\\
&< \frac{\varepsilon}{4} + \frac{\varepsilon}{4} = \frac{\varepsilon}{2},
\end{aligned}$$

since $\delta_-\big(\gamma_n', \alpha_n'\big) = \tau_n'$, and we obtain

$$\begin{aligned}
&\delta_+\big(\delta_-(\alpha_m,\alpha_p), \delta_-(\gamma_p,\gamma_m)\big)\\
&= \delta_+\big(\delta_-(\gamma_p,\alpha_p), \delta_-(\alpha_m,\gamma_m)\big) = \delta_+\big(\tau_p, \delta_-(\tau_m,t_0)\big) = \delta_-(\tau_m,\tau_p) \in E\left\{\frac{\varepsilon}{2}, f, S_0\right\}.
\end{aligned}$$

Hence, we obtain

$$\begin{aligned}
&D_\infty\big(f\big(\delta_+(\alpha_p,t),x\big), f\big(\delta_+(\alpha_m,t),x\big)\big)\\
\leq\ &\sup_{(t,x)\in\mathbb{T}^*\times S_0} D_\infty\big(f\big(\delta_+(\alpha_p,t),x\big), f\big(\delta_+(\alpha_m,t),x\big)\big)\\
\leq\ &\sup_{(t,x)\in\mathbb{T}^*\times S_0} D_\infty\big(f\big(\delta_+(\delta_-(\alpha_m,\alpha_p),t),x\big), f(t,x)\big)\\
\leq\ &\sup_{(t,x)\in\mathbb{T}^*\times S_0} D_\infty\big(f\big(\delta_+(\delta_-(\alpha_m,\alpha_p),t),x\big), f\big(\delta_+(\delta_-(\gamma_m,\gamma_p),t),x\big)\big)\\
+\ &\sup_{(t,x)\in\mathbb{T}^*\times S_0} D_\infty\big(f\big(\delta_+(\delta_-(\gamma_m,\gamma_p),t),x\big), f(t,x)\big) < \frac{\varepsilon}{2} + \frac{\varepsilon}{2} = \varepsilon.
\end{aligned}$$

Thus, we can take sequences $\alpha^{(k)} = \{\alpha_n^{(k)}\}$, $k = 1, 2, \ldots$, and $\alpha^{(k+1)} \subset \alpha^{(k)} \subset \alpha$ such that for any integers m, p, and all $(t, x) \in \mathbb{T}^* \times S_0$, the following holds:

$$D_\infty\big(f\big(\delta_+(\alpha_p^{(k)},t),x\big), f\big(\delta_+(\alpha_m^{(k)},t),x\big)\big) < \frac{1}{k},\ k = 1, 2, \ldots.$$

For all sequences $\alpha^{(k)}$, $k = 1, 2, \ldots$, we can take a sequence $\beta = \{\beta_n\}$, $\beta_n = \alpha_n^{(n)}$, and then it follows that $\big\{f\big(\delta_+(\beta_n,t),x\big)\big\} \subset \big\{f\big(\delta_+(\alpha_n,t),x\big)\big\}$ for any integers p, m with $p < m$ and all $(t, x) \in \mathbb{T}^* \times S_0$, the following holds:

$$D_\infty\big(f\big(\delta_+(\beta_p,t),x\big), f\big(\delta_+(\beta_m,t),x\big)\big) < \frac{1}{p}.$$

Therefore, $\big\{f\big(\delta_+(\beta_n,t),x\big)\big\}$ converges uniformly on $\mathbb{T}^* \times S_0$, i.e., $T_\beta^S f(t,x) = g(t,x)$ holds uniformly on $\mathbb{T}^* \times S_0$, where $\beta = \{\beta_n\} \subset \alpha$.

Next, we show that $g(t, x)$ is continuous on $\mathbb{T}^* \times D$. If this is not true, there will exist $(t_0, x_0) \in \mathbb{T}^* \times D$ such that $g(t, x)$ is not continuous at this point. Then there exist $\varepsilon_0 > 0$ and sequences $\{\delta_m^*\}$, $\{t_m\}$, $\{x_m\}$, where $\delta_m^* > 0$, $\delta_m^* \to 0$ as $m \to +\infty$, $|t_0 - t_m| + |x_0 - x_m| < \delta_m^*$ and

$$D_\infty\big(g(t_0, x_0), g(t_m, x_m)\big) \geq \varepsilon_0. \tag{4.22}$$

Let $X = \{x_m\} \bigcup \{x_0\}$, and obviously, X is a compact subset of D. Hence, there exists positive integer $N = N(\varepsilon_0, X)$ such that $n > N$ implies

$$D_\infty\big(f\big(\delta_+(\beta_n, t_m), x_m\big), g(t_m, x_m)\big) < \frac{\varepsilon_0}{3} \quad \text{for all} \quad m \in \mathbb{Z}^+ \tag{4.23}$$

and

$$D_\infty\big(f\big(\delta_+(\beta_n, t_0), x_0\big), g(t_0, x_0)\big) < \frac{\varepsilon_0}{3}. \tag{4.24}$$

According to the uniform continuity of $f(t, x)$ on $\mathbb{T}^* \times D$, for sufficiently large m, we have

$$D_\infty\big(f\big(\delta_+(\beta_n, t_0), x_0\big), f\big(\delta_+(\beta_n, t_m), x_m\big)\big) < \frac{\varepsilon_0}{3}. \tag{4.25}$$

From (4.23)–(4.25), we get $D_\infty\big(g(t_0, x_0), g(t_m, x_m)\big) < \varepsilon_0$, which contradicts (4.22). Therefore, $g(t, x)$ is continuous on $\mathbb{T}^* \times D$.

Finally, for any compact set $S_0 \subset D$ and given $\varepsilon > 0$, one can select $\tau \in E\{\varepsilon, f, S_0\}$, and then for all $(t, x) \in \mathbb{T}^* \times S_0$, we have

$$D_\infty\big(f\big(\delta_+(\beta_n, \delta_+(\tau, t)), x\big), f\big(\delta_+(\beta_n, t), x\big)\big) < \varepsilon.$$

Let $n \to +\infty$, for all $(t, x) \in \mathbb{T}^* \times S_0$, and we have

$$D_\infty\big(g\big(\delta_+(\tau, t), x\big), g(t, x)\big) \leq \varepsilon,$$

which implies that $E\{\varepsilon, g, S_0\}$ is relatively dense. Therefore, $g(t, x)$ is S-almost periodic in t uniformly for $x \in D$ under shifts $\delta_\pm$. This completes the proof. □

Next, we will give a sequentially compact criterion of S-almost periodic fuzzy vector-valued functions through the shift operator T^S.

Theorem 4.5 *Let $f \in C(\mathbb{T} \times D, \mathbb{R}^n_{\mathcal{F}})$. If for any sequence $\alpha' \subset \Pi$, there exists $\alpha \subset \alpha'$ such that $T^S_\alpha f(t, x)$ exists uniformly on $\mathbb{T}^* \times S_0$, then $f(t, x)$ is S-almost periodic in t uniformly for $x \in D$ under shifts $\delta_\pm$.*

Proof (*By contradiction*). If this is not true, then there exist $\varepsilon_0 > 0$ and $S_0 \subset D$ such that for any sufficiently large $l > 0$, there is an interval with length of l such that there is no ε_0-shift numbers of $f(t, x)$ in this interval, that is, every point in this interval is not in the set $E\{\varepsilon_0, f, S_0\}$.

One can select a number $\alpha_1' \in \Pi$ to obtain an interval (a_1, b_1) with $b_1 - a_1 > 2|\alpha_1'|$, where $a_1, b_1 \in \Pi$ such that there is no ε_0-shift numbers of $f(t, x)$ in this interval. Next, taking $\alpha_2' \in (\delta_+(\alpha_1', a_1), \delta_+(\alpha_1', b_1))$, obviously, $\delta_-(\alpha_1', \alpha_2') \in (a_1, b_1)$, so $\delta_-(\alpha_1', \alpha_2') \notin E\{\varepsilon_0, f, S_0\}$; then one can obtain an interval (a_2, b_2) with $b_2 - a_2 > 2(|\alpha_1'| + |\alpha_2'|)$, where $a_2, b_2 \in \Pi$ such that there is no ε_0-shift numbers of $f(t, x)$ in this interval. Now, selecting $\alpha_3' \in (\delta_+(\alpha_2', a_2), \delta_+(\alpha_2', b_2))$, $\alpha_3' \in (\delta_+(\alpha_1', a_2), \delta_+(\alpha_1', b_2))$, obviously, $\delta_-(\alpha_2', \alpha_3')$, $\delta_-(\alpha_1', \alpha_3') \notin E\{\varepsilon_0, f, S_0\}$. One can repeat this process again to get the numbers α_4', α_5', $\ldots$, such that $\delta_-(\alpha_j', \alpha_i') \notin E\{\varepsilon_0, f, S_0\}$, $i > j$. Hence, for any $i \neq j$, $i, j = 1, 2, \ldots$, without loss of generality, let $i > j$, for $x \in S_0$ we obtain

$$\begin{aligned}
&\sup_{(t,x)\in\mathbb{T}^*\times S_0} D_\infty\big(f(\delta_+(\alpha_i', t), x), f(\delta_+(\alpha_j', t), x)\big)\\
&= \sup_{(t,x)\in\mathbb{T}^*\times S_0} D_\infty\big(f(\delta_+(\delta_-(\alpha_j', \alpha_i'), t), x), f(t, x)\big) \geq \varepsilon_0.
\end{aligned}$$

Therefore, there is no uniformly convergent subsequence of $\{f(\delta_+(\alpha_n', t), x)\}$ for $(t, x) \in \mathbb{T}^* \times S_0$, and this is a contradiction. Thus, $f(t, x)$ is S-almost periodic in t uniformly for $x \in D$ under shifts $\delta_\pm$. This completes the proof. □

From Theorems 4.4 and 4.5, we obtain the following equivalent definition of uniformly S-almost periodic fuzzy vector-valued functions.

Definition 4.10 Let $f \in C(\mathbb{T} \times D, \mathbb{R}^n_{\mathcal{F}})$. If for any given sequence $\alpha' \subset \Pi$, there exists a subsequence $\alpha \subset \alpha'$ such that $T^S_\alpha f(t, x)$ exists uniformly on $\mathbb{T}^* \times S$, then $f(t, x)$ is called a S-almost periodic fuzzy vector-valued function in t uniformly for $x \in D$ under shifts $\delta_\pm$.

Theorem 4.6 *If $f \in C(\mathbb{T} \times D, \mathbb{R}^n_{\mathcal{F}})$ is S-almost periodic in t uniformly for $x \in D$ and $\varphi(t)$ is S-almost periodic with $\{\varphi(t) : t \in \mathbb{T}\} \subset S_0$, then $f(t, \varphi(t))$ is S-almost periodic.*

Proof For any given sequence $\alpha' \subset \Pi$, there exists $\alpha \subset \alpha'$, $\psi(t)$, $g(t, x)$ such that $T^S_\alpha \varphi(t) = \psi(t)$ exists uniformly on $\mathbb{T}^*$ and $T^S_\alpha f(t, x) = g(t, x)$ exists uniformly on $\mathbb{T}^* \times S_0$, where $\psi(t)$ is S-almost periodic and $g(t, x)$ is S-almost periodic in t uniformly for $x \in D$. Therefore, $g(t, x)$ is uniformly continuous on $\mathbb{T}^* \times S_0$, then for any given $\varepsilon > 0$, there exists $\delta^*\left(\frac{\varepsilon}{2}\right) > 0$ such that for any $x_1, x_2 \in S_0$ and all $t \in \mathbb{T}^*$, $|x_1 - x_2| < \delta^*\left(\frac{\varepsilon}{2}\right)$ implies

$$D_\infty\big(g(t, x_1), g(t, x_2)\big) < \frac{\varepsilon}{2}.$$

For a sufficiently large $N_0(\varepsilon) > 0$ such that $n \geq N_0(\varepsilon)$, we obtain

$$D_\infty\big(f(\delta_+(\alpha_n, t), x), g(t, x)\big) < \frac{\varepsilon}{2}, \quad \forall (t, x) \in \mathbb{T}^* \times S_0,$$

$$D_\infty\big(\varphi\big(\delta_+(\alpha_n,t)\big),\psi(t)\big)<\delta^*\Big(\frac{\varepsilon}{2}\Big),\quad \forall t\in\mathbb{T}^*,$$

where $\{\varphi(\delta_+(\alpha_n,t)):t\in\mathbb{T}^*\}\subset S_0$, $\{\psi(t):t\in\mathbb{T}^*\}\subset S_0$. Therefore, when $n\geq N_0(\varepsilon)$, we have

$$\begin{aligned}&D_\infty\big(T_\alpha^S f\big(t,\varphi(t)\big),g\big(t,\psi(t)\big)\big)\\&=D_\infty\big(f\big(\delta_+(\alpha_n,t),\varphi\big(\delta_+(\alpha_n,t)\big)\big),g\big(t,\psi(t)\big)\big)\\&\leq D_\infty\big(f\big(\delta_+(\alpha_n,t),\varphi\big(\delta_+(\alpha_n,t)\big)\big),g\big(t,\varphi(\delta_+(\alpha_n,t))\big)\big)\\&\quad+D_\infty\big(g\big(t,\varphi\big(\delta_+(\alpha_n,t)\big)\big),g\big(t,\psi(t)\big)\big)<\varepsilon.\end{aligned}$$

Note that

$$\lim_{n\to\infty}f\big(\delta_+(\alpha_n,t),\varphi\big(\delta_+(\alpha_n,t)\big)\big)=T_\alpha^S\big(f\big(t,\varphi(t)\big),$$

and thus, $T_\alpha^S f\big(t,\varphi(t)\big)=g\big(t,\psi(t)\big)$ exists uniformly on $\mathbb{T}^*\times S_0$. Thus, $f\big(t,\varphi(t)\big)$ is S-almost periodic under shifts $\delta_\pm$. The proof is completed. □

Definition 4.11 Let $f\in C(\mathbb{T}\times D,\mathbb{R}^n_{\mathcal{F}})$. Then $H_S(f)=\{g(t,x):\mathbb{T}\times D\to\mathbb{R}^n_{\mathcal{F}}|$ there exits $\alpha\in\Pi$ such that $T_\alpha^S f(t,x)=g(t,x)$ exists uniformly on $\mathbb{T}^*\times S_0\}$ is called the S-hull of $f(t,x)$ under shifts $\delta_\pm$.

Theorem 4.7 *$H_S(f)$ is compact if and only if $f(t,x)$ is S-almost periodic in t uniformly for $x\in D$.*

Proof If $H_S(f)$ is compact, for any given $\alpha'=\{\alpha'_n\}\subseteq\Pi$, there exists subsequence $\{f\big(\delta_+(\alpha_{n_k},t),x\big)\}_{k\in\mathbb{N}}$ of the sequence $\{f\big(\delta_+(\alpha'_n,t),x\big)\}_{n\in\mathbb{N}}$ such that

$$f\big(\delta_+(\alpha_{n_k},t),x\big)\to g(t,x)\ (k\to\infty),\forall(t,x)\in\mathbb{T}^*\times S_0.$$

Note that $\alpha=\{\alpha_{n_k}\}$, obviously, $\alpha\subset\alpha'$, so $T_\alpha^S f(t,x)$ exists uniformly on $\mathbb{T}^*\times S_0$.

Conversely, if $f(t,x)$ is S-almost periodic in t uniformly for $x\in D$ and $\{g_n(t,x)\}\subset H_S(f)$, then we can choose $\alpha'=\{\alpha'_n\}$ such that

$$D_\infty\big(f\big(\delta_+(\alpha'_n,t),x\big),g_n(t,x)\big)<\frac{1}{n},\quad\forall(t,x)\in\mathbb{T}^*\times S_0.$$

One can choose $\alpha\subset\alpha'$ so that $T_\alpha^S f(t,x)$ exists uniformly. Let $\beta\subset\gamma=\{n\}$ such that β and α are common subsequences, then

$$D_\infty\big(f\big(\delta_+(\alpha_n,t),x\big),g_{\beta_n}(t,x)\big)\to 0\ \ (n\to\infty),\quad\forall(t,x)\in\mathbb{T}^*\times S_0,$$

so that

$$\lim_{n\to\infty}g_{\beta_n}(t,x)=T_\alpha^S f(t,x)\in H_S(f),\forall(t,x)\in\mathbb{T}^*\times S_0.$$

Thus, $H_S(f)$ is compact. The proof is completed. □

Theorem 4.8 *If $f \in C(\mathbb{T} \times D, \mathbb{R}^n_{\mathcal{F}})$ is S-almost periodic in t uniformly for $x \in D$ under shifts $\delta_{\pm}$, then for any $g(t,x) \in H_S(f)$, $H_S(f) = H_S(g)$.*

Proof For any $h(t,x) \in H_S(g)$ there exists $\alpha' \subseteq \Pi$ such that $T^S_{\alpha'} g(t,x) = h(t,x)$. Since $f(t,x)$ is S-almost periodic in t uniformly for $x \in D$, from the sequence $\{\alpha'_n\} \subseteq \Pi$, one can extract a sequence $\{\alpha_n\}$ such that $T^S_\alpha f(t,x) = \lim_{n\to\infty} f\big(\delta_+(\alpha_n, t), x\big)$ exists uniformly on $\mathbb{T}^* \times S_0$.

For $g(t,x) \in H_S(f)$, there exists $\alpha^{(1)} \subseteq \Pi$ such that

$$\lim_{n\to+\infty} f\big(\delta_+(\alpha^{(1)}_n, t), x\big) = g(t,x), \quad \forall (t,x) \in \mathbb{T}^* \times S_0,$$

so we have

$$\lim_{n\to+\infty} f\big(\delta_+(\delta_+(\alpha^{(1)}_n, \alpha_n), t), x\big) = g\big(\delta_+(\alpha_n, t), x\big), \quad \forall (t,x) \in \mathbb{T}^* \times S_0,$$

and then we can take $\beta = \{\beta_n\} = \{\delta_+(\alpha^{(1)}_n, \alpha_n)\}$ such that

$$D_\infty\Big(f\big(\delta_+(\beta_n, t), x\big), g\big(\delta_+(\alpha_n, t), x\big)\Big) < \frac{1}{n}, \quad \forall (t,x) \subset \mathbb{T}^* \times S_0.$$

It follows that $T^S_\beta f(t,x) = T^S_\alpha g(t,x) = T^S_{\alpha'} g(t,x) = h(t,x)$. Hence $h(t,x) \in H_S(f)$. Thus, $H_S(g) \subseteq H_S(f)$.

On the other hand, for any $g(t,x) \in H_S(f)$, there exists α such that $T^S_\alpha f(t,x) = g(t,x)$, then

$$D_\infty\big(f\big(\delta_+(\alpha_n, t), x\big), g(t,x)\big) \to 0, \quad n \to \infty, \forall (t,x) \in \mathbb{T}^* \times S_0,$$

so making the change of variable $\delta_+(\alpha_n, t) = s$, one has

$$D_\infty\Big(f(s,x), g\big(\delta_-(\alpha_n, s), x\big)\Big) \to 0, \quad \forall (s,x) \in \mathbb{T}^* \times S_0,$$

that is

$$\lim_{n\to\infty} g\big(\delta_-(\alpha_n, t), x\big) =: T^S_{\alpha^{-1}} g(t,x) = f(t,x).$$

Thus, $f(t,x) \in H_S(g)$, then by what was shown above, $H_S(f) \subseteq H_S(g)$.

According to the above, it follows that $H_S(f) = H_S(g)$. The proof is completed. □

From Definition 4.11 and Theorem 4.8, one can directly obtain the following theorem.

Theorem 4.9 *If $f \in C(\mathbb{T} \times D, \mathbb{R}^n_{\mathcal{F}})$ is S-almost periodic in t uniformly for $x \in D$ under shifts $\delta_{\pm}$, then for any $g(t,x) \in H_S(f)$, $g(t,x)$ is S-almost periodic in t uniformly for $x \in D$ under shifts $\delta_{\pm}$.*

Now, we establish the following theorems to guarantee four fundamental operations among S-almost periodic fuzzy vector-valued functions.

Theorem 4.10 *Let $f \in C(\mathbb{T} \times D, \mathbb{R}^n_{\mathcal{F}})$ be S-almost periodic in t uniformly for $x \in D$ under shifts $\delta_\pm$. Then for any $\varepsilon > 0$, there exists a positive constant $L = L(\varepsilon, S)$, for any $a \in \Pi$, there exist a constant $\eta > 0$ and $\alpha \in \Pi$ such that $[\alpha, \delta_+(\alpha, \eta)]_{\mathbb{T}^*} \subset [a, \delta_+(a, L)]_{\mathbb{T}^*}$ and $[\alpha, \delta_+(\alpha, \eta)]_{\mathbb{T}^*} \subset E(\varepsilon, f, S)$.*

Proof Since $f(t, x)$ is uniformly continuous on $\mathbb{T}^* \times S_0$, given any $\varepsilon > 0$, there exists $\delta^*(\varepsilon_1, S_0) > 0$ so that $|t_1 - t_2| < \delta^*(\varepsilon_1, S_0)$ implies

$$D_\infty\big(f(t_1, x), f(t_2, x)\big) < \varepsilon_1, \quad \forall x \in S_0,$$

where $\varepsilon_1 = \dfrac{\varepsilon}{2}$.

We take $\eta = \delta^*\left(\dfrac{\varepsilon}{2}, S_0\right) = \delta^*(\varepsilon_1, S_0)$, and $L = \delta_+\big(l(\varepsilon_1, S_0), \eta\big)$, where $l(\varepsilon_1, S_0)$ is the inclusion length of $E(\varepsilon_1, f, S_0)$.

For any $a \in \Pi$, consider an interval $\big[a, \delta_+(a, L)\big]_{\mathbb{T}^*}$, and take

$$\tau \in E(f, \varepsilon_1, S) \cap \Big[\delta_+(\eta_*, a), \delta_-\big(\eta_*, \delta_+\big(a, l(\varepsilon_1, S_0)\big)\big)\Big]_{\mathbb{T}^*},$$

where $\eta_* > t_0$ (t_0 is the initial point) satisfying $\delta_+(\eta_*, \eta_*) \le \frac{\eta}{2}$ (in fact, we also have $\eta_* - \delta_-(\eta_*, t_0) \le \eta_* < \delta_+(\eta_*, \eta_*) \le \frac{\eta}{2}$), so we obtain

$$\Big[\delta_+\big(\tau, \delta_-(\eta_*, t_0)\big), \delta_+(\tau, \eta_*)\Big]_{\mathbb{T}^*} \subset \big[a, \delta_+(a, L)\big]_{\mathbb{T}^*}.$$

Hence, for all $\xi \in \Big[\delta_+\big(\tau, \delta_-(\eta_*, t_0)\big), \delta_+(\tau, \eta_*)\Big]_{\mathbb{T}^*}$, we can obtain

$$|\xi - \tau| \le |\delta_+\big(\tau, \delta_-(\eta_*, t_0)\big) - \delta_+(\tau, \eta_*)| \le 2\delta_+(\eta_*, \eta_*) \le \eta.$$

Therefore, for any $(t, x) \in \mathbb{T}^* \times S_0$,

$$\begin{aligned} D_\infty\big(f\big(\delta_+(\xi, t), x\big), f(t, x)\big) \le\ & D_\infty\big(f\big(\delta_+(\xi, t), x\big), f\big(\delta_+(\tau, t), x\big)\big) \\ & + D_\infty\big(f\big(\delta_+(\tau, t), x\big), f(t, x)\big) \le \varepsilon. \end{aligned}$$

We let $\alpha = \delta_-(\eta_*, \tau)$, then $\big[\alpha, \delta_+(\alpha, \eta)\big]_{\mathbb{T}^*} \subset E(\varepsilon, f, S_0)$. This completes the proof. □

In the following theorem, for $\eta \in \mathbb{T}^*$, we will use notations $\delta^+_\eta = \delta_+(\eta, t_0) = \eta$, $\delta^+_{2\eta} = \delta_+(\eta, \delta^+_\eta)$, $\delta^+_{3\eta} = \delta_+(\eta, \delta^+_{2\eta}), \ldots, \delta^+_{m\eta} := \delta_+(\eta, \delta^+_{(m-1)\eta})$ for simplicity.

Theorem 4.11 *If $f, g \in C(\mathbb{T} \times D, \mathbb{R}^n_{\mathcal{F}})$ are S-almost periodic in t uniformly for $x \in D$ under shifts $\delta_\pm$, then for any $\varepsilon > 0$, $E(f, \varepsilon, S_0) \cap E(g, \varepsilon, S_0)$ is a nonempty relatively dense set in Π.*

Proof Since f, g are S-almost periodic in t uniformly for $x \in D$ under shifts $\delta_{\pm}$, they are uniformly continuous on $\mathbb{T}^* \times S_0$. For any given $\varepsilon > 0$, one can take $\delta_i^* = \delta_i^*\left(\frac{\varepsilon}{2}, S_0\right)$ $(i = 1, 2)$; and $l_1 = l_1\left(\frac{\varepsilon}{2}, S_0\right)$, $l_2 = l_2\left(\frac{\varepsilon}{2}, S_0\right)$ are inclusion lengths of $E(f, \frac{\varepsilon}{2}, S_0)$, $E(g, \frac{\varepsilon}{2}, S_0)$, respectively.

According to Theorem 4.7, we choose

$$\eta = \eta(\varepsilon, S) = \min(\delta_1^*, \delta_2^*) \in \Pi,\ \ L_i = \delta_+(l_i, \eta)\ (i = 1, 2),\ \ L = \max(L_1, L_2).$$

Hence, one can select $\frac{\varepsilon}{2}$-shift numbers of $f(t, x)$ and $g(t, x)$: $\tau_1 = \delta_{m\eta}^+$ and $\tau_2 = \delta_{n\eta}^+$, respectively, where $\tau_1, \tau_2 \in \left[a, \delta_+(a, L)\right]_{\mathbb{T}^*}$, m, n are integers. Since

$$\delta_-(\tau_2, t_0) \in \left[\delta_-\left(a, \delta_-(L, t_0)\right), \delta_-(a, t_0)\right]_{\mathbb{T}^*},$$

then it follows that

$$\begin{aligned}\delta_+\left(\delta_-\left(a, \delta_-(L, t_0)\right), a\right) &\le \delta_+\left(\tau_1, \delta_-\left(a, \delta_-(L, t_0)\right)\right)\\ &\le \delta_-(\tau_2, \tau_1) \le \delta_-(a, \tau_1) \le \delta_-\left(a, \delta_+(a, L)\right),\end{aligned}$$

thus, $\Lambda\left(\delta_-(\tau_2, \tau_1)\right) \le L$. Let $m - n = s$, then s can only be selected from a finite number set $\{s_1, s_2, \ldots, s_p\}$. When $m - n = s_j$, $j = 1, 2, \ldots, p$, then the $\frac{\varepsilon}{2}$-shift numbers of $f(t)$ and $g(t)$ can be denoted by τ_1^j, τ_2^j, respectively, i.e., $\delta_-\left(\tau_2^j, \tau_1^j\right) = \delta_{s_j\eta}^+$, $j = 1, 2, \ldots, p$, and we choose $T = \max\limits_j\{\Lambda(\tau_1^j), \Lambda(\tau_2^j)\}$.

For any $a \in \Pi$, on the interval $\left[\delta_+(a, T), \delta_+\left(\delta_+(a, T), L\right)\right]_{\mathbb{T}^*}$, we can take $\frac{\varepsilon}{2}$-shift numbers of $f(t, x)$ and $g(t, x)$ as τ_1 and τ_2, respectively, and there exists some integer s_j such that

$$\delta_-(\tau_2, \tau_1) = \delta_{s_j\eta}^+ = \delta_-\left(\tau_2^j, \tau_1^j\right).$$

Let $\tau(\varepsilon, S_0) = \delta_-\left(\tau_1^j, \tau_1\right) = \delta_-\left(\tau_2^j, \tau_2\right)$, then $\tau(\varepsilon, S_0) \in \left[a, \delta_+\left(a, \delta_+\left(L, \delta_+(T, T)\right)\right)\right]_{\mathbb{T}^*}$, and for any $(t, x) \in \mathbb{T}^* \times S_0$, we have

$$\begin{aligned}D_\infty\left(f\left(\delta_+(\tau, t), x\right), f(t, x)\right) \le\ & D_\infty\left(f\left(\delta_+(\delta_-(\tau_1^j, \tau_1), t), x\right), f\left(\delta_-(\tau_1^j, t), x\right)\right)\\ &+ D_\infty\left(f\left(\delta_-(\tau_1^j, t), x\right), f(t, x)\right) < \varepsilon\end{aligned}$$

and

$$\begin{aligned}D_\infty\left(g\left(\delta_+(\tau, t), x\right), g(t, x)\right) \le\ & D_\infty\left(g\left(\delta_+(\delta_-(\tau_2^j, \tau_2), t), x\right), g\left(\delta_-(\tau_2^j, t), x\right)\right)\\ &+ D_\infty\left(g\left(\delta_-(\tau_2^j, t), x\right), g(t, x)\right) < \varepsilon.\end{aligned}$$

Therefore, there exists at least a $\tau = \tau(\varepsilon, S)$ on any interval

$$\big[a, \delta_+\big(a, \delta_+\big(L, \delta_+(T, T)\big)\big)\big]_{\mathbb{T}^*}$$

with the length $\delta_+\big(L, \delta_+(T, T)\big)$ such that $\tau \in E(f, \varepsilon, S_0) \cap E(g, \varepsilon, S_0)$. The proof is completed. □

From Definition 4.9, one can easily prove the following theorem.

Theorem 4.12 *If $f \in C(\mathbb{T} \times D, \mathbb{R}^n_{\mathscr{F}})$ is S-almost periodic in t uniformly for $x \in D$ under shifts $\delta_\pm$, then for any $\alpha \in \mathbb{R}$, $b \in \Pi$, the functions $\alpha \cdot f(t, x)$, $f\big(\delta_\pm(b, t), x\big)$ are S-almost periodic in t uniformly for $x \in D$ under shifts $\delta_\pm$.*

Theorem 4.13 *Let $f, g \in C(\mathbb{T} \times D, \mathbb{R}^n_{\mathscr{F}})$ be S-almost periodic in t uniformly for $x \in D$ under shifts $\delta_\pm$. If $u \tilde{-}_{gH_{(I)}} v$ exists and one of the following conditions is satisfied,*

(i) If $I^{\alpha,(I)}_{f_i,\omega_i}, I^{\alpha,(I)}_{g_i,\omega_i} > 0$ or $I^{\alpha,(I)}_{f_i,\omega_i}, I^{\alpha,(I)}_{g_i,\omega_i} < 0$,
(ii) If $I^{\alpha,(II)}_{f_i,\omega_i}, I^{\alpha,(II)}_{g_i,\omega_i} > 0$ or $I^{\alpha,(II)}_{f_i,\omega_i}, I^{\alpha,(II)}_{g_i,\omega_i} < 0$,

then $f * g$ and $f \circledast g$ are S-almost periodic in t uniformly for $x \in D$.

Proof From Theorem 4.11, for any $\varepsilon > 0$, $E(f, \frac{\varepsilon}{2}, S_0) \cap E(g, \frac{\varepsilon}{2}, S_0)$ is a nonempty relatively dense set. Without loss of generality, we assume that condition (i) is satisfied (the proof of the other case when condition (ii) is satisfied is similar). According to Theorem 2.6 (ii), we have

$$\begin{aligned} &g\big(\delta_+(\tau, t), x\big) * f\big(\delta_+(\tau, t), x\big) + g\big(\delta_+(\tau, t), x\big) * f(t, x) \\ &= g\big(\delta_+(\tau, t), x\big) * \big(f\big(\delta_+(\tau, t), x\big) + f(t, x)\big), \end{aligned}$$

$$g(\delta_+(\tau, t), x) * f(t, x) + g(t, x) * f(t, x) = \big(g(\delta_+(\tau, t), x) + g(t, x)\big) * f(t, x).$$

Denote $\sup\limits_{(t,x)\in\mathbb{T}^*\times S_0} \|f(t, x)\|_{\mathscr{F}} = M_1$, $\sup\limits_{(t,x)\in\mathbb{T}^*\times S_0} \|g(t, x)\|_{\mathscr{F}} = M_2$, take

$$\tau \in E(f, \frac{\varepsilon}{2}, S_0) \cap E(g, \frac{\varepsilon}{2}, S_0),$$

then for all $(t, x) \in \mathbb{T}^* \times S_0$ we have

$$\begin{aligned} &D_\infty\big(f\big(\delta_+(\tau, t), x\big) * g\big(\delta_+(\tau, t), x\big), f(t, x) * g(t, x)\big) \\ &\le \big\|g\big(\delta_+(\tau, t), x\big)\big\|_{\mathscr{F}} D_\infty\big(f\big(\delta_+(\tau, t), x\big), f(t, x)\big) \\ &+\|f(t, x)\|_{\mathscr{F}} D_\infty\big(g(\delta_+(\tau, t), x), g(t, x)\big) \\ &\le (M_1 + M_2)\varepsilon \equiv \varepsilon_1. \end{aligned}$$

Therefore, $\tau \in E(f * g, \varepsilon_1, S_0)$ and $E(f * g, \varepsilon_1, S_0)$ is a relatively dense set, so $f * g$ is S-almost periodic in t uniformly for $x \in D$. This completes the proof. □

Theorem 4.14 *If $f, g \in C(\mathbb{T} \times D, \mathbb{R}^n_{\mathcal{F}})$ are S-almost periodic in t uniformly for $x \in D$ under shifts $\delta_\pm$, then $f \tilde{+} g$ are S-almost periodic in t uniformly for $x \in D$.*

Proof From Theorem 4.11, for any $\varepsilon > 0$, $E\left(f, \frac{\varepsilon}{2}, S_0\right) \cap E\left(g, \frac{\varepsilon}{2}, S_0\right)$ is a nonempty relatively dense set. It is easy to observe that if $\tau \in E\left(f, \frac{\varepsilon}{2}, S_0\right) \cap E\left(g, \frac{\varepsilon}{2}, S_0\right)$, then $\tau \in E\left(f \tilde{+} g, \varepsilon, S_0\right)$. Hence

$$\left(E(f, \frac{\varepsilon}{2}, S_0) \cap E(g, \frac{\varepsilon}{2}, S_0)\right) \subset E(f \tilde{+} g, \varepsilon, S_0).$$

Therefore, $E(f \tilde{+} g, \varepsilon, S_0)$ is a relatively dense set, so $f \tilde{+} g$ is S-almost periodic in t uniformly for $x \in D$. The proof is completed. □

In what follows, we will establish a convergence theorem of S-almost periodic function sequences.

Theorem 4.15 *If $f_n \in C(\mathbb{T} \times D, \mathbb{R}^n_{\mathcal{F}})$, $n = 1, 2, \ldots$ are S-almost periodic in t for $x \in D$, and the sequence $\{f_n(t, x)\}$ uniformly converges to $f(t, x)$ on $\mathbb{T}^* \times S_0$, then $f(t, x)$ is S-almost periodic in t uniformly for $x \in D$.*

Proof For any $\varepsilon > 0$, there exists sufficiently large $\tilde{n}_0$ such that for all $(t, x) \in \mathbb{T}^* \times S_0$,

$$D_\infty(f(t, x), f_{\tilde{n}_0}(t, x)) < \frac{\varepsilon}{3}.$$

Take $\tau \in E\big\{f_{\tilde{n}_0}, \frac{\varepsilon}{3}, S_0\big\}$, and then for all $(t, x) \in \mathbb{T}^* \times S_0$, we obtain

$$\begin{aligned}
&D_\infty\big(f\big(\delta_+(\tau, t), x\big), f(t, x)\big) \\
&\le D_\infty\big(f\big(\delta_\tau(t), x\big), f_{\tilde{n}_0}\big(\delta_\tau(t), x\big)\big) + D_\infty\big(f_{\tilde{n}_0}\big(\delta_\tau(t), x\big), f_{\tilde{n}_0}(t, x)\big) \\
&\quad + D_\infty(f_{\tilde{n}_0}(t, x), f(t, x)) < \varepsilon,
\end{aligned}$$

that is, $\tau \in E(f, \varepsilon, S_0)$. Therefore, $E(f, \varepsilon, S_0)$ is also a relatively dense set, $f(t, x)$ is S-almost periodic in t uniformly for $x \in D$. This completes the proof. □

Theorem 4.16 *If $f \in C(\mathbb{T} \times D, \mathbb{R}^n_{\mathcal{F}})$ is S-almost periodic in t uniformly for $x \in D$, $F(\cdot)$ is uniformly continuous on the value field of $f(t, x)$, then $F \circ f$ is S-almost periodic in t uniformly for $x \in D$ under shifts $\delta_\pm$.*

Proof For F is uniformly continuous on the value field of $f(t,x)$, and $f(t,x)$ is S-almost periodic in t uniformly for $x \in D$, then there exists a real sequence $\alpha = \{\alpha_n\} \subseteq \Pi$ such that

$$\begin{aligned} T_\alpha^S(F \circ f) = T_\alpha^S\big(F\big(f(t,x)\big)\big) &= \lim_{n\to+\infty} F\big(f\big(\delta_+(\alpha_n,t),x\big)\big) \\ &= F\big(\lim_{n\to+\infty} f\big(\delta_+(\alpha_n,t),x\big)\big) = F(T_\alpha^S f) \end{aligned}$$

holds uniformly on $\mathbb{T}^* \times S_0$. Hence, $F \circ f$ is S-almost periodic in t uniformly for $x \in D$. □

Theorem 4.17 *Let $f \in C(\mathbb{T} \times D, \mathbb{R}^n_{\mathcal{F}})$ and j be an embedding mapping in Theorem 2.5. Then (i) $j \circ f$ is continuous on $\mathbb{T}$ if and only if f is continuous on $\mathbb{T}$. (ii) $j \circ f$ is S-almost periodic if and only if f is S-almost periodic. (iii) If f is gH-Δ-differentiable on $\mathbb{T}$, then $j \circ f$ is Δ-differentiable on $\mathbb{T}$ and $(j \circ f)^\Delta(t) = \big(j \circ f^\Delta\big)(t)$ for $t \in \mathbb{T}$.*

Proof From the property of the support function of the fuzzy set and Theorem 2.5, one can obtain

$$D_\infty\big(f(t_1), f(t_2)\big) = \|(j \circ f)(t_1) - (j \circ f)(t_2)\|_{\times_{i=1}^n(\bar{C}\times\bar{C})},$$

thus, we can obtain (i) immediately.

Next, we prove (ii). From the property of the support function of the fuzzy set and Theorem 2.5, one obtains

$$D_\infty\big(f(\delta_+(\tau,t)), f(t)\big) = \|(j \circ f)(\delta_+(\tau,t)) - (j \circ f)(t)\|_{\times_{i=1}^n(\bar{C}\times\bar{C})}. \tag{4.26}$$

Assume that f is S-almost periodic. For any $\varepsilon > 0$, there exists a constant $l(\varepsilon) > 0$ such that each interval of length $l(\varepsilon)$ contains a $\tau \in E\{\varepsilon, f\}$ such that

$$D_\infty\big(f(\delta_+(\tau,t)), f(t)\big) < \varepsilon.$$

From (4.26), one obtains

$$\big\|(j \circ f)(\delta_+(\tau,t)) - (j \circ f)(t)\big\|_{\times_{i=1}^n(\bar{C}\times\bar{C})} < \varepsilon,$$

which implies that $j \circ f$ is S-almost periodic. If $j \circ f$ is S-almost periodic, similarly, according to (4.26), one can obtain f is S-almost periodic immediately.

Now we show (iii). If f is gH-Δ-differentiable, then there exists a neighborhood U of t such that

$$\begin{aligned}&\big\|(j\circ f)(\sigma(t))-(j\circ f)(t+h)-(j\circ f^{\Delta})(t)(\mu(t)-h)\big\|_{\times_{i=1}^{n}(\bar{C}\times\bar{C})}\\&=\big\|j\circ\big[\big(f(\sigma(t))\tilde{-}_{gH}f(t+h)\big)\tilde{-}_{gH}f^{\Delta}(t)\big(\mu(t)-h\big)\big]\big\|_{\times_{i=1}^{n}(\bar{C}\times\bar{C})}\\&\leq D_{\infty}\big(f(\sigma(t))\tilde{-}_{gH}f(t+h),\,f^{\Delta}(t)(\mu(t)-h)\big)<\varepsilon|\mu(t)-h|\end{aligned}$$

for all $t+h\in U$ with $|h|<\delta$. The proof is completed. □

Theorem 4.18 *If $f\in C(\mathbb{T}\times D,\mathbb{R}^n_{\mathscr{F}})$ is shift-Δ-almost periodic in t uniformly for $x\in D$ under shifts $\delta_{\pm}$, denote*

$$F(t,x)=\int_{t_0}^{t}f(s,x)\Delta s,\;\; t_0\in\mathbb{T}^*,$$

then $F(t,x)$ is S-almost periodic in t uniformly for $x\in D$ under shifts $\delta_{\pm}$ if and only if $F(t,x)$ is bounded on $\mathbb{T}^*\times S_0$, where S_0 is any compact subset of D.

Proof For any $\alpha\in[0,1]$, we have

$$\begin{aligned}\big[F(t,x)\big]^{\alpha}&=\times_{i=1}^{n}\big[F_i^-(t,x),F_i^+(t,x)\big]_{\alpha}=\left[\int_{t_0}^{t}f(s,x)\Delta s\right]^{\alpha}=\int_{t_0}^{t}\big[f(s,x)\big]^{\alpha}\Delta s\\&=\times_{i=1}^{n}\left[\int_{t_0}^{t}f_i^-(s,x)\Delta s,\int_{t_0}^{t}f_i^+(s,x)\Delta s\right]_{\alpha}.\end{aligned}$$

According to Definition 4.9 (iii), for $i=1,2,\ldots,n$, we obtain

$$D_{\infty}^{(i)}\big(f_i(\delta_{\pm}(\tau,t),x)\delta_{\pm}^{\Delta}(\tau,t),f_i(t,x)\big)=\big\|f_i(\delta_{\pm}(\tau,t),x)\delta_{\pm}^{\Delta}(\tau,t)\tilde{-}_{gH}f_i(t,x)\big\|_{\mathscr{F}_0}<\varepsilon,$$

which implies that for each $i\in\{1,2,\ldots,n\}$, we have

$$\max\big\{\big|f_i^-(\delta_{\pm}(\tau,t),x)\delta_{\pm}^{\Delta}(\tau,t)-f_i^-(t,x)\big|,\big|f_i^+(\delta_{\pm}(\tau,t),x)\delta_{\pm}^{\Delta}(\tau,t)-f_i^+(t,x)\big|\big\}<\varepsilon.$$

Hence, $f_i^+,f_i^-\in C(\mathbb{T}\times D,\mathbb{R})$ is shift-Δ-almost periodic in t uniformly for $x\in D$ under shifts $\delta_{\pm}$. Without loss of generality, we will prove that $F_i^-(t,x)=\int_{t_0}^{t}f_i^-(s,x)\Delta s$ is S-almost periodic in t uniformly for $x\in D$ under shifts $\delta_{\pm}$ if and only if $F_i^-(t,x)$ is bounded on $\mathbb{T}^*\times S_0$ (the proof of the case $F_i^+(t,x)=\int_{t_0}^{t}f_i^+(s,x)\Delta s$ is similar).

First, if $F_i^-(t,x)$ is S-almost periodic in t uniformly for $x\in D$, it is easy to see that $F_i^-(t,x)$ is bounded on $\mathbb{T}^*\times S_0$.

Suppose $F_i^-(t,x)$ is bounded. Denote

$$G^-:=\sup_{(t,x)\in\mathbb{T}^*\times S_0}F_i^-(t,x)>g^-:=\inf_{(t,x)\in\mathbb{T}^*\times S_0}F_i^-(t,x),$$

for any $\varepsilon>0$, there exist t_1 and t_2 such that

$$F_i^-(t_1,x)<g^-+\frac{\varepsilon}{6},\qquad F_i^-(t_2,x)>G^--\frac{\varepsilon}{6},\quad\forall x\in S_0.$$

Let $l = l(\varepsilon_1, S_0)$ be an inclusion length of $E(f, \varepsilon_1, S_0)$, where $\varepsilon_1 = \dfrac{\varepsilon}{6d}$, $d = |t_1 - t_2|$. For any $\alpha \in \Pi$, take $\tau \in E(f, \varepsilon_1, S_0)$ such that

$$\delta_+(\tau, t_1) \in \big[\delta_+(\alpha, t_0), \delta_+\big(\delta_+(\alpha, l), t_0\big)\big]_{\mathbb{T}}.$$

Denote $s_i = \delta_+(\tau, t_i)$, $(i = 1, 2)$, $L = \delta_+(l, \tilde{d})$, where $\tilde{d} \in \Pi$ and $\tilde{d} > d$, so $s_1, s_2 \in \big[\delta_+(\alpha, t_0), \delta_+\big(\delta_+(\alpha, L), t_0\big)\big]_{\mathbb{T}}$, for all $x \in S_0$,

$$\begin{aligned}
&F_i^-(s_2, x) - F_i^-(s_1, x)\\
&= F_i^-(t_2, x) - F_i^-(t_1, x) - \int_{t_1}^{t_2} f_i^-(t, x)\Delta t + \int_{\delta_+(\tau,t_1)}^{\delta_+(\tau,t_2)} f_i^-(t, x)\Delta t\\
&= F_i^-(t_2, x) - F_i^-(t_1, x) + \int_{t_1}^{t_2} \big[f_i^-\big(\delta_+(\tau, t), x\big)\delta_+^{\Delta}(\tau, t) - f_i^-(t, x)\big]\Delta t\\
&> G^- - g^- - \frac{\varepsilon}{3} - \varepsilon_1 d = G^- - g^- - \frac{\varepsilon}{2},
\end{aligned}$$

that is

$$\big(F_i^-(s_1, x) - g^-\big) + \big(G^- - F_i^-(s_2, x)\big) < \frac{\varepsilon}{2}.$$

Since

$$F_i^-(s_1, x) - g^- \geq 0, \qquad G^- - F_i^-(s_2, x) \geq 0,$$

in any interval with length L, there exist s_1, s_2 such that

$$F_i^-(s_1, x) < g^- + \frac{\varepsilon}{2}, \qquad F_i^-(s_2, x) > G^- - \frac{\varepsilon}{2}.$$

Next, note that $L \neq t_0$, we have $\inf |\delta_L(t) - t| > q > 0$ for all $t \in \mathbb{T}^*$, where q is some positive constant. Let $\varepsilon_2 = \dfrac{\varepsilon}{2q}$, and we will show that if $\tau \in E(f, \varepsilon_2, S)$, then $\tau \in E(F, \varepsilon, S)$. In fact, for any $(t, x) \in \mathbb{T}^* \times S$, we can choose $s_1, s_2 \in [t, \delta_+(L, t)]_{\mathbb{T}}$ such that

$$F_i^-(s_1, x) < g^- + \frac{\varepsilon}{2}, \qquad F_i^-(s_2, x) > G^- - \frac{\varepsilon}{2}.$$

Thus, for $\tau \in E(f, \varepsilon_2, S)$, one obtains

$$\begin{aligned}
&F_i^-\big(\delta_+(\tau, t), x\big) - F_i^-(t, x)\\
&= F_i^-\big(\delta_+(\tau, s_1), x\big) - F_i^-(s_1, x)\\
&\quad + \int_t^{s_1} f_i^-(t, x)\Delta t - \int_{\delta_+(\tau,t)}^{\delta_+(\tau,s_1)} f_i^-(t, x)\Delta t\\
&> g^- - (g^- + \frac{\varepsilon}{2}) - \int_t^{s_1} \big[f_i^-\big(\delta_+(\tau, t), x\big)\big)\delta_+^{\Delta}(\tau, t) - f_i^-(t, x)\big]\Delta t\\
&> -\frac{\varepsilon}{2} - \varepsilon_2 q = -\varepsilon
\end{aligned}$$

since $|t - s_1| = |\delta_+(\tau, t_1) - t| \leq |\delta_+(L, t) - t|$. Moreover, we also have

$$\begin{aligned} & F_i^-(\delta_+(\tau, t), x) - F_i^-(t, x) \\ &= F_i^-(\delta_+(\tau, s_2), x) - F_i^-(s_2, x) + \int_t^{s_2} f_i^-(t, x)\Delta t \\ &- \int_{\delta_+(\tau,t)}^{\delta_+(\tau,s_2)} f_i^-(t, x)\Delta t < G^- - \left(G^- - \frac{\varepsilon}{2}\right) + \varepsilon_2 q = \varepsilon. \end{aligned}$$

Similar to the proof above, we can also obtain F_i^+ is S-almost periodic in t uniformly for $x \in D$ under shifts $\delta_\pm$ if and only if $F_i^+(t, x)$ is bounded on $\mathbb{T}^* \times S_0$.

Therefore, for $\tau \in E(f, \varepsilon_2, S_0)$, we have $\tau \in E(F, \varepsilon, S_0)$, thus $F(t, x)$ is S-almost periodic in t uniformly for $x \in D$. The proof is completed. □

Finally, we will establish a sufficient and necessary criterion for S-almost periodic functions.

Theorem 4.19 *A function $f \in C(\mathbb{T} \times D, \mathbb{R}^n_{\mathscr{F}})$ is S-almost periodic in t uniformly for $x \in D$ under shifts $\delta_\pm$ if and only if for every pair of sequences $\alpha', \beta' \subseteq \Pi$, there exist common subsequences $\alpha \subset \alpha', \beta \subset \beta'$ such that*

$$T^S_{\delta_+(\alpha,\beta)} f(t, x) = T^S_\alpha T^S_\beta f(t, x). \tag{4.27}$$

Proof If $f(t, x)$ is S–almost periodic in t uniformly for $x \in D$, for any two sequences $\alpha', \beta' \subseteq \Pi$, there exists subsequence $\beta'' \subset \beta'$ such that $T^S_{\beta''} f(t, x) = g(t, x)$ holds uniformly on $\mathbb{T}^* \times S_0$ and $g(t, x)$ is S-almost periodic in t uniformly for $x \in D$.

Choose $\alpha'' \subset \alpha'$ and α'', β'' are the common subsequences of α', β', respectively, then there exists $\alpha''' \subset \alpha''$ such that $T^S_{\alpha'''} g(t, x) = h(t, x)$ holds uniformly on $\mathbb{T}^* \times S_0$.

Similarly, take $\beta''' \subset \beta''$ such that β''', α''' are the common subsequences of β'', α'', respectively, then there exist common subsequence $\alpha \subset \alpha''', \beta \subset \beta'''$ such that $T^S_{\delta_+(\alpha,\beta)} f(t, x) = k(t, x)$ holds uniformly on $\mathbb{T}^* \times S_0$. From the above, one can easily observe that

$$T^S_\beta f(t, x) = g(t, x), \quad T^S_\alpha g(t, x) = h(t, x)$$

hold uniformly on $\mathbb{T}^* \times S_0$. Thus, for all $\varepsilon > 0$, if n is sufficiently large, then for any $(t, x) \in \mathbb{T}^* \times S_0$, we have

$$D_\infty\big(f\big(\delta_+(\delta_+(\alpha_n, \beta_n), t), x\big), k(t, x)\big) < \frac{\varepsilon}{3},$$

$$D_\infty\big(g(t, x), f\big(\delta_+(\beta_n, t), x\big)\big) < \frac{\varepsilon}{3}, \quad D_\infty\big(h(t, x), g\big(\delta_{\alpha_n}(t), x\big)\big) < \frac{\varepsilon}{3}.$$

Therefore

$$\begin{aligned} D_\infty(h(t,x),k(t,x)) \le\ & D_\infty\big(h(t,x), g\big(\delta_+(\alpha_n,t),x\big)\big) \\ & + D_\infty\big(g\big(\delta_+(\alpha_n,t),x\big), f\big(\delta_+(\delta_+(\alpha_n,\beta_n),t),x\big)\big) \\ & + D_\infty\big(f\big(\delta_+(\delta_+(\alpha_n,\beta_n),t),x\big),k(t,x)\big) < \varepsilon \end{aligned}$$

holds for all $(t,x) \in \mathbb{T}^* \times S_0$. Since $\varepsilon > 0$ is arbitrary, we have $h(t,x) \equiv k(t,x)$, that is, $T^S_{\delta_+(\alpha,\beta)} f(t,x) = T^S_\alpha T^S_\beta f(t,x)$ holds uniformly on $\mathbb{T}^* \times S_0$.

On the other hand, if (4.27) holds, then for any sequence $\gamma' \subseteq \Pi$, there exists subsequence $\gamma \subset \gamma'$, such that $T^S_\gamma f(t,x)$ exists uniformly on $\mathbb{T}^* \times S_0$.

In the following, we will show that $f(t,x)$ is S-almost periodic in t uniformly for $x \in D$.

If this is not true, i.e., $T^S_\gamma f(t,x)$ does not converge uniformly on $\mathbb{T}^* \times S_0$, then there exist $\varepsilon_0 > 0$ and $t_0 \in \mathbb{T}$, subsequences $\alpha' \subset \gamma, \beta' \subset \gamma, s' \subseteq \Pi$ and $\alpha' = \{\alpha'_n\}, \beta' = \{\beta'_n\}, s = \{s'_n\}$ such that

$$D_\infty\big(f\big(\delta_+(\delta_+(s'_n,\alpha'_n),t_0),x\big), f\big(\delta_+(\delta_+(s'_n,\beta'_n),t_0),x\big)\big) \ge \varepsilon_0 > 0. \tag{4.28}$$

According to (4.27), there exist common subsequences $\alpha'' \subset \alpha', s'' \subset s'$ such that for all $(t,x) \in \mathbb{T}^* \times S_0$, we have

$$T^S_{\delta_+(s'',\alpha'')} f(t,x) = T^S_{s''} T^S_{\alpha''} f(t,x). \tag{4.29}$$

Selecting $\beta'' \subset \beta'$ such that β'', α'', s'' are common subsequences of β', α', s', respectively, such that for all $(t,x) \in \mathbb{T}^* \times S_0$, one has

$$T^S_{\delta_+(s,\beta)} f(t,x) = T^S_s T^S_\beta f(t,x). \tag{4.30}$$

Similarly, taking $\alpha \subset \alpha''$ satisfying α, β, s are common subsequences of α'', β'', s'', respectively, according to (4.29), for all $(t,x) \in \mathbb{T}^* \times S_0$, we get

$$T^S_{\delta_+(s,\alpha)} f(t,x) = T^S_s T^S_\alpha f(t,x). \tag{4.31}$$

Since $T^S_\alpha f(t,x) = T^S_\beta f(t,x) = T^S_\gamma f(t,x)$, from (4.30) and (4.31), for all $(t,x) \in \mathbb{T}^* \times S_0$, we obtain

$$T^S_{\delta_+(s,\beta)} f(t,x) = T^S_{\delta_+(s,\alpha)} f(t,x),$$

that is, for all $(t,x) \in \mathbb{T}^* \times S_0$,

$$\lim_{n\to+\infty} f\big(\delta_+(\delta_+(s_n,\beta_n),t),x\big) = \lim_{n\to+\infty} f\big(\delta_+(\delta_+(s_n,\alpha_n),t),x\big).$$

Letting $t = t_0$, this contradicts (4.28). Therefore, $f(t,x)$ is S-almost periodic in t uniformly for $x \in D$ under shifts $\delta_\pm$. The proof is completed. □

5 Division of Fuzzy Vector-Valued Functions Depending on Determinant Algorithm

5.1 Basic Results of Fuzzy Multidimensional Spaces

This section is mainly concerned with an embedding theorem in a fuzzy multidimensional space.

Now we introduce an equivalent norm $\|\cdot\|_{\mathscr{L}}$ for any

$$\big((f_1, g_1), (f_2, g_2), \ldots, (f_n, g_n)\big) \in \times_{i=1}^{n}\big(\bar{C}[0,1] \times \bar{C}[0,1]\big)$$

as follows:

$$\big\|\big((f_1, g_1), (f_2, g_2), \ldots, (f_n, g_n)\big)\big\|_{\mathscr{L}} = \left\{\sum_{i=1}^{n}\int_0^1 \Big(f_i^2(x) + g_i^2(x)\Big)\mathrm{d}x\right\}^{\frac{1}{2}}.$$

In fact, there exists two positive constants c_1, c_2 such that $c_1\|\cdot\|_{\mathscr{L}} \le \|\cdot\|_{\times_{i=1}^{n}(\bar{C}\times\bar{C})} \le c_2\|\cdot\|_{\mathscr{L}}$.

Theorem 5.1 *The space* $\big(\times_{i=1}^{n}\big(\bar{C}[0,1] \times \bar{C}[0,1]\big), \langle\cdot,\cdot\rangle\big)$ *is an inner product space equipped with the inner product*

$$\langle x, y\rangle = \sum_{i=1}^{n}\int_0^1 \big(x_i^{(1)}(\hat{t})y_i^{(1)}(\hat{t}) + x_i^{(2)}(\hat{t})y_i^{(2)}(\hat{t})\big)\mathrm{d}\hat{t},$$

where

$$x = \big((x_1^{(1)}, x_1^{(2)}), (x_2^{(1)}, x_2^{(2)}), \ldots, (x_n^{(1)}, x_n^{(2)})\big)$$

and

$$y = \big((y_1^{(1)}, y_1^{(2)}), (y_2^{(1)}, y_2^{(2)}), \ldots, (y_n^{(1)}, y_n^{(2)})\big).$$

C. Wang and R. P. Agarwal, *Dynamic Equations and Almost Periodic Fuzzy Functions on Time Scales*, Synthesis Lectures on Mathematics & Statistics,
https://doi.org/10.1007/978-3-031-11236-2_5

Proof For any $x, y, z \in \times_{i=1}^{n}(\bar{C}[0,1] \times \bar{C}[0,1])$ and $\alpha \in \mathbb{R}$, we can obtain (i). $\langle x, x\rangle \geq 0$ and $\langle x, x\rangle = 0 \Leftrightarrow x = 0$; (ii). $\langle x, y\rangle = \langle y, x\rangle$; (iii). $\langle \alpha x, y\rangle = \alpha\langle x, y\rangle$ for $\alpha \in \mathbb{R}$; (iv). we can also obtain

$$\begin{aligned}
&\langle x+y, z\rangle \\
&= \sum_{i=1}^{n}\int_0^1 \left((x_i^{(1)}(\hat{t}) + y_i^{(1)}(\hat{t}))z_i^{(1)}(\hat{t}) + (x_i^{(2)}(\hat{t}) + y_i^{(2)}(\hat{t}))z_i^{(2)}(\hat{t})\right)\mathrm{d}\hat{t} \\
&= \sum_{i=1}^{n}\int_0^1 \left(x_i^{(1)}(\hat{t})z_i^{(1)}(\hat{t}) + y_i^{(1)}(\hat{t})z_i^{(1)}(\hat{t}) + x_i^{(2)}(\hat{t})z_i^{(2)}(\hat{t}) + y_i^{(2)}(\hat{t})z_i^{(2)}(\hat{t})\right)\mathrm{d}\hat{t} \\
&= \sum_{i=1}^{n}\int_0^1 (x_i^{(1)}(\hat{t})z_i^{(1)}(\hat{t}) + x_i^{(2)}(\hat{t})z_i^{(2)}(\hat{t}))\mathrm{d}\hat{t} \\
&\quad + \sum_{i=1}^{n}\int_0^1 (y_i^{(1)}(\hat{t})z_i^{(1)}(\hat{t}) + y_i^{(2)}(\hat{t})z_i^{(2)}(\hat{t}))\mathrm{d}\hat{t} \\
&= \langle x, z\rangle + \langle y, z\rangle.
\end{aligned}$$

Therefore, $\left(\times_{i=1}^{n}(\bar{C}[0,1] \times \bar{C}[0,1]), \langle\cdot,\cdot\rangle\right)$ is an inner product space. The proof is completed. □

Remark 5.1 Notice that in the inner product space $\left(\times_{i=1}^{n}(\bar{C}[0,1] \times \bar{C}[0,1]), \langle\cdot,\cdot\rangle\right)$, the inner product will generate the norm $\|\cdot\|_{\mathscr{L}}$, i.e.,

$$\sqrt{\langle x, x\rangle} = \|x\|_{\mathscr{L}} = \left(\sum_{i=1}^{n}\int_0^1 ((x_i^{(1)}(\hat{t}))^2 + (x_i^{(2)}(\hat{t}))^2)\mathrm{d}(\hat{t})\right)^{\frac{1}{2}}$$

which is equivalent to the norm $\|\cdot\|_{\times_{i=1}^{n}(\bar{C}\times\bar{C})}$.

Remark 5.2 From Theorem 2.4, we obtain that $\left(\times_{i=1}^{n}(\bar{C}[0,1] \times \bar{C}[0,1]), \langle\cdot,\cdot\rangle\right)$ is a Hilbert space.

5.2 A New Division of Multidimensional Intervals and Fuzzy Vectors

In this subsection, using the algorithm determinant of the multiplication of Sect. 2.2, we introduce the corresponding division of multidimensional intervals and fuzzy vectors.

Definition 5.1 Let an interval be $H = [h^-, h^+]$ and $0 \notin H$. The "inverse" of H (it is not the inverse in the algebraic sense) is defined by $H^{-1} = [1/h^+, 1/h^{-1}]$.

Remark 5.3 If $0 \in (h^-, h^+)$, the inverse of H is nonexistent. However, for intervals $H = [0, h^+]$ or $H = [h^-, 0]$, the general inverse of H can be defined by letting $H = [\varepsilon, H^+]$ or $H = [h^-, -\varepsilon]$ and we obtain the result by the limit as $\varepsilon \to 0^+$. For example, consider $H = [0, 3]$, and we have $H^{-1} = [1/3, \lim_{\varepsilon\to 0^+} 1/\varepsilon] = [1/3, +\infty)$. Similarly, for $H = [-3, 0]$, we have $H^{-1} = [\lim_{\varepsilon\to 0^+} 1/-\varepsilon, -1/3] = (-\infty, -1/3]$. Notice that the concept of inverse and general inverse of an interval H was introduced in the literature [75] by Stefanini and it will be useful to establish the transformational relation between the multiplication and the division of fuzzy vectors under a determinant algorithm.

For any $[a^-, a^+] \subseteq [u^-, u^+]$ and $[b^-, b^+] \subseteq [v^-, v^+]$ and $0 \notin [v^-, v^+]$, we introduce the following division by a determinant algorithm:

$$\textbf{Type I.}\quad [a^-, a^+] \div_I [b^-, b^+] = \left\{a \circledast b : a \in [a^-, a^+],\ b \in \left[1/b^+, 1/b^-\right]\right\}, \tag{5.1}$$

where if $I^{(I)}_{u,v^{-1}} \le 0$, then

$$a \circledast b = \begin{cases} ab, & ab \in [u^-/v^-, u^+/v^+], \\ u^-/v^-, & ab < u^-/v^-, \\ u^+/v^+, & ab > u^+/v^+; \end{cases}$$

if $I^{(I)}_{u,v^{-1}} \ge 0$, then

$$a \circledast b = \begin{cases} ab, & ab \in [u^+/v^+, u^-/v^-], \\ u^+/v^+, & ab < u^+/v^+, \\ u^-/v^-, & ab > u^-/v^-. \end{cases}$$

$$\textbf{Type II.}\quad [a^-, a^+] \div_{II} [b^-, b^+] = \left\{a \circledast b : a \in [a^-, a^+],\ b \in [1/b^+, 1/b^-]\right\}, \tag{5.2}$$

where if $I^{(II)}_{u,v^{-1}} \le 0$, then

$$a \circledast b = \begin{cases} ab, & ab \in [u^+/v^-, u^-/v^+], \\ u^+/v^-, & ab < u^+/v^-, \\ u^-/v^+, & ab > u^-/v^+; \end{cases}$$

if $I^{(II)}_{u,v^{-1}} \ge 0$, then

$$a \circledast b = \begin{cases} ab, & ab \in [u^-/v^+, u^+/v^-], \\ u^-/v^+, & ab < u^-/v^+, \\ u^+/v^-, & ab > u^+/v^-. \end{cases}$$

Type III. $[a^-, a^+] \div_{III} [b^-, b^+] = \{a \ast b : a \in [a^-, a^+],\ b \in [1/b^+, 1/b^-]\}$, (5.3)

where if $I^{(III)}_{u,v^{-1}} \le 0$, then

$$a \ast b = \begin{cases} ab, & ab \in [u^-/v^-, u^-/v^+], \\ u^-/v^-, & ab < u^-/v^-, \\ u^-/v^+, & ab > u^-/v^+; \end{cases}$$

if $I^{(III)}_{u,v^{-1}} \ge 0$, then

$$a \ast b = \begin{cases} ab, & ab \in [u^-/v^+, u^-/v^-], \\ u^-/v^+, & ab < u^-/v^+, \\ u^-/v^-, & ab > u^-/v^-. \end{cases}$$

Type IV. $[a^-, a^+] \div_{IV} [b^-, b^+] = \{a \ast b : a \in [a^-, a^+],\ b \in [1/b^+, 1/b^-]\}$, (5.4)

where if $I^{(IV)}_{u,v^{-1}} \le 0$, then

$$a \ast b = \begin{cases} ab, & ab \in [u^+/v^-, u^+/v^+], \\ u^+/v^-, & ab < u^+/v^-, \\ u^+/v^+, & ab > u^+/v^+; \end{cases}$$

if $I^{(IV)}_{u,v^{-1}} \ge 0$, then

$$a \ast b = \begin{cases} ab, & ab \in [u^+/v^+, u^+/v^-], \\ u^+/v^+, & ab < u^+/v^+, \\ u^+/v^-, & ab > u^+/v^-. \end{cases}$$

Type V. $[a^-, a^+] \div_{V} [b^-, b^+] = \{a \ast b : a \in [a^-, a^+],\ b \in [1/b^+, 1/b^-]\}$, (5.5)

where if $I^{(V)}_{u,v^{-1}} \le 0$, then

$$a \ast b = \begin{cases} ab, & ab \in [u^-/v^+, u^+/v^+], \\ u^-/v^+, & ab < u^-/v^+, \\ u^+/v^+, & ab > u^+/v^+; \end{cases}$$

if $I^{(V)}_{u,v^{-1}} \ge 0$, then

$$a \ast b = \begin{cases} ab, & ab \in [u^+/v^+, u^-/v^+], \\ u^+/v^+, & ab < u^+/v^+, \\ u^-/v^+, & ab > u^-/v^+. \end{cases}$$

Type VI. $[a^-, a^+] \div_{VI} [b^-, b^+] = \{a \ast b : a \in [a^-, a^+],\ b \in [1/b^+, 1/b^-]\}$, (5.6)

where if $I^{(VI)}_{u,v^{-1}} \le 0$, then

$$a \ast b = \begin{cases} ab, & ab \in [u^-/v^-, u^+/v^-], \\ u^-/v^-, & ab < u^-/v^-, \\ u^+/v^-, & ab > u^+/v^-; \end{cases}$$

if $I^{(VI)}_{u,v^{-1}} \ge 0$, then

$$a \ast b = \begin{cases} ab, & ab \in [u^+/v^-, u^-/v^-], \\ u^+/v^-, & ab < u^+/v^-, \\ u^-/v^-, & ab > u^-/v^-. \end{cases}$$

Next, we will introduce six types of the division of fuzzy vectors induced by the division of compact intervals defined by (5.1)–(5.6). For any $\alpha \in [0, 1]$ and $i = 1, 2, \ldots, n$, we define the following types $I - VI$ (see (5.7)–(5.12)) with the (compact box) α-level set:

Type I. $$\left[\left(\frac{u}{v}\right)_{\mathring{I}}\right]^{\alpha} := [u \div_{\mathring{I}} v]^{\alpha} = \times_{i=1}^{n}\left([u^-_{i,\alpha}, u^+_{i,\alpha}] \div_I [v^-_{i,\alpha}, v^+_{i,\alpha}]\right),$$

$$\text{where } [u^-_{i,\alpha}, u^+_{i,\alpha}] \div_I [v^-_{i,\alpha}, v^+_{i,\alpha}] = \begin{cases} [u^-_{i,\alpha}/v^-_{i,\alpha}, u^+_{i,\alpha}/v^+_{i,\alpha}] \text{ if } I^{\alpha,(I)}_{u_i,v_i^{-1}} \le 0, \\ [u^+_{i,\alpha}/v^+_{i,\alpha}, u^-_{i,\alpha}/v^-_{i,\alpha}] \text{ if } I^{\alpha,(I)}_{u_i,v_i^{-1}} \ge 0; \end{cases} \tag{5.7}$$

Type II. $$\left[\left(\frac{u}{v}\right)_{\mathring{I}I}\right]^{\alpha} := [u \div_{\mathring{I}I} v]^{\alpha} = \times_{i=1}^{n}\left([u^-_{i,\alpha}, u^+_{i,\alpha}] \div_{II} [v^-_{i,\alpha}, v^+_{i,\alpha}]\right),$$

$$\text{where } [u^-_{i,\alpha}, u^+_{i,\alpha}] \div_{II} [v^-_{i,\alpha}, v^+_{i,\alpha}] = \begin{cases} [u^+_{i,\alpha}/v^-_{i,\alpha}, u^-_{i,\alpha}/v^+_{i,\alpha}] \text{ if } I^{\alpha,(II)}_{u_i,v_i^{-1}} \le 0, \\ [u^-_{i,\alpha}/v^+_{i,\alpha}, u^+_{i,\alpha}/v^-_{i,\alpha}] \text{ if } I^{\alpha,(II)}_{u_i,v_i^{-1}} \ge 0; \end{cases} \tag{5.8}$$

Type III. $$\left[\left(\frac{u}{v}\right)_{I\mathring{I}I}\right]^{\alpha} := [u \div_{I\mathring{I}I} v]^{\alpha} = \times_{i=1}^{n}\left([u^-_{i,\alpha}, u^+_{i,\alpha}] \div_{III} [v^-_{i,\alpha}, v^+_{i,\alpha}]\right),$$

$$\text{where } [u^-_{i,\alpha}, u^+_{i,\alpha}] \div_{III} [v^-_{i,\alpha}, v^+_{i,\alpha}] = \begin{cases} [u^-_{i,\alpha}/v^-_{i,\alpha}, u^-_{i,\alpha}/v^+_{i,\alpha}] \text{ if } I^{\alpha,(III)}_{u_i,v_i^{-1}} \le 0, \\ [u^-_{i,\alpha}/v^+_{i,\alpha}, u^-_{i,\alpha}/v^-_{i,\alpha}] \text{ if } I^{\alpha,(III)}_{u_i,v_i^{-1}} \ge 0; \end{cases} \tag{5.9}$$

Type IV. $$\left[\left(\frac{u}{v}\right)_{I\mathring{V}}\right]^{\alpha} := [u \div_{I\mathring{V}} v]^{\alpha} = \times_{i=1}^{n}\left([u^-_{i,\alpha}, u^+_{i,\alpha}] \div_{IV} [v^-_{i,\alpha}, v^+_{i,\alpha}]\right),$$

$$\text{where } [u^-_{i,\alpha}, u^+_{i,\alpha}] \div_{IV} [v^-_{i,\alpha}, v^+_{i,\alpha}] = \begin{cases} [u^+_{i,\alpha}/v^-_{i,\alpha}, u^+_{i,\alpha}/v^+_{i,\alpha}] \text{ if } I^{\alpha,(IV)}_{u_i,v_i^{-1}} \leq 0, \\ [u^+_{i,\alpha}/v^+_{i,\alpha}, u^+_{i,\alpha}/v^-_{i,\alpha}] \text{ if } I^{\alpha,(IV)}_{u_i,v_i^{-1}} \geq 0; \end{cases} \tag{5.10}$$

Type V. $$\left[\left(\frac{u}{v}\right)_{\mathring{V}}\right]^\alpha := [u \div_{\mathring{V}} v]^\alpha = \times^n_{i=1}\big([u^-_{i,\alpha}, u^+_{i,\alpha}] \div_V [v^-_{i,\alpha}, v^+_{i,\alpha}]\big),$$

$$\text{where } [u^-_{i,\alpha}, u^+_{i,\alpha}] \div_V [v^-_{i,\alpha}, v^+_{i,\alpha}] = \begin{cases} [u^-_{i,\alpha}/v^+_{i,\alpha}, u^+_{i,\alpha}/v^+_{i,\alpha}] \text{ if } I^{\alpha,(V)}_{u_i,v_i^{-1}} \leq 0, \\ [u^+_{i,\alpha}/v^+_{i,\alpha}, u^-_{i,\alpha}/v^+_{i,\alpha}] \text{ if } I^{\alpha,(V)}_{u_i,v_i^{-1}} \geq 0; \end{cases} \tag{5.11}$$

Type VI. $$\left[\left(\frac{u}{v}\right)_{\mathring{V}I}\right]^\alpha := [u \div_{\mathring{V}I} v]^\alpha = \times^n_{i=1}\big([u^-_{i,\alpha}, u^+_{i,\alpha}] \div_{VI} [v^-_{i,\alpha}, v^+_{i,\alpha}]\big),$$

$$\text{where } [u^-_{i,\alpha}, u^+_{i,\alpha}] \div_{VI} [v^-_{i,\alpha}, v^+_{i,\alpha}] = \begin{cases} [u^-_{i,\alpha}/v^-_{i,\alpha}, u^+_{i,\alpha}/v^-_{i,\alpha}] \text{ if } I^{\alpha,(VI)}_{u_i,v_i^{-1}} \leq 0, \\ [u^+_{i,\alpha}/v^-_{i,\alpha}, u^-_{i,\alpha}/v^-_{i,\alpha}] \text{ if } I^{\alpha,(VI)}_{u_i,v_i^{-1}} \geq 0. \end{cases} \tag{5.12}$$

For convenience, we introduce some notations. Let $u \div_{\mathring{I}} v := {}^I w$; $u \div_{\mathring{I}I} v := {}^{II} w$; $u \div_{\mathring{I}II} v := {}^{III} w$; $u \div_{\mathring{I}V} v := {}^{IV} w$; $u \div_{\mathring{V}} v := {}^V w$; $u \div_{\mathring{V}I} v := {}^{VI} w$; and $[{}^{\hat{I}} w]^\alpha := \times^n_{i=1}[{}^{\hat{I}} w^-_{i,\alpha}, {}^{\hat{I}} w^+_{i,\alpha}]$, where $\hat{I} \in \{I, II, \ldots, VI\}$. The following properties are immediate.

Theorem 5.2

(*i*) *If* $I^{\alpha,(I)}_{u_i,v_i^{-1}} \leq 0$, *then*

$$\times^n_{i=1}[u^-_{i,\alpha}, u^+_{i,\alpha}] = \times^n_{i=1}\{[v^-_{i,\alpha}, v^+_{i,\alpha}] \circledcirc [{}^I w^-_{i,\alpha}, {}^I w^+_{i,\alpha}]\};$$

$$\times^n_{i=1}[v^-_{i,\alpha}, v^+_{i,\alpha}] = \times^n_{i=1}\{[u^-_{i,\alpha}, u^+_{i,\alpha}] \circ [{}^I w^-_{i,\alpha}, {}^I w^+_{i,\alpha}]^{-1}\};$$

(*ii*) *if* $I^{\alpha,(I)}_{u_i,v_i^{-1}} \geq 0$, *then*

$$\times^n_{i=1}[u^-_{i,\alpha}, u^+_{i,\alpha}] = \times^n_{i=1}\{[v^-_{i,\alpha}, v^+_{i,\alpha}] \circ [{}^I w^-_{i,\alpha}, {}^I w^+_{i,\alpha}]\};$$

$$\times^n_{i=1}[v^-_{i,\alpha}, v^+_{i,\alpha}] = \times^n_{i=1}\{[u^-_{i,\alpha}, u^+_{i,\alpha}] \circledcirc [{}^I w^-_{i,\alpha}, {}^I w^+_{i,\alpha}]^{-1}\};$$

(*iii*) *if* $I^{\alpha,(II)}_{u_i,v_i^{-1}} \leq 0$, *then*

$$\times^n_{i=1}[u^-_{i,\alpha}, u^+_{i,\alpha}] = \times^n_{i=1}\{[v^-_{i,\alpha}, v^+_{i,\alpha}] \circledcirc [{}^{II} w^-_{i,\alpha}, {}^{II} w^+_{i,\alpha}]\};$$

$$\times^n_{i=1}[v^-_{i,\alpha}, v^+_{i,\alpha}] = \times^n_{i=1}\{[u^-_{i,\alpha}, u^+_{i,\alpha}] \circledcirc [{}^{II} w^-_{i,\alpha}, {}^{II} w^+_{i,\alpha}]^{-1}\};$$

(iv) *if* $I^{\alpha,(II)}_{u_i,v_i^{-1}} \geq 0$, *then*

$$\times_{i=1}^{n}[u_{i,\alpha}^{-}, u_{i,\alpha}^{+}] = \times_{i=1}^{n}\{[v_{i,\alpha}^{-}, v_{i,\alpha}^{+}] \circ [{}^{II}w_{i,\alpha}^{-}, {}^{II}w_{i,\alpha}^{+}]\};$$

$$\times_{i=1}^{n}[v_{i,\alpha}^{-}, v_{i,\alpha}^{+}] = \times_{i=1}^{n}\{[u_{i,\alpha}^{-}, u_{i,\alpha}^{+}] \circ [{}^{II}w_{i,\alpha}^{-}, {}^{II}w_{i,\alpha}^{+}]^{-1}\};$$

(v) *if* $I^{\alpha,(III)}_{u_i,v_i^{-1}} \leq 0$ *or* $I^{\alpha,(III)}_{u_i,v_i^{-1}} \geq 0$, *then*

$$\times_{i=1}^{n}[v_{i,\alpha}^{-}, v_{i,\alpha}^{+}] = \times_{i=1}^{n}\{[u_{i,\alpha}^{-}, u_{i,\alpha}^{+}] \boxtimes [{}^{III}w_{i,\alpha}^{-}, {}^{III}w_{i,\alpha}^{+}]^{-1}\};$$

(vi) *if* $I^{\alpha,(IV)}_{u_i,v_i^{-1}} \leq 0$ *or* $I^{\alpha,(IV)}_{u_i,v_i^{-1}} \geq 0$, *then*

$$\times_{i=1}^{n}[v_{i,\alpha}^{-}, v_{i,\alpha}^{+}] = \times_{i=1}^{n}\{[u_{i,\alpha}^{-}, u_{i,\alpha}^{+}] \boxdot [{}^{IV}w_{i,\alpha}^{-}, {}^{IV}w_{i,\alpha}^{+}]^{-1}\};$$

(vii) *if* $I^{\alpha,(V)}_{u_i,v_i^{-1}} \leq 0$ *or* $I^{\alpha,(V)}_{u_i,v_i^{-1}} \geq 0$, *then*

$$\times_{i=1}^{n}[u_{i,\alpha}^{-}, u_{i,\alpha}^{+}] = \times_{i=1}^{n}\{[v_{i,\alpha}^{-}, v_{i,\alpha}^{+}] \boxdot [{}^{V}w_{i,\alpha}^{-}, {}^{V}w_{i,\alpha}^{+}]\};$$

(viii) *if* $I^{\alpha,(VI)}_{u_i,v_i^{-1}} \leq 0$ *or* $I^{\alpha,(VI)}_{u_l,v_i^{-1}} \geq 0$, *then*

$$\times_{i=1}^{n}[u_{i,\alpha}^{-}, u_{i,\alpha}^{+}] = \times_{i=1}^{n}\{[v_{i,\alpha}^{-}, v_{i,\alpha}^{+}] \boxtimes [{}^{VI}w_{i,\alpha}^{-}, {}^{VI}w_{i,\alpha}^{+}]\}.$$

Proof (i). From $I^{\alpha,(I)}_{u_i,v_i^{-1}} \leq 0$, we have

$$\times_{i=1}^{n}\big([u_{i,\alpha}^{-}, u_{i,\alpha}^{+}] \div_I [v_{i,\alpha}^{-}, v_{i,\alpha}^{+}]\big) = \times_{i=1}^{n}\big[u_{i,\alpha}^{-}/v_{i,\alpha}^{-}, u_{i,\alpha}^{+}/v_{i,\alpha}^{+}\big],$$

so we obtain $\times_{i=1}^{n}[{}^{I}w_{i,\alpha}^{-}, {}^{I}w_{i,\alpha}^{+}] = \times_{i=1}^{n}\big[u_{i,\alpha}^{-}/v_{i,\alpha}^{-}, u_{i,\alpha}^{+}/v_{i,\alpha}^{+}\big]$. Then we have

$$\begin{vmatrix} v_{i,\alpha}^{+} & v_{i,\alpha}^{-} \\ u_{i,\alpha}^{-}/v_{i,\alpha}^{-} & u_{i,\alpha}^{+}/v_{i,\alpha}^{+} \end{vmatrix} = u_{i,\alpha}^{+} - u_{i,\alpha}^{-} \geq 0,$$

which implies that

$$\times_{i=1}^{n}\{[v_{i,\alpha}^{-}, v_{i,\alpha}^{+}] \circledcirc [{}^{I}w_{i,\alpha}^{-}, {}^{I}w_{i,\alpha}^{+}]\} = \times_{i=1}^{n}[u_{i,\alpha}^{-}, u_{i,\alpha}^{+}].$$

On the other hand, we have $\times_{i=1}^{n}[{}^{I}w_{i,\alpha}^{-}, {}^{I}w_{i,\alpha}^{+}]^{-1} = \times_{i=1}^{n}\big[v_{i,\alpha}^{+}/u_{i,\alpha}^{+}, v_{i,\alpha}^{-}/u_{i,\alpha}^{-}\big]$, so we have

$$\begin{vmatrix} u_{i,\alpha}^{-} & u_{i,\alpha}^{+} \\ v_{i,\alpha}^{+}/u_{i,\alpha}^{+} & v_{i,\alpha}^{-}/u_{i,\alpha}^{-} \end{vmatrix} = v_{i,\alpha}^{-} - v_{i,\alpha}^{+} \leq 0,$$

which implies that

$$\times_{i=1}^{n}\{[u_{i,\alpha}^{-}, u_{i,\alpha}^{+}] \circledcirc [{}^{I}w_{i,\alpha}^{-}, {}^{I}w_{i,\alpha}^{+}]^{-1}\} = \times_{i=1}^{n}[v_{i,\alpha}^{-}, v_{i,\alpha}^{+}].$$

By using a similar way, one can also check $(ii) - (viii)$ and we will not repeat their proof process here. □

Theorem 5.3 *Let* $u, v \in [\mathbb{R}_{\mathscr{F}}^{n}]$ *(here* 1 *is the same as* $\{1\}$*). Then*

(i) *if* $I_{u_i,v_i^{-1}}^{\alpha,(I)} \leq 0$ *and* $u \div_{\mathring{I}} v := {}^{I}w$*, then* $u = v \circledast ({}^{I}w)$ *and* $v = u * ({}^{I}w)^{-1}$*;*

(ii) *if* $I_{u_i,v_i^{-1}}^{\alpha,(I)} \geq 0$ *and* $u \div_{\mathring{I}} v := {}^{I}w$*, then* $u = v * ({}^{I}w)$ *and* $v = u \circledast ({}^{I}w)^{-1}$*;*

(iii) *if* $I_{u_i,v_i^{-1}}^{\alpha,(II)} \leq 0$ *and* $u \div_{\mathring{II}} v := {}^{II}w$*, then* $u = v \circledast ({}^{II}w)$ *and* $v = u \circledast ({}^{II}w)^{-1}$*;*

(iv) *if* $I_{u_i,v_i^{-1}}^{\alpha,(II)} \geq 0$ *and* $u \div_{\mathring{II}} v := {}^{II}w$*, then* $u = v * ({}^{II}w)$*;* $v = u * (w^{II})^{-1}$*;*

(v) *if* $I_{u_i,v_i^{-1}}^{\alpha,(III)} \leq 0$ *or* $I_{u_i,v_i^{-1}}^{\alpha,(III)} \geq 0$ *and* $u \div_{\mathring{III}} v := {}^{III}w$*, then* $v = u \hat{*} (w^{III})^{-1}$*;*

(vi) *if* $I_{u_i,v_i^{-1}}^{\alpha,(IV)} \leq 0$ *or* $I_{u_i,v_i^{-1}}^{\alpha,(IV)} \geq 0$ *and* $u \div_{\mathring{IV}} v := {}^{IV}w$*, then* $v = u \hat{\circledast} ({}^{IV}w)^{-1}$*;*

(vii) *if* $I_{u_i,v_i^{-1}}^{\alpha,(V)} \leq 0$ *or* $I_{u_i,v_i^{-1}}^{\alpha,(V)} \geq 0$ *and* $u \div_{\mathring{V}} v := {}^{V}w$*, then* $u = v \hat{\circledast} ({}^{V}w)$*;*

$(viii)$ *if* $I_{u_i,v_i^{-1}}^{\alpha,(VI)} \leq 0$ *or* $I_{u_i,v_i^{-1}}^{\alpha,(VI)} \geq 0$ *and* $u \div_{\mathring{VI}} v := {}^{VI}w$*, then* $u = v \hat{*} ({}^{VI}w)$*.*

Proof By Theorem 5.2, the results $(i) - (viii)$ are immediate. □

5.3 Basic Results of Calculus of Fuzzy Vector-Valued Functions on Time Scales

In this section, we will provide some determinant algorithm properties of the gH-Δ-derivatives of the product of two fuzzy vector-valued functions on time scales which belong to the transformations of Theorems 3.10 and 3.11 from the literature [104]. For convenience, we adopt the notation $f(\sigma(t)) = f^{\sigma}(t)$ in some statements. Using the following properties, we will deduce the determinant algorithm properties of the gH-Δ-derivatives of the **division** of two fuzzy vector-valued functions on time scales.

Theorem 5.4 *Let* $f, g^{-1} : \mathbb{T} \to [\mathbb{R}_{\mathscr{F}}^{n}]$ *be* (I)*-*gH*-*Δ*-differentiable and* $\mathbf{0} \notin [g^{-1}]^{\alpha}$ *for* $\alpha \in [0, 1]$*. Then*

(i) *if* $I_{f_i,g_i^{-1}}^{\alpha,(I)} \leq 0$*,* $I_{f_i,(g_i^{-1})^{\Delta_I}}^{\alpha,(I)} \leq 0$*,* $I_{f_i^{\Delta_I},(g_i^{-1})^{\sigma}}^{\alpha,(I)} \leq 0$ *and* $f \div_I g$ *is* (I)*-*gH*-*Δ*-differentiable, then*

$$(f \div_I g)^{\Delta_I} = (f * g^{-1})^{\Delta_I} = f * (g^{-1})^{\Delta_I} \tilde{+} f^{\Delta_I} * (g^{-1})^{\sigma}.$$

(ii) *if* $I^{\alpha,(I)}_{f_i,g_i^{-1}} \le 0$, $I^{\alpha,(I)}_{f_i,(g_i^{-1})^{\Delta_I}} \ge 0$, $I^{\alpha,(I)}_{f_i^{\Delta_I},(g_i^{-1})^{\sigma}} \ge 0$ *and* $f \div_I g$ *is* (II)-gH-Δ-*differentiable, then*

$$(f \div_I g)^{\Delta_{II}} = (f * g^{-1})^{\Delta_{II}} = f * (g^{-1})^{\Delta_I} \widetilde{+} f^{\Delta_I} * (g^{-1})^{\sigma}.$$

(iii) *if* $I^{\alpha,(II)}_{f_i,g_i^{-1}} \le 0$, $I^{\alpha,(II)}_{f_i,(g_i^{-1})^{\Delta_I}} \le 0$, $I^{\alpha,(II)}_{f_i^{\Delta_I},(g_i^{-1})^{\sigma}} \le 0$ *and* $f \div_{II} g$ *is* (I)-gH-Δ-*differentiable, then*

$$(f \div_{II} g)^{\Delta_I} = (f \circledast g^{-1})^{\Delta_I} = f \circledast (g^{-1})^{\Delta_I} \widetilde{+} (g^{-1})^{\sigma} \circledast f^{\Delta_I}.$$

(iv) *if* $I^{\alpha,(II)}_{f_i,g_i^{-1}} \le 0$, $I^{\alpha,(II)}_{f_i,(g_i^{-1})^{\Delta_I}} \ge 0$, $I^{\alpha,(II)}_{f_i^{\Delta_I},(g_i^{-1})^{\sigma}} \ge 0$ *and* $f \div_{II} g$ *is* (II)-gH-Δ-*differentiable, then*

$$(f \div_{II} g)^{\Delta_{II}} = (f \circledast g^{-1})^{\Delta_{II}} = f \circledast (g^{-1})^{\Delta_I} \widetilde{+} f^{\Delta_I} \circledast (g^{-1})^{\sigma}.$$

(v) *if* $I^{\alpha,(I)}_{f_i,g_i^{-1}} \ge 0$, $I^{\alpha,(I)}_{f_i,(g_i^{-1})^{\Delta_I}} \ge 0$, $I^{\alpha,(I)}_{f_i^{\Delta_I},(g_i^{-1})^{\sigma}} \ge 0$ *and* $f \div_I g$ *is* (I)-gH-Δ-*differentiable, then*

$$(f \div_I g)^{\Delta_I} = (f * g^{-1})^{\Delta_I} = f * (g^{-1})^{\Delta_I} \widetilde{+} f^{\Delta_I} * (g^{-1})^{\sigma}.$$

(vi) *if* $I^{\alpha,(I)}_{f_i,g_i^{-1}} \ge 0$, $I^{\alpha,(I)}_{f_i,(g_i^{-1})^{\Delta_I}} \le 0$, $I^{\alpha,(I)}_{f_i^{\Delta_I},(g_i^{-1})^{\sigma}} \le 0$ *and* $f \div_I g$ *is* (II)-gH-Δ-*differentiable, then*

$$(f \div_I g)^{\Delta_{II}} = (f * g^{-1})^{\Delta_{II}} = f * (g^{-1})^{\Delta_I} \widetilde{+} f^{\Delta_I} * (g^{-1})^{\sigma}.$$

(vii) *if* $I^{\alpha,(II)}_{f_i,g_i^{-1}} \ge 0$, $I^{\alpha,(II)}_{f_i,(g_i^{-1})^{\Delta_I}} \ge 0$, $I^{\alpha,(II)}_{f_i^{\Delta_I},(g_i^{-1})^{\sigma}} \ge 0$ *and* $f \div_{II} g$ *is* (I)-gH-Δ-*differentiable, then*

$$(f \div_{II} g)^{\Delta_I} = (f \circledast g^{-1})^{\Delta_I} = f \circledast (g^{-1})^{\Delta_I} \widetilde{+} (g^{-1})^{\sigma} \circledast f^{\Delta_I}.$$

(viii) *if* $I^{\alpha,(II)}_{f_i,g_i^{-1}} \ge 0$, $I^{\alpha,(II)}_{f_i,(g_i^{-1})^{\Delta_I}} \le 0$, $I^{\alpha,(II)}_{f_i^{\Delta_I},(g_i^{-1})^{\sigma}} \le 0$ *and* $f \div_{II} g$ *is* (II)-gH-Δ-*differentiable, then*

$$(f \div_{II} g)^{\Delta_{II}} = (f \circledast g^{-1})^{\Delta_{II}} = f \circledast (g^{-1})^{\Delta_I} \widetilde{+} f^{\Delta_I} \circledast (g^{-1})^{\sigma}.$$

Proof Let $[f(t)]^{\alpha} = \times_{i=1}^{n}[f_{i,\alpha}^{-}(t), f_{i,\alpha}^{+}(t)]$ and $[g(t)]^{\alpha} = \times_{i=1}^{n}[g_{i,\alpha}^{-}(t), g_{i,\alpha}^{+}(t)]$, then $[g^{-1}(t)]^{\alpha} = ([g(t)]^{\alpha})^{-1} = \times_{i=1}^{n}[\frac{1}{g_{i,\alpha}^{+}(t)}, \frac{1}{g_{i,\alpha}^{-}(t)}]$.

Since $I^{\alpha,(I)}_{f_i,g_i^{-1}} \leq 0$, then

$$\begin{vmatrix} f^-_{i,\alpha} & f^+_{i,\alpha} \\ 1/g^+_{i,\alpha} & 1/g^-_{i,\alpha} \end{vmatrix} = \frac{f^-_{i,\alpha}}{g^-_{i,\alpha}} - \frac{f^+_{i,\alpha}}{g^+_{i,\alpha}} \leq 0,$$

which implies that $[f * g^{-1}]^\alpha = \times_{i=1}^n [\frac{f^-_{i,\alpha}}{g^-_{i,\alpha}}, \frac{f^+_{i,\alpha}}{g^+_{i,\alpha}}]$.

Thus we obtain

$$\begin{aligned}\left[\frac{f^-_{i,\alpha}}{g^-_{i,\alpha}}, \frac{f^+_{i,\alpha}}{g^+_{i,\alpha}}\right]^{\Delta_I} &= \left[\left(\frac{f^-_{i,\alpha}}{g^-_{i,\alpha}}\right)^\Delta, \left(\frac{f^+_{i,\alpha}}{g^+_{i,\alpha}}\right)^\Delta\right] \\ &= \times_{i=1}^n \left[\frac{(f^-_{i,\alpha})^\Delta g^-_{i,\alpha} - f^-_{i,\alpha}(g^-_{i,\alpha})^\Delta}{g^-_{i,\alpha}(g^-_{i,\alpha})^\sigma}, \frac{(f^+_{i,\alpha})^\Delta g^+_{i,\alpha} - f^+_{i,\alpha}(g^+_{i,\alpha})^\Delta}{g^+_{i,\alpha}(g^+_{i,\alpha})^\sigma}\right].\end{aligned}$$

On the other hand, from $I^{\alpha,(I)}_{f_i,(g_i^{-1})^{\Delta_I}} \leq 0$ and $I^{\alpha,(I)}_{f_i^{\Delta_I},(g_i^{-1})^\sigma} \leq 0$, we have

$$\begin{vmatrix} f^-_{i,\alpha} & f^+_{i,\alpha} \\ (1/g^+_{i,\alpha})^\Delta & (1/g^-_{i,\alpha})^\Delta \end{vmatrix} = -\frac{f^-_{i,\alpha}(g^-_{i,\alpha})^\Delta}{g^-_{i,\alpha}(g^-_{i,\alpha})^\sigma} + \frac{f^+_{i,\alpha}(g^+_{i,\alpha})^\Delta}{g^+_{i,\alpha}(g^+_{i,\alpha})^\sigma} \leq 0,$$

$$\begin{vmatrix} (f^-_{i,\alpha})^\Delta & (f^+_{i,\alpha})^\Delta \\ ((g^+_{i,\alpha})^{-1})^\sigma & ((g^-_{i,\alpha})^{-1})^\sigma \end{vmatrix} = \frac{(f^-_{i,\alpha})^\Delta}{(g^-_{i,\alpha})^\sigma} - \frac{(f^+_{i,\alpha})^\Delta}{(g^+_{i,\alpha})^\sigma} \leq 0,$$

then

$$[f * (g^{-1})^{\Delta_I}]^\alpha = \times_{i=1}^n \left[-\frac{f^-_{i,\alpha}(g^-_{i,\alpha})^\Delta}{g^-_{i,\alpha}(g^-_{i,\alpha})^\sigma}, -\frac{f^+_{i,\alpha}(g^+_{i,\alpha})^\Delta}{g^+_{i,\alpha}(g^+_{i,\alpha})^\sigma}\right]$$

and

$$[f^{\Delta_I} * (g^{-1})^\sigma]^\alpha = \times_{i=1}^n \left[\frac{(f^-_{i,\alpha})^\Delta}{(g^-_{i,\alpha})^\sigma}, \frac{(f^+_{i,\alpha})^\Delta}{(g^+_{i,\alpha})^\sigma}\right],$$

so we have

$$\begin{aligned}&[f * (g^{-1})^{\Delta_I} + f^{\Delta_I} * (g^{-1})^\sigma]^\alpha \\ &= \times_{i=1}^n \left[\frac{(f^-_{i,\alpha})^\Delta g^-_{i,\alpha} - f^-_{i,\alpha}(g^-_{i,\alpha})^\Delta}{g^-_{i,\alpha}(g^-_{i,\alpha})^\sigma}, \frac{(f^+_{i,\alpha})^\Delta g^+_{i,\alpha} - f^+_{i,\alpha}(g^+_{i,\alpha})^\Delta}{g^+_{i,\alpha}(g^+_{i,\alpha})^\sigma}\right],\end{aligned}$$

which implies that $(f * g^{-1})^{\Delta_I} = f * (g^{-1})^{\Delta_I} \tilde{+} f^{\Delta_I} * (g^{-1})^\sigma$.

By applying a similar analysis as in (i) to $(ii) - (viii)$, one can obtain the results $(ii) - (viii)$ immediately. □

Theorem 5.5 *Let $f, g^{-1} : \mathbb{T} \to [\mathbb{R}^n_{\mathscr{F}}]$ be (II)-gH-Δ-differentiable and $\mathbf{0} \notin [g^{-1}]^\alpha$ for $\alpha \in [0, 1]$. Then*

(i) *if* $I^{\alpha,(II)}_{f_i,g_i^{-1}} \leq 0$, $I^{\alpha,(I)}_{f_i,(g_i^{-1})^{\Delta_{II}}} \geq 0$, $I^{\alpha,(I)}_{f_i^{\Delta_{II}},(g_i^{-1})^{\sigma}} \leq 0$ *and* $f \div_{II} g$ *is* (I)-gH-Δ-*differentiable, then*

$$(f \div_{II} g)^{\Delta_I} = (f \circledast g^{-1})^{\Delta_I} = f * (g^{-1})^{\Delta_{II}} \widetilde{\mp} f^{\Delta_{II}} * (g^{-1})^{\sigma}.$$

(ii) *if* $I^{\alpha,(II)}_{f_i,g_i^{-1}} \leq 0$, $I^{\alpha,(I)}_{f_i,(g_i^{-1})^{\Delta_{II}}} \leq 0$, $I^{\alpha,(I)}_{f_i^{\Delta_{II}},(g_i^{-1})^{\sigma}} \geq 0$ *and* $f \div_{II} g$ *is* (II)-gH-Δ-*differentiable, then*

$$(f \div_{II} g)^{\Delta_{II}} = (f \circledast g^{-1})^{\Delta_{II}} = f * (g^{-1})^{\Delta_I} \widetilde{\mp} f^{\Delta_I} * (g^{-1})^{\sigma}.$$

(iii) *if* $I^{\alpha,(II)}_{f_i,g_i^{-1}} \geq 0$, $I^{\alpha,(I)}_{f_i,(g_i^{-1})^{\Delta_{II}}} \leq 0$, $I^{\alpha,(I)}_{f_i^{\Delta_{II}},(g_i^{-1})^{\sigma}} \geq 0$ *and* $f \div_{II} g$ *is* (I)-gH-Δ-*differentiable, then*

$$(f \div_{II} g)^{\Delta_I} = (f \circledast g^{-1})^{\Delta_I} = f * (g^{-1})^{\Delta_{II}} \widetilde{\mp} (g^{-1})^{\sigma} * f^{\Delta_{II}}.$$

(iv) *if* $I^{\alpha,(II)}_{f_i,g_i^{-1}} \geq 0$, $I^{\alpha,(I)}_{f_i,(g_i^{-1})^{\Delta_{II}}} \geq 0$, $I^{\alpha,(I)}_{f_i^{\Delta_{II}},(g_i^{-1})^{\sigma}} \leq 0$ *and* $f \div_{II} g$ *is* (II)-gH-Δ-*differentiable, then*

$$(f \div_{II} g)^{\Delta_{II}} = (f \circledast g^{-1})^{\Delta_{II}} = f * (g^{-1})^{\Delta_{II}} \widetilde{\mp} f^{\Delta_{II}} * (g^{-1})^{\sigma}.$$

(v) *if* $I^{\alpha,(I)}_{f_i,g_i^{-1}} \leq 0$, $I^{\alpha,(II)}_{f_i,(g_i^{-1})^{\Delta_{II}}} \geq 0$, $I^{\alpha,(II)}_{f_i^{\Delta_{II}},(g_i^{-1})^{\sigma}} \leq 0$ *and* $f \div_I g$ *is* (I)-gH-Δ-*differentiable, then*

$$(f \div_I g)^{\Delta_I} = (f * g^{-1})^{\Delta_I} = f \circledast (g^{-1})^{\Delta_{II}} \widetilde{\mp} f^{\Delta_{II}} \circledast (g^{-1})^{\sigma}.$$

(vi) *if* $I^{\alpha,(I)}_{f_i,g_i^{-1}} \leq 0$, $I^{\alpha,(II)}_{f_i,(g_i^{-1})^{\Delta_{II}}} \leq 0$, $I^{\alpha,(II)}_{f_i^{\Delta_{II}},(g_i^{-1})^{\sigma}} \geq 0$ *and* $f \div_I g$ *is* (II)-gH-Δ-*differentiable, then*

$$(f \div_I g)^{\Delta_{II}} = (f * g^{-1})^{\Delta_{II}} = f \circledast (g^{-1})^{\Delta_{II}} \widetilde{\mp} f^{\Delta_{II}} \circledast (g^{-1})^{\sigma}.$$

(vii) *if* $I^{\alpha,(I)}_{f_i,g_i^{-1}} \geq 0$, $I^{\alpha,(II)}_{f_i,(g_i^{-1})^{\Delta_{II}}} \leq 0$, $I^{\alpha,(II)}_{f_i^{\Delta_{II}},(g_i^{-1})^{\sigma}} \geq 0$ *and* $f \div_I g$ *is* (I)-gH-Δ-*differentiable, then*

$$(f \div_I g)^{\Delta_I} = (f * g^{-1})^{\Delta_I} = f \circledast (g^{-1})^{\Delta_{II}} \widetilde{\mp} (g^{-1})^{\sigma} \circledast f^{\Delta_{II}}.$$

(viii) *if* $I^{\alpha,(I)}_{f_i,g_i^{-1}} \geq 0$, $I^{\alpha,(II)}_{f_i,(g_i^{-1})^{\Delta_{II}}} \geq 0$, $I^{\alpha,(II)}_{f_i^{\Delta_{II}},(g_i^{-1})^{\sigma}} \leq 0$ *and* $f \div_I g$ *is* (II)-gH-Δ-*differentiable, then*

$$(f \div_I g)^{\Delta_{II}} = (f * g^{-1})^{\Delta_{II}} = f \circledast (g^{-1})^{\Delta_{II}} \widetilde{\mp} f^{\Delta_{II}} \circledast (g^{-1})^{\sigma}.$$

Proof Denote $[f(t)]^{\alpha} = \times_{i=1}^{n}[f_{i,\alpha}^{-}(t), f_{i,\alpha}^{+}(t)]$ and $[g(t)]^{\alpha} = \times_{i=1}^{n}[g_{i,\alpha}^{-}(t), g_{i,\alpha}^{+}(t)]$, then $[g^{-1}(t)]^{\alpha} = ([g(t)]^{\alpha})^{-1} = \times_{i=1}^{n}[\frac{1}{g_{i,\alpha}^{+}(t)}, \frac{1}{g_{i,\alpha}^{-}(t)}]$.

Since $I^{\alpha,(II)}_{f_i,g_i^{-1}} \le 0$, then

$$\begin{vmatrix} f^+_{i,\alpha} & f^-_{i,\alpha} \\ 1/g^+_{i,\alpha} & 1/g^-_{i,\alpha} \end{vmatrix} = \frac{f^+_{i,\alpha}}{g^-_{i,\alpha}} - \frac{f^-_{i,\alpha}}{g^+_{i,\alpha}} \le 0,$$

which implies that $[f \circledast g^{-1}]^\alpha = \times_{i=1}^n [\frac{f^+_{i,\alpha}}{g^-_{i,\alpha}}, \frac{f^-_{i,\alpha}}{g^+_{i,\alpha}}]$.

Thus we obtain

$$\begin{aligned} \left[\frac{f^+_{i,\alpha}}{g^-_{i,\alpha}}, \frac{f^-_{i,\alpha}}{g^+_{i,\alpha}}\right]^{\Delta_I} &= \left[\left(\frac{f^+_{i,\alpha}}{g^-_{i,\alpha}}\right)^\Delta, \left(\frac{f^-_{i,\alpha}}{g^+_{i,\alpha}}\right)^\Delta\right] \\ &= \left[\frac{(f^+_{i,\alpha})^\Delta g^-_{i,\alpha} - f^+_{i,\alpha}(g^-_{i,\alpha})^\Delta}{g^-_{i,\alpha}(g^-_{i,\alpha})^\sigma}, \frac{(f^-_{i,\alpha})^\Delta g^+_{i,\alpha} - f^-_{i,\alpha}(g^+_{i,\alpha})^\Delta}{g^+_{i,\alpha}(g^+_{i,\alpha})^\sigma}\right]. \end{aligned}$$

On the other hand, from $I^{\alpha,(I)}_{f_i,(g_i^{-1})^{\Delta_{II}}} \ge 0$, $I^{\alpha,(I)}_{f_i^{\Delta_{II}},(g_i^{-1})^\sigma} \le 0$, we have

$$\begin{vmatrix} f^-_{i,\alpha} & f^+_{i,\alpha} \\ (1/g^-_{i,\alpha})^\Delta & (1/g^+_{i,\alpha})^\Delta \end{vmatrix} = -\frac{f^-_{i,\alpha}(g^+_{i,\alpha})^\Delta}{g^+_{i,\alpha}(g^+_{i,\alpha})^\sigma} + \frac{f^+_{i,\alpha}(g^-_{i,\alpha})^\Delta}{g^-_{i,\alpha}(g^-_{i,\alpha})^\sigma} \ge 0,$$

$$\begin{vmatrix} (f^+_{i,\alpha})^\Delta & (f^-_{i,\alpha})^\Delta \\ ((g^+_{i,\alpha})^{-1})^\sigma & ((g^-_{i,\alpha})^{-1})^\sigma \end{vmatrix} = \frac{(f^+_{i,\alpha})^\Delta}{(g^-_{i,\alpha})^\sigma} - \frac{(f^-_{i,\alpha})^\Delta}{(g^+_{i,\alpha})^\sigma} \le 0,$$

then

$$[f * (g^{-1})^{\Delta_I}]^\alpha = \times_{i=1}^n \left[-\frac{f^+_{i,\alpha}(g^-_{i,\alpha})^\Delta}{g^-_{i,\alpha}(g^-_{i,\alpha})^\sigma}, -\frac{f^-_{i,\alpha}(g^+_{i,\alpha})^\Delta}{g^+_{i,\alpha}(g^+_{i,\alpha})^\sigma}\right]$$

and

$$[f^{\Delta_I} * (g^{-1})^\sigma]^\alpha = \times_{i=1}^n \left[\frac{(f^+_{i,\alpha})^\Delta}{(g^-_{i,\alpha})^\sigma}, \frac{(f^-_{i,\alpha})^\Delta}{(g^+_{i,\alpha})^\sigma}\right],$$

so we have

$$\begin{aligned} &[f * (g^{-1})^{\Delta_I} + f^{\Delta_I} * (g^{-1})^\sigma]^\alpha \\ &= \times_{i=1}^n \left[\frac{(f^+_{i,\alpha})^\Delta g^-_{i,\alpha} - f^+_{i,\alpha}(g^-_{i,\alpha})^\Delta}{g^-_{i,\alpha}(g^-_{i,\alpha})^\sigma}, \frac{(f^-_{i,\alpha})^\Delta g^+_{i,\alpha} - f^-_{i,\alpha}(g^+_{i,\alpha})^\Delta}{g^+_{i,\alpha}(g^+_{i,\alpha})^\sigma}\right], \end{aligned}$$

which implies that $(f \circledast g^{-1})^{\Delta_I} = f * (g^{-1})^{\Delta_{II}} \widetilde{\mp} f^{\Delta_{II}} * (g^{-1})^\sigma$.

By applying a similar analysis as in (i) to $(ii) - (viii)$, one can obtain the results $(ii) - (viii)$ immediately. □

Almost Periodic Generalized Fuzzy Multidimensional Dynamic Equations and Applications

6

6.1 Almost Periodic Generalized Fuzzy Multidimensional Dynamic Equations

In the literature [116], the authors established a theory of fuzzy multidimensional dynamic equations on time scales. In this section, we introduce some new concepts related to fuzzy dynamic equations under shifts on time scales, and then we establish some basic results on almost periodic generalized fuzzy multidimensional dynamic equations .

Consider the following nonlinear fuzzy multidimensional dynamic equation

$$x^{\Delta}(t) = f(t, x), \tag{6.1}$$

where $f \in C(\mathbb{T} \times [\mathbb{R}^n_{\mathscr{F}}], [\mathbb{R}^n_{\mathscr{F}}])$, and let $\Omega = \{x \in [\mathbb{R}^n_{\mathscr{F}}] : x(t)$ is a bounded solution to (6.1)$\}$.

Definition 6.1 If $\Omega \neq \emptyset$, then $\lambda = \inf_{x\in\Omega} \|x\|_{\mathscr{F}}$ exists, and λ is called the least-value of solutions to (6.1). If there exists $\varphi(t) \in \Omega$ such that $\|\varphi\|_{\infty} = \lambda$, then $\varphi(t)$ is called a minimum norm solution to (6.1), where $\|\cdot\|_{\mathscr{F}} = D_{\infty}(\cdot, \tilde{\mathbf{0}})$ and $\|\cdot\|_{\infty} = \sup_{t\in\mathbb{T}} \|\cdot\|_{\mathscr{F}}$.

A similar proof to that in Theorem 5.1 in [26] gives the following.

Lemma 6.1 *If $f \in C(\mathbb{T} \times S, [\mathbb{R}^n_{\mathscr{F}}])$ is bounded on $\mathbb{T}^* \times S_0$ and* (6.1) *has a bounded solution $\varphi(t)$ such that $\{\varphi(t), t \in \mathbb{T}\} \subset S_0$ and $\tilde{\mathbf{0}} \in S_0$, then there is a minimum norm solution to* (6.1).

Lemma 6.2 *If $f \in C(\mathbb{T} \times [\mathbb{R}^n_{\mathscr{F}}], [\mathbb{R}^n_{\mathscr{F}}])$ is S-almost periodic in t uniformly for $x \in [\mathbb{R}^n_{\mathscr{F}}]$ under shifts $\delta_{\pm}$, $S_0 = \overline{\{\varphi(t) : t \geq t_0\}}$ and* (6.1) *has a bounded solution $\varphi(t)$ on $[t_0, \infty)_{\mathbb{T}}$, then* (6.1) *has a S-almost periodic solution $\psi(t)$ satisfying $\{\psi(t), t \in \mathbb{T}\} \subset S_0$.*

C. Wang and R. P. Agarwal, *Dynamic Equations and Almost Periodic Fuzzy Functions on Time Scales*, Synthesis Lectures on Mathematics & Statistics,
https://doi.org/10.1007/978-3-031-11236-2_6

Proof First, we may choose $\alpha' = \{\alpha'_k\} \subset \Pi$ such that $\lim_{k\to+\infty} \alpha'_k = +\infty$ implies

$$T^S_{\alpha'} f(t,x) = \lim_{k\to\infty} f\big(\delta_+(\alpha'_k, t), x\big) = f(t,x)$$

holds uniformly on $\mathbb{T}^* \times S_0$. For any fixed $a \in \mathbb{T}^*$, consider the interval $(a,\infty)_{\mathbb{T}}$ and $\varphi_k(t) = \varphi\big(\delta_+(\alpha'_k, t)\big)$.

Now, we show that for k sufficiently large, $\{\varphi_k\}$ is defined on $(a,\infty)_{\mathbb{T}}$ and is a solution to $x^\Delta(t) = f(t,x)$. Because $\varphi(t)$ is a solution to (6.1), then it follows that $T^S_\alpha\big(\varphi^\Delta(t)\big) = T^S_\alpha\big(f(t,x)\big)$, i.e.,

$$\lim_{k\to\infty} \big(\varphi\big(\delta_+(\alpha'_k, t)\big)\big)^\Delta = \big(\lim_{k\to\infty} \big(\varphi\big(\delta_+(\alpha'_k, t)\big)\big)\big)^\Delta = \lim_{k\to\infty} f\big(\delta_+(\alpha'_k, t), x\big) = f(t,x).$$

One can easily observe that $\{\varphi_k(t)\}$ is uniformly bounded and equicontinuous on $(a,\infty)_{\mathbb{T}}$. Then let α be a sequence which goes to $+\infty$, and it follows from Corollary 3.4 of [85] that there will exist $\alpha = \{\alpha'_{k_n}\} \subset \alpha'$ such that $T^S_\alpha \varphi(t) = \lim_{n\to\infty} \varphi\big(\delta_+(\alpha'_{k_n}, t)\big) = \psi(t)$ holds uniformly on $\mathbb{T}^*$. Thus, for all $t \in \mathbb{T}^*$, one can obtain $\psi(\cdot) \in S_0$. Since $T^S_\alpha f(t,x) = \lim_{n\to\infty} f\big(\delta_+(\alpha'_{k_n}, t), x\big) = f(t,x)$, then $\psi(t)$ is a S-almost periodic solution to (6.1). This completes the proof. □

Lemma 6.3 *Let $f \in C(\mathbb{T} \times [\mathbb{R}^n_{\mathcal{F}}], [\mathbb{R}^n_{\mathcal{F}}])$ be S-almost periodic in t uniformly for $x \in [\mathbb{R}^n_{\mathcal{F}}]$ under shifts $\delta_\pm$. If* (6.1) *has a minimum norm solution, then for any $g \in H_S(f)$, the following equation*

$$x^\Delta(t) = g(t,x) \tag{6.2}$$

has the same least-value of solutions as that to (6.1).

Proof Let $\varphi(t)$ be the minimum norm solution to (6.1) and λ is the least-value. Since $g \in H_S(f)$, then there exists a sequence $\alpha' \in \Pi$ such that $T^S_{\alpha'} f(t,x) = g(t,x)$ holds uniformly on $\mathbb{T}^* \times S_0$. From Corollary 3.4 in [85], there exists $\alpha \subset \alpha'$ such that $T^S_\alpha \varphi(t) = \psi(t)$ holds uniformly on $\mathbb{T}^*$. Hence, $\psi(t)$ is a solution to (6.2). For $D_\infty(\varphi(t), \tilde{\mathbf{0}}) \le \lambda$, one has $D_\infty(\psi(t), \tilde{\mathbf{0}}) \le \lambda$, thus, $\lambda' = \|\psi\|_\infty \le \lambda$. Since $\varphi(t) = T^S_{\alpha^{-1}} \psi(t)$ and $D_\infty(\psi(t), \tilde{\mathbf{0}}) \le \lambda'$, it follows that $D_\infty(\varphi(t), \tilde{\mathbf{0}}) \le \lambda'$, thus, $\lambda = \|\varphi\|_\infty \le \lambda'$. Therefore, $\lambda = \lambda'$, where $\|\cdot\|_\infty = \sup_{t\in\mathbb{T}} D_\infty(\cdot, \tilde{\mathbf{0}})$. The proof is completed. □

From the proof process of Lemma 6.3, one can obtain the following lemma immediately.

Lemma 6.4 *If $\varphi(t)$ is a minimum norm solution to* (6.1) *then there exists a sequence $\alpha' \subseteq \Pi$ such that $T^S_{\alpha'} f(t,x) = g(t,x)$ exists uniformly on $\mathbb{T}^* \times S_0$. Moreover, if there exists* *a subsequence* $\alpha \subset \alpha'$ *such that* $T_\alpha^S \varphi(t) = \psi(t)$ *holds uniformly on* $\mathbb{T}^*$, *then* $\psi(t)$ *is a minimum norm solution to* (6.2).

Lemma 6.5 *If* $f \in C(\mathbb{T} \times [\mathbb{R}^n_{\mathscr{F}}], [\mathbb{R}^n_{\mathscr{F}}])$ *is S-almost periodic in t uniformly for* $x \in [\mathbb{R}^n_{\mathscr{F}}]$ *and for every* $g \in H_S(f)$, (6.2) *has a unique minimum norm solution, then these minimum norm solutions are S-almost periodic on* $\mathbb{T}^*$.

Proof For a fixed $g \in H_S(f)$, we obtain that (6.2) has the unique minimum norm solution $\psi(t)$. Because $g(t,x)$ is S-almost periodic in t uniformly for $x \in [\mathbb{R}^n_{\mathscr{F}}]$, it follows that for any sequences $\alpha', \beta' \subseteq \Pi$, there exist common subsequences $\alpha \subset \alpha', \beta \subset \beta'$ such that

$$T^S_{\delta_+(\alpha,\beta)} g(t,x) = T^S_\alpha T^S_\beta g(t,x)$$

holds uniformly on $\mathbb{T}^* \times S_0$ and $T^S_\alpha T^S_\beta \psi(t)$, $T^S_{\delta_+(\alpha,\beta)} \psi(t)$ exist uniformly on $\mathbb{T}^*$. It follows from Lemmas 6.3 and 6.4 that $T^S_\alpha T^S_\beta \psi(t)$ and $T^S_{\delta_+(\alpha,\beta)} \psi(t)$ are minimum norm solutions to the following equation:

$$x^\Delta(t) = T^S_{\delta_+(\alpha,\beta)} g(t,x).$$

From the uniqueness of the minimum norm solution, one has

$$T^S_\alpha T^S_\beta \psi(t) = T^S_{\delta_+(\alpha,\beta)} \psi(t).$$

Therefore, $\psi(t)$ is S-almost periodic. The proof is completed. □

We will now discuss the linear S-almost periodic dynamic equation on $\mathbb{T}$ as follows:

$$x^\Delta(t) = A(t)x \tilde{+} f(t) \tag{6.3}$$

and its associated homogeneous equation

$$x^\Delta(t) = A(t)x, \tag{6.4}$$

where $A : \mathbb{T} \to \mathbb{R}^{n\times n}$ is an S-almost periodic matrix-valued function and $f : \mathbb{T} \to [\mathbb{R}^n_{\mathscr{F}}]$ is an S-almost periodic fuzzy vector-valued function.

Definition 6.2 If $B \in H_S(A)$, we say that

$$y^\Delta(t) = B(t)y \tag{6.5}$$

is a homogeneous equation in the hull of (6.3).

Definition 6.3 If $B \in H_S(A)$ and $g \in H_S(f)$, we say that

$$y^{\Delta}(t) = B(t)y\widetilde{+}g(t) \tag{6.6}$$

is an equation in the hull of (6.3).

Definition 6.4 We say the vector functions $x_1, x_2, \dots, x_n \in C(\mathbb{T}, [\mathbb{R}^n_{\mathscr{F}}])$ are linearly independent if there exist constants $c_1, c_2, \dots, c_n$ which are not complete zeroes such that

$$c_1 x_1(t) + c_2 x_2(t) + \cdots + c_n x_n(t) = \tilde{0}.$$

Otherwise, we say $x_1, x_2, \dots, x_n \in C(\mathbb{T}, [\mathbb{R}^n_{\mathscr{F}}])$ are linearly dependent.

Definition 6.5 We say $X(t) = \big(x_1(t), x_2(t), \dots, x_n(t)\big)$ is a fundamental solution matrix for (6.4) if $X^{\Delta}(t) = A(t)X(t)$ for all $t \in \mathbb{T}$ and $x_1, x_2, \dots, x_n \in C(\mathbb{T}, \mathbb{R}^n_{\mathscr{F}})$ are linearly independent. In particular, for any initial point $t_0 \in \mathbb{T}$, we say $X(t) = e_A(t, t_0)$ is a matrix-valued nontrivial solution for (6.4).

Remark 6.1 We let $[\mathbb{R}^{n\times n}_{\mathscr{F}}] := \underbrace{[\mathbb{R}^n_{\mathscr{F}}] \times [\mathbb{R}^n_{\mathscr{F}}] \times \cdots \times [\mathbb{R}^n_{\mathscr{F}}]}_{n \text{ terms}}$. Notice that a matrix-valued nontrivial solution $X(t) = e_A(t, t_0)$ satisfies $X(t) \in \mathbb{R}^{n\times n} \subset [\mathbb{R}^{n\times n}_{\mathscr{F}}]$, where

$$\mathbb{R}^{n\times n} := \underbrace{\big\{\chi_{\{x_1\}}, x_1 \in \mathbb{R}^n\big\} \times \big\{\chi_{\{x_2\}}, x_2 \in \mathbb{R}^n\big\} \times \cdots \times \big\{\chi_{\{x_n\}}, x_n \in \mathbb{R}^n\big\}}_{n \text{ terms}},$$

where $x_1, x_2, \dots, x_n$ are real vectors which are linearly independent. Notice that if $X(t)$ is invertible, then $X^{-1}(t) \in \mathbb{R}^{n\times n}$. By considering the embedding mapping from Theorem 2.5, if we let $(j \circ X)(t) = \big((j \circ x_1)(t), (j \circ x_2)(t) \dots, (j \circ x_n)(t)\big)$, from $X(t) \in \mathbb{R}^{n\times n}$, we have

$$\begin{aligned}
(j \circ X)(t) &= \big(j \circ X)(t) = \big((j \circ x_1)(t), (j \circ x_2)(t) \dots, (j \circ x_n)(t)\big) \\
&= \big(j \circ \chi_{\{x_1(t)\}}, j \circ \chi_{\{x_2(t)\}} \dots, j \circ \chi_{\{x_n(t)\}}\big) \\
&= \begin{pmatrix} x_{11}(t) & x_{21}(t) & \cdots & x_{n1}(t) \\ x_{12}(t) & x_{22}(t) & \cdots & x_{n2}(t) \\ \vdots & \vdots & \ddots & \vdots \\ x_{1n}(t) & x_{2n}(t) & \cdots & x_{nn}(t) \end{pmatrix} = X(t)
\end{aligned}$$

where $x_i(t) = \big(x_{i1}(t), x_{i2}(t), \dots, x_{in}(t)\big)^T \in \mathbb{R}^n$, $i = 1, 2, \dots, n$. Hence, we can obtain that $(j \circ X)^{-1}(t) = X^{-1}(t) = (j \circ X^{-1})(t)$ for all invertible $X(t) \in \mathbb{R}^{n\times n}$. For the definition of $[\mathbb{R}^{n\times n}_{\mathscr{F}}]$ and $\mathscr{A} \in [\mathbb{R}^{n\times n}_{\mathscr{F}}]$, where $\mathscr{A} = (x_{ij})_{n\times n}$ and $x_{ij} \in \mathbb{R}_{\mathscr{F}}$, we equip the space $[\mathbb{R}^{n\times n}_{\mathscr{F}}]$ with the function $\|\mathscr{A}\|_{\hat{\mathscr{F}}} = (\sum_{j=1}^{n}\sum_{i=1}^{n}\|x_{ij}\|^2_{\mathscr{F}_0})^{\frac{1}{2}}$.

From Remark 6.1, one can obtain the following lemma directly.

Lemma 6.6 $(j \circ X)^{-1}(t) = X^{-1}(t) = (j \circ X^{-1})(t)$ *for all invertible* $X(t) \in \mathbb{R}^{n \times n}$.

In what follows, we introduce the following exponential dichotomy which will be used later.

Definition 6.6 Let $A(t)$ be an $n \times n$ rd-continuous matrix-valued function on $\mathbb{T}$. The linear system

$$x^{\Delta}(t) = A(t)x(t) \tag{6.7}$$

is said to admit an exponential dichotomy on $\mathbb{T}$ if there exist positive constants K, α, projection P and the fundamental solution matrix $X(t)$ of (6.7), satisfying

$$\begin{cases} \|X(t)PX^{-1}(s)\|_{\hat{\mathscr{F}}} \leq Ke_{\ominus\alpha}\big(t, \rho(s)\big), s, t \in \mathbb{T}, t \geq s, \\ \|X(t)(I - P)X^{-1}(s)\|_{\hat{\mathscr{F}}} \leq Ke_{\ominus\alpha}\big(s, \rho(t)\big), s, t \in \mathbb{T}, t \leq s. \end{cases}$$

In the following, we give sufficient conditions to guarantee that (6.7) admits exponential dichotomy.

Lemma 6.7 *For any solution* $x(t, x_0) = x(t)$, *of* (6.7) *and* $t \geq t_0$, *the following inequality holds*

$$\begin{aligned} &\left(\exp\left\{\int_{t_0}^{t} \xi_{\mu(s)}\big(2\lambda(s) + \lambda_0(s)\mu(s)\big)\Delta s\right\}\right)^{\frac{1}{2}} \|x_0\|_{\mathscr{F}} \leq \|x(t, x_0)\|_{\mathscr{F}} \\ &\leq \left(\exp\left\{\int_{t_0}^{t} \xi_{\mu(s)}\big(2\Lambda(s) + \Lambda_0(s)\mu(s)\big)\Delta s\right\}\right)^{\frac{1}{2}} \|x_0\|_{\mathscr{F}}, \end{aligned} \tag{6.8}$$

where $\lambda(t)$ *and* $\Lambda(t)$ *are the smallest and the largest eigenvalue of the matrix* $\hat{A}(t) = \frac{1}{2}\big(A(t) + A^T(t)\big)$ *respectively, and* $\lambda_0(t)$ *and* $\Lambda_0(t)$ *are the smallest and the largest eigenvalue of the matrix* $\tilde{A}(t) = A(t)A^T(t)$, $A^T(t)$ *is the transpose of the matrix* $A(t)$, $\|x\|_{\mathscr{F}} = D_{\infty}(x, \tilde{\mathbf{0}})$.

Proof Let $z(t) = (j \circ x)(t)$. Then (6.7) turns into

$$(j \circ x)^{\Delta}(t) = A(t)(j \circ x)(t), \text{ i.e., } z^{\Delta}(t) = A(t)z(t). \tag{6.9}$$

If $z(t, t_0) = 0$, i.e., $x(t, t_0) = \tilde{0}$, then (6.8) holds. Let $z(t, t_0) = z(t)$ be a nontrivial solution of the fuzzy system (6.9), by Theorem 5.1, $\|z\|^2_{\mathscr{L}} = \langle z, z\rangle$. Then for $t \in \mathbb{T}$, we have

$$\begin{aligned}\left(\|z(t)\|^2_{\mathscr{L}}\right)^{\Delta} &= \left(\langle z(t), z(t)\rangle\right)^{\Delta} = \langle z^{\Delta}(t), z(t)\rangle + \langle z^{\sigma}(t), z^{\Delta}(t)\rangle \\ &= \langle z^{\Delta}(t), z(t)\rangle + \langle z^{\Delta}(t)\mu(t) + z(t), z^{\Delta}(t)\rangle \\ &= \langle z^{\Delta}(t), z(t)\rangle + \mu(t)\langle z^{\Delta}(t), z^{\Delta}(t)\rangle + \langle z(t), z^{\Delta}(t)\rangle \\ &= 2\langle A(t)z(t), z(t)\rangle + \mu(t)\langle A(t)z(t), A(t)z(t)\rangle \\ &= 2\langle \hat{A}(t)z(t), z(t)\rangle + \mu(t)\langle A(t)A^T(t)z(t), z(t)\rangle.\end{aligned}$$

Since $\hat{A}(t)$, $\tilde{A}(t)$ are symmetric, then we have

$$\lambda(t)\langle z(t), z(t)\rangle \le \langle \hat{A}(t)z(t), z(t)\rangle \le \Lambda(t)\langle z(t), z(t)\rangle, \tag{6.10}$$

where $\lambda(t)$ and $\Lambda(t)$ are the smallest and the largest eigenvalue of the matrix $\hat{A}(t)$. We also have

$$\lambda_0(t)\langle z(t), z(t)\rangle \le \langle A(t)A^T(t)z(t), z(t)\rangle \le \Lambda_0(t)\langle z(t), z(t)\rangle. \tag{6.11}$$

Then from (6.10) and (6.11), we obtain

$$\left(2\lambda(t) + \lambda_0(t)\mu(t)\right)\langle z(t), z(t)\rangle \le \left(\|z(t)\|^2_{\mathscr{L}}\right)^{\Delta} \le \left(2\Lambda(t) + \Lambda_0(t)\mu(t)\right)\langle z(t), z(t)\rangle. \tag{6.12}$$

Thus, for $t \ge t_0$, from (6.12) and by Lemmas 3.4 and 3.6 from [82], we have

$$\begin{aligned}&\left(\exp\left\{\int_{t_0}^{t} \xi_{\mu(s)}\left(2\lambda(s) + \lambda_0(s)\mu(s)\right)\Delta s\right\}\right)^{\frac{1}{2}} \|z_0\|_{\mathscr{L}} \le \|z(t, t_0)\|_{\mathscr{L}} \\ \le &\left(\exp\left\{\int_{t_0}^{t} \xi_{\mu(s)}\left(2\Lambda(s) + \Lambda_0(s)\mu(s)\right)\Delta s\right\}\right)^{\frac{1}{2}} \|z_0\|_{\mathscr{L}}.\end{aligned}$$

Thus, from $\|z\|_{\mathscr{L}} = \|j \circ x\|_{\mathscr{L}} = \|j \circ x - j \circ \tilde{0}\|_{\mathscr{L}} = D_{\infty}(x, \tilde{\mathbf{0}})$, the estimate (6.8) follows from this inequality. This completes the proof. □

Theorem 6.1 *For (6.7), if $\Lambda : \mathbb{T} \to \mathbb{R}$ is the largest eigenvalue of matrix $\hat{A}(t)$ and $\lambda_0 : \mathbb{T} \to \mathbb{R}$ is the smallest eigenvalue of matrix $\tilde{A}(t)$, there exists a positive constant $c_\mu > 0$ and $M > 0$ such that*

(*i*) $p(t) \le c_\mu m[p] < 0$ *and* $c_\mu \mu \le M$ *for all* $t \in \mathbb{T}$;
(*ii*) $\frac{\nu}{\mu} \le M$ *for all right scattered points of* $\mathbb{T}$;
(*iii*) $Mm[p] \in \mathscr{R}^+$;

where $\hat{A}(t) = \frac{1}{2}\left(A(t) + A^T(t)\right)$ *and* $\tilde{A}(t) = A(t)A^T(t)$, $A^T(t)$ *is the transpose of the matrix* $A(t)$, $i = 1, 2, \ldots, n$ *and*

$$p(t) = \Lambda(t) + \frac{1}{2}\Lambda_0(t)\mu(t), \quad p \in \mathscr{R}^+.$$

Assume $|m[p]| < +\infty$. *Then the linear system* (6.7) *admits an exponential dichotomy on* $\mathbb{T}$, *where*

$$m[p] = \lim_{T\to\infty} \frac{1}{T} \int_t^{\delta_+(T,t)} p(s)\Delta s$$

denotes the mean-value of the function p under the shift δ_+.

Proof Let $X(t,\tau)$ be the solution the matrix Cauchy problem

$$X^{\Delta} = A(t)X, \quad X(\tau,\tau) = I,$$

i.e., a matriciant of (6.4). Then any nontrivial matrix-valued solution $X(t)$ of (6.4) can be represented as $X(t) = X(t,s)X(s)$, i.e., $X(t)X^{-1}(s) = X(t,s)$. The linear system (6.4) has a unique solution $x(t) = X(t,s)x(s)$. Using Lemma 6.7, we obtain

$$\begin{aligned}
&\|x(t)\|_{\mathscr{F}} = \big\|X(t)X^{-1}(s)x(s)\big\|_{\mathscr{F}} = \|X(t,s)x(s)\|_{\mathscr{F}} \\
&\le \left(\exp\left\{\int_s^t \xi_{\mu(s)}\big(2\Lambda(s) + \Lambda_0(s)\mu(s)\big)\Delta s\right\}\right)^{\frac{1}{2}} \|x(s)\|_{\mathscr{F}} \\
&\le \exp\left\{\int_s^t \xi_{\mu(s)}\big(p(s)\big)\Delta s\right\}\|x(s)\|_{\mathscr{F}} \le \exp\left\{\int_s^t \xi_{\mu(s)}\big(cm[p]\big)\Delta s\right\}\|x(s)\|_{\mathscr{F}} \\
&= e_{cm[p]}(t,s)\|x(s)\| = e_{cm[p]}\big(t,\rho(s)\big)e_{cm[p]}\big(\rho(s),s\big)\|x(s)\| \\
&\le k_0 e_{cm[p]}\big(t,\rho(s)\big)\|x(s)\|_{\mathscr{F}},
\end{aligned}$$

where $k_0 = e_{cm[p]}\big(\rho(s),s\big) \ge 1$. In fact, if $\rho(s)$ is a right scatted point, then

$$\begin{aligned}
e_{cm[p]}\big(\rho(s),s\big) &= \exp\Big\{-\int_{\rho(s)}^{\sigma(\rho(s))} \frac{\ln(1 + c_{\mu}\mu(\tau)m[p])}{\mu(\tau)}\Delta\tau\Big\} \\
&= \exp\left\{-\nu(s)\cdot\frac{\ln(1 + c_{\mu}\mu(s)m[p])}{\mu(s)}\right\} \\
&= \exp\left\{\frac{\nu}{\mu}\cdot\ln\frac{1}{1 + c_{\mu}\mu m[p]}\right\} \le \exp\left\{M\cdot\ln\frac{1}{1 + Mm[p]}\right\} := k_0,
\end{aligned}$$

if $\rho(s)$ is a right dense point, then $\sigma\big(\rho(s)\big) = \rho(s) = s$, and then $k_0 = e_{cm[p]}\big(\rho(s),s\big) = 1$.

Thus, we have

$$\frac{\|X(t,x)x(s)\|_{\mathscr{F}}}{\|x(s)\|_{\mathscr{F}}} \le k_0 e_{cm[p]}\big(t,\rho(s)\big),$$

which implies that $\|X(t,s)\|_{\hat{\mathscr{F}}} \le k_0 e_{cm[p]}\big(t,\rho(s)\big)$, i.e., (6.7) admits an exponential dichotomy on $\mathbb{T}$ with the projection $P = I$. This completes the proof. □

Corollary 6.1 *Let* $\mathbb{R}_i^n \subset \mathbb{R}^n$ *be the vector subspace in which all the elements of the vector are equal to zero except the ith one and*

$$A(t) = \left(\chi_{\{c_1(t)\}}, \chi_{\{c_2(t)\}}, \ldots, \chi_{\{c_n(t)\}}\right), \quad c_i(t) \in \mathbb{R}_i^n,$$

where $c_i \in C(\mathbb{T}, \mathbb{R}^-)$ are bounded functions and $c_i(t) \in \mathscr{R}^+$, $i = 1, 2, \ldots, n$. Assume there are positive constants c_{μ_0}, M such that

(i) $c_\mu \mu \leq M$ *for all* $t \in \mathbb{T}$;
(ii) $\frac{\nu}{\mu} \leq M$ *for all right scattered points of* $\mathbb{T}$;
(iii) $m\left[c^{\mathcal{M}}(t) + \frac{1}{2}\left(c_i^m(t)\right)^2\mu\right] = \lim_{T\to\infty} \frac{1}{T}\int_t^{\delta_+(T,t)} \left(c^{\mathcal{M}}(t) + \frac{1}{2}\left(c_i^m(t)\right)^2\mu\right)\Delta t < 0$;
(iv) $|m[c_i^{\mathcal{M}} + \frac{1}{2}(c_i^m)^2\mu]| < \infty$ *and* $Mm[c_i^{\mathcal{M}} + \frac{1}{2}(c_i^m)^2\mu] \in \mathscr{R}^+$;

where $c^{\mathcal{M}}(t) = \max_i\left(c_i(t)\right)$, $c^m(t) = \min_i\left(c_i(t)\right)$. *Then (6.7) admits an exponential dichotomy on* $\mathbb{T}$.

Proof From Corollary 6.1, one can easily obtain $\Lambda(t) = c^{\mathcal{M}}(t)$ and $\Lambda_0 = \left(c_i^m(t)\right)^2$. According to Theorem 6.1, (6.7) admits an exponential dichotomy. This completes the proof. □

In the following, we will establish Favard's theorem for homogeneous linear fuzzy dynamic equation.

Lemma 6.8 *If $A(t)$ is an S-almost periodic fuzzy matrix-valued function and $x(t)$ is an S-almost periodic solution of the homogeneous linear fuzzy dynamic equation*

$$x^\Delta(t) = A(t)x,$$

then $\inf_{t\in\mathbb{T}} \|x(t)\|_{\mathscr{F}} = \inf_{t\in\mathbb{T}} D_\infty\left(x(t), \tilde{\mathbf{0}}\right) > 0$ *or* $x(t) \equiv \tilde{\mathbf{0}}$.

Proof If $\inf_{t\in\mathbb{T}} \|x(t)\|_{\mathscr{F}} = 0$, then there exists $\{t_n\} \subset \mathbb{T}$ such that $\|x(t_n)\|_{\mathscr{F}} \to 0$ as $n \to \infty$. It follows from $x \in H_S(x)$ that there exists $\alpha' \subset \Pi$ such that

$$\lim_{n\to+\infty} x\left(\delta_+(\alpha_n', t)\right) = x(t)$$

for all $t \in \mathbb{T}^*$, which implies that for any $\varepsilon > 0$, there exists $N > 0$ so that $n > N$ implies $\left\|x\left(\delta_+(\alpha_n', t_n)\right)\tilde{-}_{gH}x(t_n)\right\|_{\mathscr{F}} < \frac{\varepsilon}{2}$. Furthermore, since $x(t)$ is S-almost periodic on $\mathbb{T}$, it is uniformly continuous on $\mathbb{T}$, one can choose $t_0 \in \mathbb{T}$ such that $|t_0 - t_n| < \delta^*$ implies

$$\left\|x\left(\delta_{\alpha_n'}(t_0)\right)\tilde{-}_{gH}x\left(\delta_{\alpha_n'}(t_n)\right)\right\|_{\mathscr{F}} < \frac{\varepsilon}{2}.$$

Therefore, for sufficiently large $n \in \mathbb{N}$, we have

$$\|x(\delta_+(\alpha_n', t_0))\tilde{-}_{gH}x(t_n)\|_{\mathscr{F}} \le \|x(\delta_+(\alpha_n', t_0))\tilde{-}_{gH}x(\delta_+(\alpha_n', t_n))\|_{\mathscr{F}}$$
$$+\|x(\delta_+(\alpha_n', t_n))\tilde{-}_{gH}x(t_n)\|_{\mathscr{F}} < \frac{\varepsilon}{2} + \frac{\varepsilon}{2} < \varepsilon.$$

Hence, we can easily see that $\|x(\delta_+(\alpha_n', t_0))\|_{\mathscr{F}} \to 0$ as $n \to +\infty$. Since $A(t)$ is S-almost periodic on $\mathbb{T}$, there exists sequence $\alpha \subset \alpha'$ such that

$$T_\alpha^S A(t) = B(t), \qquad T_\alpha^S x(t) = y(t),$$

$$T_{\alpha^{-1}}^S B(t) = A(t), \qquad T_{\alpha^{-1}}^S y(t) = x(t)$$

hold uniformly on $\mathbb{T}^*$, then one obtains

$$y^\Delta(t) = (T_\alpha^S x(t))^\Delta = \left(\lim_{n\to\infty} x(\delta_+(\alpha_n, t))\right)^\Delta = \lim_{n\to\infty}\left(x(\delta_+(\alpha_n, t))\right)^\Delta$$
$$= \lim_{n\to\infty} A(\delta_+(\alpha_n, t))x(\delta_+(\alpha_n, t)) = \lim_{n\to\infty} A(\delta_+(\alpha_n, t)) \cdot \lim_{n\to\infty} x(\delta_+(\alpha_n, t)),$$

that is, $y(t)$ is a solution to the following equation:

$$y^\Delta = B(t)y$$

satisfying the initial condition:

$$y(t_0) = T_\alpha^S x(t_0) = \lim_{n\to+\infty} x(\delta_+(\alpha_n, t_0)) = \tilde{\mathbf{0}}.$$

Hence, $y(t) = y(t_0)e_B(t, t_0) \equiv \tilde{\mathbf{0}}$, and therefore, $x(t) = T_{\alpha^{-1}}^S y(t) \equiv \tilde{\mathbf{0}}$. The proof is completed. □

Lemma 6.9 *Suppose that* (6.4) *has an S-almost periodic solution* $x(t)$ *and*

$$\inf_{t\in\mathbb{T}} \|x(t)\|_{\mathscr{F}} > 0.$$

If (6.3) *has bounded solution on* $[t_0, \infty)_{\mathbb{T}}$, *then* (6.3) *has an S-almost periodic solution.*

Proof From Lemmas 6.1 and 6.2, we know that there is a minimum norm solution to (6.3) on $\mathbb{T}^*$ and for every pair of $T_\alpha^S A(t) = B(t)$ and $T_\alpha^S f(t) = g(t)$, (6.6) has a minimum norm solution.

Next, we show that the minimum norm solution to (6.6) is unique.

For a fixed pair of $B(t)$ and $g(t)$, we consider (6.6). If $x(t)$ is a minimum norm solution with least-value λ then

$$\left\|(j\circ x)(t)\right\|_{\mathscr{L}} = \left|s(p,[x(t)]^\alpha)\right| \le \|[x(t)]^\alpha\|_*\|p\| \le \sup_{\alpha\in[0,1]} \|[x(t)]^\alpha\|_*$$
$$= \|x(t)\|_{\mathscr{F}} \le \|x\|_\infty = \lambda.$$

Now, let (6.6) have two different minimum norm solutions $x_1(t)$ and $x_2(t)$, and their least-values are equal to λ. Since $\frac{1}{2}\big[x_1(t)\tilde{-}_{gH}x_2(t)\big]$ is a bounded non-trivial solution to (6.5), from the condition of the lemma, there exists a real number $\rho > 0$ such that

$$\begin{aligned}\inf_{t\in\mathbb{T}}\frac{1}{2}\big\|x_1(t)\tilde{-}_{gH}x_2(t)\big\|_{\mathscr{F}} &= \inf_{t\in\mathbb{T}}\frac{1}{2}D_\infty\big(x_1(t),x_2(t)\big)\\ &= \inf_{t\in\mathbb{T}}\frac{1}{2}\|(j\circ x_1)(t)-(j\circ x_2)(t)\|_{\mathscr{L}} \ge \rho > 0.\end{aligned}$$

Now by the parallelogram law, we have

$$\begin{aligned}&\left\|\frac{1}{2}((j\circ x_1)(t)+(j\circ x_2)(t))\right\|^2_{\mathscr{L}} + \left\|\frac{1}{2}((j\circ x_1)(t)-(j\circ x_2)(t))\right\|^2_{\mathscr{L}}\\ &= \frac{1}{2}\big(\|(j\circ x_1)(t)\|^2 + \|(j\circ x_2)(t)\|^2_{\mathscr{L}}\big) \le \lambda^2,\end{aligned}$$

and note that

$$\begin{aligned}\Big(\frac{1}{2}\big(x_1(t)\tilde{+}x_2(t)\big)\Big)^\Delta &= \frac{1}{2}x_1^\Delta\tilde{+}\frac{1}{2}x_2^\Delta\\ &= \frac{1}{2}\big(B(t)x_1\tilde{+}g(t)\big)\tilde{+}\frac{1}{2}\big(B(t)x_2\tilde{+}g(t)\big)\\ &= B(t)\Big(\frac{1}{2}(x_1\tilde{+}x_2)\Big)\tilde{+}g(t),\end{aligned}$$

then $\frac{1}{2}\big(x_1(t)\tilde{+}x_2(t)\big)$ is a solution to (6.6). Hence, we obtain

$$\begin{aligned}\left\|\Big(j\circ\Big(\frac{1}{2}(x_1\tilde{+}x_2)\Big)\Big)(t)\right\|_{\mathscr{L}} &= \left\|\frac{1}{2}((j\circ x_1)(t)+(j\circ x_2)(t))\right\|_{\mathscr{L}}\\ &< \sqrt{\lambda^2-\rho^2} < \lambda.\end{aligned}$$

This is a contradiction. The proof is completed. □

Lemma 6.10 *If every bounded solution of a homogeneous equation in the hull of* (6.3) *is S-almost periodic, then all bounded solutions of* (6.3) *are S-almost periodic.*

Proof From Lemma 6.8, we know that every non-trivial bounded solution of equations in the hull of (6.3) satisfies $\inf_{t\in\mathbb{T}}\|x(t)\|_{\mathscr{F}} > 0$. From Lemma 6.9 it follows that if (6.3) has bounded solutions on $\mathbb{T}$, then (6.3) will have an S-almost periodic solution $\psi(t)$. If $\varphi(t)$ is an arbitrary bounded solution of (6.3), then $\eta(t) = \psi(t)\tilde{-}_{gH}\varphi(t)$ is a bounded solution of its associated homogeneous equation (6.4), and it is S-almost periodic. Thus, $\varphi(t)$ is S-almost periodic. This completes the proof. □

Lemma 6.11 *If a homogeneous equation in the hull of* (6.3) *has the unique bounded solution* $x(t) \equiv \tilde{\mathbf{0}}$, *then* (6.3) *has a unique S-almost periodic solution .*

Proof Let $\psi(t)$, $\varphi(t)$ be two bounded solutions to (6.3), and then $x(t) = \varphi(t) \tilde{-}_{gH} \psi(t)$ is a solution of a homogeneous equation in the hull of (6.3), since $x(t) \equiv \tilde{\mathbf{0}}$, we have that $\varphi(t) \equiv \psi(t)$. Thus, by Lemma 6.10, (6.3) has a unique S-almost periodic solution. This completes the proof. □

Using a proof similar to that in Lemmas 7.4–7.5 in [36], one can easily prove Lemmas 6.12–6.13 (we omit their proofs).

Lemma 6.12 *Let P be a projection and X a differentiable invertible matrix such that XPX^{-1} is bounded on $\mathbb{T}$. Then there exists a differentiable matrix S such that $XPX^{-1} = SPS^{-1}$ for all $t \in \mathbb{T}$ and S, S^{-1} are bounded on $\mathbb{T}$. In fact, there is an S of the form $S = XQ^{-1}$, where Q commutes with P.*

Lemma 6.13 *If* (6.4) *has an exponential dichotomy and $X(t)$ is the fundamental solution matrix of* (6.4), *C non-singular, then $X(t)C$ has an exponential dichotomy with the same projection P if and only if $CP = PC$.*

Lemma 6.14 *Suppose that $A(t)$ is an S-almost periodic matrix function under shifts $\delta_{\pm}$ and* (6.4) *has an exponential dichotomy, then for every $B(\cdot) \in H_S(A)$,* (6.5) *has an exponential dichotomy with the same projection P and the same constants K, α.*

Proof Let X be the fundamental solution matrix satisfying (6.6). Let Q and S be given as in Lemma 6.12 and $T_\alpha^S A = B$ uniformly on $\mathbb{T}^*$. For any given $t_0 \in \mathbb{T}^*$, let $X_n(t) = X\big(\delta_+(\alpha_n, t)\big)Q^{-1}\big(\delta_+(\alpha_n, t_0)\big)$, it follows from (6.4) that

$$X^\Delta(t) = A(t)X(t), \tag{6.13}$$

Now, replacing t with $\delta_+(\alpha_n, t)$ in (6.13), we can obtain

$$X^\Delta\big(\delta_+(\alpha_n, t)\big) = A\big(\delta_+(\alpha_n, t)\big)X\big(\delta_+(\alpha_n, t)\big),$$

i.e., $X_n^\Delta(t) = A\big(\delta_+(\alpha_n, t)\big)X_n(t)$, then $X_n(t)$ is a fundamental solution matrix to $x^\Delta(t) = A\big(\delta_+(\alpha_n, t)\big)x$, by Lemma 6.13, it has an exponential dichotomy with the same projection P and the same constants. This is true since Q^{-1} commutes with P. One may select subsequences so that $X_n(t_0)$ and $X_n^{-1}(t_0)$ converge since their forms are $S\big(\delta_+(\alpha_n, t_0)\big)$ and

$S^{-1}\big(\delta_+(\alpha_n, t_0)\big)$ and they are bounded. Without changing notation we may assume that $X_n(t_0) \to Y_0$, then $X_n^{-1}(t_0) \to Z_0$, where $Z_0 = Y_0^{-1}$. But now for a suitable subsequence $X_n(t)$ converges to a solution of $y^\Delta = By$ uniformly on $\mathbb{T}^*$. Denote this solution by Y. Then $Y(t_0) = Y_0$ is non-singular and clearly Y satisfies (6.6) since X_n does for all n. The proof is completed. □

Lemma 6.15 *If the homogeneous equation* (6.4) *has an exponential dichotomy , then* (6.4) *has only one bounded solution* $x(t) \equiv \tilde{\mathbf{0}}$.

Proof Let $X(t)$ be the fundamental solution matrix to (6.4). For any sequence $\alpha \subset \Pi$, we have $A_n = A\big(\delta_+(\alpha_n, t)\big)$, $X_n(t) = X\big(\delta_+(\alpha_n, t)\big)$. Since the homogeneous equation (6.4) has an exponential dichotomy, it is easy to observe that there exists a constant M such that $\|X_n(t)\|_{\hat{\mathscr{F}}} \leq M$ and $\|X_n^\Delta(t)\|_{\hat{\mathscr{F}}} = \|A_n(t)X_n(t)\|_{\hat{\mathscr{F}}} \leq \bar{A}M$, where $\bar{A} = \sup_{t\in\mathbb{T}} \|A(t)\|_{\hat{\mathscr{F}}}$. Therefore, from Corollary 3.4 in [85], there exists $\{\alpha_{n_k}\} := \alpha' \subset \alpha$ such that $\{X_{n_k}\}$ converges uniformly on $\mathbb{T}^*$ and $\lim\limits_{n\to+\infty} X\big(\delta_+(\alpha_n, t)\big)$ exists uniformly on $\mathbb{T}^*$. Thus $X(t)$ is S-almost periodic. Since the homogeneous equation (6.4) has an exponential dichotomy, then $\inf_{t\in\mathbb{T}} \|x(t)\|_{\mathscr{F}} = 0$, from Lemma 6.8, $x(t) \equiv \tilde{\mathbf{0}}$. This completes the proof. □

Lemma 6.16 *If the homogeneous equation* (6.4) *has an exponential dichotomy, then all equations in the hull of* (6.4) *have only one bounded solution* $x(t) \equiv \tilde{\mathbf{0}}$.

Proof From Lemma 6.14, all equations in the hull of (6.4) have an exponential dichotomy, according to Lemma 6.15, all equations in the hull of (6.4) have only one bounded solution $x(t) \equiv \tilde{\mathbf{0}}$. This completes the proof. □

Lemma 6.17 *Let* $A \in \mathscr{R}$ *and* $x_i : \mathbb{T} \to [\mathbb{R}^n_{\mathscr{F}}]$. *Now* (6.4) *has a fundamental solution matrix* $X(t) = \big(x_1(t), x_2(t), \ldots, x_n(t)\big)$ *if and only if* $\big((j \circ x_1)(t), (j \circ x_2)(t) \ldots, (j \circ x_n)(t)\big)$ *is a fundamental solution for the dynamic equations:*

$$Z^\Delta(t) = A(t)Z(t), \tag{6.14}$$

where $Z(t) = (j \circ X)(t) = \big((j \circ x_1)(t), (j \circ x_2)(t) \ldots, (j \circ x_n)(t)\big)$.

Proof If $X(t)$ is the fundamental solution matrix for (6.4), one obtains that

$$j\big(X^\Delta(t)\big) = j\big(A(t)X(t)\big).$$

From Theorem 2.5 and Theorem 4.17 (iii), since $A(t)$ is a matrix-valued function, we can obtain

$$(j \circ X)^\Delta(t) = A(t)\big(j \circ X\big)(t). \tag{6.15}$$

On the other hand, if $Z(t)$ is the fundamental matrix solution for (6.14), because j is a bijective mapping, by employing the mapping j^{-1} on the both sides of (6.15), one gets the desired result immediately. This completes the proof. □

Theorem 6.2 *Let $A : \mathbb{T} \to \mathbb{R}^{n\times n}$ be an S-almost periodic matrix-valued function and $f : \mathbb{T} \to [\mathbb{R}^n_{\mathcal{F}}]$ be an S-almost periodic fuzzy vector-valued function. If* (6.4) *admits an exponential dichotomy, then* (6.3) *has a unique S-almost periodic fuzzy vector-valued solution*

$$x(t) = \int_{-\infty}^{t} X(t)PX^{-1}(\sigma(s))f(s)\Delta s \tilde{-}_{gH} \int_{t}^{+\infty} X(t)(I-P)X^{-1}(\sigma(s))f(s)\Delta s,$$

where $X(t)$ is the fundamental trivial solution matrix of (6.4).

Proof First, by Lemma 6.17, let $Z(t) = (j \circ X)(t)$, we prove that $z(t) = (j \circ x)(t)$ is a bounded solution of the following vector-valued system:

$$z(t) = A(t)z(t) + \tilde{f}(t), \text{ where } \tilde{f}(t) = (j \circ f)(t). \tag{6.16}$$

In fact,

$$\begin{aligned}
& z^{\Delta}(t) - A(t)z(t) \\
&= Z^{\Delta}(t)\int_{-\infty}^{t} PZ^{-1}(\sigma(s))\tilde{f}(s)\Delta s + Z(\sigma(t))PZ^{-1}(\sigma(t))\tilde{f}(t) \\
&\quad -Z^{\Delta}(t)\int_{t}^{+\infty} (I-P)Z^{-1}(\sigma(s))\tilde{f}(s)\Delta s + Z(\sigma(t))(I-P)Z^{-1}(\sigma(t))\tilde{f}(t) \\
&\quad -A(t)Z(t)\int_{-\infty}^{t} PZ^{-1}(\sigma(s))\tilde{f}(s)\Delta s + A(t)Z(t)\int_{t}^{+\infty} (I-P)Z^{-1}(\sigma(s))\tilde{f}(s)\Delta s \\
&= Z(\sigma(t))(P+I-P)Z^{-1}(\sigma(t))\tilde{f}(t) = \tilde{f}(t),
\end{aligned}$$

thus, we obtain (6.16) has a bounded solution $z(t)$ satisfying

$$z(t) + \int_{t}^{+\infty} Z(t)(I-P)Z^{-1}(\sigma(s))\tilde{f}(s)\Delta s = \int_{-\infty}^{t} Z(t)PZ^{-1}(\sigma(s))\tilde{f}(s)\Delta s,$$

by employing the inverse mapping j^{-1} and Lemma 6.6, and one can obtain $x(t) = (j^{-1} \circ z)(t)$ is a solution of (6.3).

Moreover, we also have

$$\begin{aligned}
\|x\|_{\mathscr{F}} &= \sup_{t\in\mathbb{T}} \bigg\| \int_{-\infty}^{t} X(t)PX^{-1}(\sigma(s))f(s)\Delta s \\
&\widetilde{-}_{gH} \int_{t}^{+\infty} X(t)(I-P)X^{-1}(\sigma(s))f(s)\Delta s \bigg\|_{\mathscr{F}} \\
&\leq \sup_{t\in\mathbb{T}} \left(\left| \int_{-\infty}^{t} e_{\ominus\alpha}(t,s)f(s)\Delta s \right| + \left| \int_{t}^{+\infty} e_{\ominus\alpha}(\sigma(s),t)f(s)\Delta s \right| \right) K\|f\|_{\mathscr{F}} \\
&\leq \frac{2K}{\alpha}\|f\|_{\infty},
\end{aligned}$$

where $\|\cdot\|_{\infty} = \sup_{t\in\mathbb{T}} D_{\infty}(\cdot, \tilde{\mathbf{0}})$.

Now, we show that the fuzzy vector-valued solution $x(t)$ is S-almost periodic. According to Lemma 6.16, we obtain that all equations in the hull of (6.4) have only one bounded solution $\bar{x}(t) \equiv \tilde{\mathbf{0}}$. Thus, for given α', β', one can choose common subsequences $\alpha \subset \alpha'$, $\beta \subset \beta'$ such that $T^S_{\delta_+(\alpha,\beta)}A = T^S_\alpha T^S_\beta A$, $T^\delta_{\delta_+(\alpha,\beta)}f = T^S_\alpha T^S_\beta f$, $y = T^S_{\delta_+(\alpha,\beta)}x$, and $z = T^S_\alpha T^S_\beta x$ exists uniformly on $\mathbb{T}^*$. However, $y \widetilde{-}_{gH} z = T^S_{\delta_+(\alpha,\beta)}x \widetilde{-}_{gH} T^S_\alpha T^S_\beta x \equiv \tilde{\mathbf{0}}$ because $y\widetilde{-}_{gH}z$ is the bounded solution to all equations in the hull of (6.4). Hence, $T^S_{\delta_+(\alpha,\beta)}x = T^S_\alpha T^S_\beta x$, then it follows from Theorem 4.19 that $x \in AP_S(\mathbb{T})$. This completes the proof. □

6.2 Applications on Fuzzy Dynamic Equations and Models on Time Scales

In this section, we provide several applications to demonstrate the S-almost periodic theory of fuzzy-vector-valued functions on irregular time scales.

Application 6.2.1 *As an application of our results obtained in the previous sections, in what follows, we consider the following S-almost periodic fuzzy delay vector-valued dynamic equation with variable delays under shifts $\delta_\pm$:*

$$x^{\Delta}(t) = A(t)x(t)\widetilde{+}\sum_{i=1}^{n} f\big(t, x\big(\delta_-(\tau_i(t),t)\big)\big), \tag{6.1}$$

where $A(t)$ is an S-almost periodic matrix-valued function on $\mathbb{T}$, $\tau_i(t) : \mathbb{T}^* \to \Pi$ is S-almost periodic on $\mathbb{T}$ for every $i = 1, 2, \ldots, n$, $f \in C(\mathbb{T}\times[\mathbb{R}^n_{\mathscr{F}}], [\mathbb{R}^n_{\mathscr{F}}])$ is S-almost periodic uniformly in t for $x \in [\mathbb{R}^n_{\mathscr{F}}]$.

Theorem 6.3 *Suppose that the following hold:*

(H_1) *$x^{\Delta}(t) = A(t)x(t)$ admits an exponential dichotomy on $\mathbb{T}$ with positive constants K and α.*

(H_2) *There exists* $M < \dfrac{\alpha}{2Kn}$ *such that* $D_\infty\big(f(t,x), f(t,y)\big) \le M D_\infty(x,y)$ *for* $t \in \mathbb{T}$, x, $y \in [\mathbb{R}^n_{\mathcal{F}}]$. *Then system* (6.1) *has a unique S-almost periodic fuzzy vector-valued solution.*

Proof For any $\varphi \in AP_S(\mathbb{T})$, consider the following equation

$$x^{\Delta}(t) = A(t)x(t)\widetilde{+}\sum_{i=1}^{n} f\big(t, \varphi\big(\delta_-(\tau_i(t), t)\big)\big). \tag{6.2}$$

From Theorem 6.2, then (6.2) has a unique solution $T\varphi \in AP_S(\mathbb{T})$ and

$$\begin{aligned} T\varphi(t) = &\int_{-\infty}^{t} X(t)PX^{-1}(\sigma(s)) \sum_{i=1}^{n} f\big(s, \varphi\big(\delta_-(\tau_i(s), s)\big)\big)\Delta s \\ &\stackrel{\sim}{-}_{gH} \int_{t}^{+\infty} X(t)(I-P)X^{-1}(\sigma(s)) \sum_{i=1}^{n} f\big(s, \varphi\big(\delta_-(\tau_i(s), s)\big)\big)\Delta s. \end{aligned}$$

Define a mapping $T : AP_S(\mathbb{T}) \to AP_S(\mathbb{T})$ by setting $(T\varphi)(t) = x_\varphi(t)$, $\forall\, x \in AP_S(\mathbb{T})$. From (H_1), we have

$$\|X(t)PX^{-1}(s)\|_{\hat{\mathcal{F}}} \le Ke_{\ominus\alpha}\big(t, \rho(s)\big), s, t \in \mathbb{T}, t \ge s,$$

$$\|X(t)(I-P)X^{-1}(s)\|_{\hat{\mathcal{F}}} \le Ke_{\ominus\alpha}\big(s, \rho(t)\big), s, t \in \mathbb{T}, t \le s.$$

For any $\varphi, \psi \in AP_S(\mathbb{T})$, we have

$$\begin{aligned} &D_\infty(T\varphi, T\psi) \\ &= D_\infty\Big(\int_{-\infty}^{t} X(t)PX^{-1}(\sigma(s)) \sum_{i=1}^{n} f\big(s, \varphi\big(\delta_-(\tau_i(s), s)\big)\big)\Delta s \\ &\quad \stackrel{\sim}{-}_{gH} \int_{t}^{+\infty} X(t)(I-P)X^{-1}(\sigma(s)) \sum_{i=1}^{n} f\big(s, \varphi\big(\delta_-(\tau_i(s), s)\big)\big)\Delta s, \\ &\quad \int_{-\infty}^{t} X(t)PX^{-1}(\sigma(s)) \sum_{i=1}^{n} f\big(s, \psi\big(\delta_-(\tau_i(s), s)\big)\big)\Delta s \\ &\quad \stackrel{\sim}{-}_{gH} \int_{t}^{+\infty} X(t)(I-P)X^{-1}(\sigma(s)) \sum_{i=1}^{n} f\big(s, \psi\big(\delta_-(\tau_i(s), s)\big)\big)\Delta s\Big) \\ &\le \Big\|\int_{-\infty}^{t} X(t)PX^{-1}(\sigma(s))\Big(\sum_{i=1}^{n}\big|f\big(s, \varphi\big(\delta_-(\tau_i(s), s)\big)\big)\stackrel{\sim}{-}_{gH} f\big(s, \psi\big(\delta_-(\tau_i(s), s)\big)\big)\big|\Big)\Delta s \end{aligned}$$

$$
\begin{aligned}
&\tilde{-}_{gH} \int_t^{+\infty} X(t)(I-P)X^{-1}(\sigma(s))\Big(\sum_{i=1}^{n} \big| f\big(s, \varphi(\delta_-(\tau_i(s), s))\big) \\
&\tilde{-}_{gH} f\big(s, \psi(\delta_-(\tau_i(s), s))\big)\big|\Big)\Delta s\Big\|_{\mathscr{F}} \\
&\leq \left[\left|\int_{-\infty}^{t} K e_{\ominus\alpha}(t, s)\Delta s\right| + \left|\int_t^{+\infty} K e_{\ominus\alpha}(\sigma(s), t)\Delta s\right|\right]\sum_{i=1}^{n} M D_\infty(\varphi, \psi) \\
&\leq \frac{2}{\alpha} K n M D_\infty(\varphi, \psi).
\end{aligned}
$$

From (H_2), T is a contraction. Therefore, (6.1) has a unique S-almost periodic fuzzy vector-valued solution. □

Application 6.2.2 *Consider an two dimensional fuzzy vector-valued delay dynamic equations :*

$$
V^\Delta(t) = A(t)V(t) + \sum_{i=1}^{2} F\big(t, V\big(\delta_-(\tau_i(t), t)\big)\big), \tag{6.3}
$$

where $A : \mathbb{T} \to \mathbb{R}^{2\times 2}$ *is a matrix-valued function and*

$$
[V(t)]^\alpha = [v_{1-}(t), v_{1+}(t)] \times [v_{2-}(t), v_{2+}(t)]
$$

and

$$
[F(t, V)]^\alpha = [f_{1-}(t, V), f_{1+}(t, V)] \times [f_{2-}(t, V), f_{2+}(t, V)].
$$

Assume that

(i) $A(t)$ *satisfies all the conditions in Theorem 6.1;*
(ii) $f_{1-}, f_{1+}, f_{2-}, f_{2+} \in C(\mathbb{T} \times [\mathbb{R}^2_{\mathscr{F}}], \mathbb{R})$ *are S-almost periodic in t uniformly for* $V \in [\mathbb{R}^2_{\mathscr{F}}]$.

From Theorem 6.1, one can obtain that $V^\Delta(t) = A(t)V(t)$ *admits an exponential dichotomy on* $\mathbb{T}$ *with* $P = I$. *For* $(t_1, V_1), (t_2, V_1) \in \mathbb{T} \times [\mathbb{R}^2_{\mathscr{F}}]$, *we can obtain*

$$
[F(t_1, V_1)]^\alpha = [f_{1-}(t_1, V_1), f_{1+}(t_1, V_1)] \times [f_{2-}(t_1, V_1), f_{2+}(t_1, V_1)],
$$

$$
[F(t_2, V_2)]^\alpha = [f_{1-}(t_2, V_2), f_{1+}(t_2, V_2)] \times [f_{2-}(t_2, V_2), f_{2+}(t_2, V_2)],
$$

and then we see that

$$
\begin{aligned}
&[F(t_1, V_1)]^\alpha \tilde{-}_{gH} [F(t_2, V_2)]^\alpha \\
&= [\min\{f_{1-}(t_1, V_1) - f_{1-}(t_2, V_2), f_{1+}(t_1, V_1) - f_{1+}(t_2, V_2)\},
\end{aligned}
$$

$$
\begin{aligned}
&\max\{f_{1-}(t_1, V_1) - f_{1-}(t_2, V_2), f_{1+}(t_1, V_1) - f_{1+}(t_2, V_2)\}] \\
&\times [\min\{f_{2-}(t_1, V_1) - f_{2-}(t_2, V_2), f_{2+}(t_1, V_1) - f_{2+}(t_2, V_2)\}, \\
&\max\{f_{2-}(t_1, V_1) - f_{2-}(t_2, V_2), f_{2+}(t_1, V_1) - f_{2+}(t_2, V_2)\}].
\end{aligned}
$$

Also we note that

$$
\begin{aligned}
&D_\infty([F(t_1, V_1)]^\alpha, [F(t_2, V_2)]^\alpha) \\
&= \max\Bigg(\sqrt{(f_{1-}(t_1, V_1) - f_{1-}(t_2, V_2))^2 + (f_{2-}(t_1, V_1) - f_{2-}(t_2, V_2))^2}, \\
&\sqrt{(f_{1-}(t_1, V_1) - f_{1-}(t_2, V_2))^2 + (f_{2+}(t_1, V_1) - f_{2+}(t_2, V_2))^2}, \\
&\sqrt{(f_{1+}(t_1, V_1) - f_{1+}(t_2, V_2))^2 + (f_{2+}(t_1, V_1) - f_{2+}(t_2, V_2))^2}, \\
&\sqrt{(f_{1+}(t_1, V_1) - f_{1+}(t_2, V_2))^2 + (f_{2-}(t_1, V_1) - f_{2-}(t_2, V_2))^2}\Bigg).
\end{aligned}
$$

Because $f_{1-}, f_{1+}, f_{2-}, f_{2+} \in C(\mathbb{T} \times [\mathbb{R}^2_{\mathscr{F}}], \mathbb{R})$ *are S-almost periodic in t uniformly for* $V \in [\mathbb{R}^2_{\mathscr{F}}]$*, we can obtain that for any* $\varepsilon > 0$*, there exists at least a* $\omega \in \Pi$ *in an interval with a length of l such that*

$$\sqrt{(f_{1-}(\delta_+(\omega, t), V) - f_{1-}(t, V))^2 + (f_{2-}(\delta_+(\omega, t), V) - f_{2-}(t, V))^2} < \varepsilon,$$

$$\sqrt{(f_{1-}(\delta_+(\omega, t), V) - f_{1-}(t, V))^2 + (f_{2+}(\delta_+(\omega, t), V) - f_{2+}(t, V))^2} < \varepsilon,$$

$$\sqrt{(f_{1+}(\delta_+(\omega, t), V) - f_{1+}(t, V))^2 + (f_{2+}(\delta_+(\omega, t), V) - f_{2+}(t, V))^2} < \varepsilon,$$

$$\sqrt{(f_{1+}(\delta_+(\omega, t), V) - f_{1+}(t, V))^2 + (f_{2-}(\delta_+(\omega, t), V) - f_{2-}(t, V))^2} < \varepsilon,$$

which implies that $D_\infty([F(\delta_+(\omega, t), V)]^\alpha, [F(t, V)]^\alpha) < \varepsilon$*. Therefore,* $F : \mathbb{T} \times [\mathbb{R}^2_{\mathscr{F}}] \to [\mathbb{R}^2_{\mathscr{F}}]$ *is S-almost periodic in t uniformly for* $V \in [\mathbb{R}^2_{\mathscr{F}}]$*. From Theorem 6.2, one can easily obtain that* (6.3) *has a S-almost periodic fuzzy solution:*

$$
\begin{aligned}
x(t) &= \int_{-\infty}^{t} e_A(t, \sigma(s)) \sum_{i=1}^{2} F(s, V(\delta_-(\tau_i(s), s)))\Delta s \\
&= \sum_{i=1}^{2} \int_{-\infty}^{t} e_A(t, \sigma(s)) F(s, V(\delta_-(\tau_i(s), s)))\Delta s. \qquad (6.4)
\end{aligned}
$$

Based on (6.4)*, we show the S-almost periodic fuzzy solutions for the following case.*

Suppose $F(t,V)=F(t)$ is a two dimensional trapezoidal fuzzy vector-valued function and $A(t)=\big(\chi_{\{c_1(t)\}},\chi_{\{c_2(t)\}}\big)$ satisfies Corollary 6.1. Let a trapezoidal fuzzy number $N=(l,m,n,r)$ be

$$N(t)=\begin{cases}0, & t\le l,\\ \frac{t-l}{m-l}, & l<t\le m,\\ 1, & m<t\le n,\\ \frac{r-t}{r-n}, & n<t\le r,\\ 0, & r<t,\end{cases}\qquad \text{where } l<m<n<r \text{ and } l,m,n,r\in\mathbb{R}.$$

One can obtain that

$$[N(t)]^1=[m,n],\ [N(t)]^0=[l,r],\ [N(t)]^\alpha=[m-(1-\alpha)(m-l),n+(1-\alpha)(r-n)].$$

Now, we construct the two dimensional S-almost periodic trapezoidal fuzzy vector-valued function

$$F(t)=\big(|\sin\sqrt{3}t|N(t),|\cos\sqrt{2}t|N(t)\big)^T.$$

We can obtain the α-level set of $F(t)$ as

$$\begin{aligned}[F(t)]^\alpha&=\big[|\sin\sqrt{3}t|\big(m-(1-\alpha)(m-l)\big),|\sin\sqrt{3}t|\big(n+(1-\alpha)(r-n)\big)\big]\\&\quad\times\big[|\cos\sqrt{2}t|\big(m-(1-\alpha)(m-l)\big),|\cos\sqrt{2}t|\big(n+(1-\alpha)(r-n)\big)\big].\end{aligned}$$

Hence, in this case, we can obtain the α-level set of the S-almost periodic fuzzy solution for (6.3) as follows.

$$\begin{aligned}&\left[\int_{-\infty}^{t}e_A\big(t,\sigma(s)\big)F(s)\Delta s\right]^\alpha\\&=\left[\int_{-\infty}^{t}e_{c_1}\big(t,\sigma(s)\big)|\sin\sqrt{3}s|\big(m-(1-\alpha)(m-l)\big)\Delta s,\right.\\&\left.\int_{-\infty}^{t}e_{c_1}\big(t,\sigma(s)\big)|\sin\sqrt{3}s|\big(n+(1-\alpha)(r-n)\big)\Delta s\right]\\&\times\left[\int_{-\infty}^{t}e_{c_2}\big(t,\sigma(s)\big)|\cos\sqrt{2}s|\big(m-(1-\alpha)(m-l)\big)\Delta s,\right.\\&\left.\int_{-\infty}^{t}e_{c_2}\big(t,\sigma(s)\big)|\cos\sqrt{2}s|\big(n+(1-\alpha)(r-n)\big)\Delta s\right]\\&:=\big([N(t)]^\alpha\times[N(t)]^\alpha\big)\cdot\begin{pmatrix}\int_{-\infty}^{t}e_{c_1}\big(t,\sigma(s)\big)|\sin\sqrt{3}s|\Delta s\\ \int_{-\infty}^{t}e_{c_2}\big(t,\sigma(s)\big)|\cos\sqrt{2}s|\Delta s\end{pmatrix}.\end{aligned}$$

Application 6.2.3 *The theory of quantum calculus plays an important role in the real world, due to its applications in several physics fields such as black holes, cosmic strings, confor-*

mal quantum mechanics, nuclear and high energy physics, fractional quantum Hall effect, high-T_c superconductors, etc. (see [46]). In applied dynamic equations, the q-difference equations which arises from quantum calculus are important dynamic systems which can be applied to model linear and nonlinear problems and play an important role in different fields of engineering and biological science (see [21]), but there is no result on almost periodic problems of fuzzy q-difference dynamic systems. With the developments of time scales, quantum time scales calculus can be extended to quantum-like time scales such as $\{(-q)^n : q > 1, n \in \mathbb{Z}\}$, $\{-q^n :, q > 1, n \in \mathbb{Z}\}$, etc.

Now, based on the general quantum time scale $\{(-q)^n : q > 1, n \in \mathbb{Z}\}$, we consider the generalized two-dimensional q-difference fuzzy dynamic equations as follows:

$$V^{\Delta}(t) = A(t)V(t) + F(t), \tag{6.5}$$

where $A : \mathbb{T} \to \mathbb{R}^{2\times 2}$ is a matrix-valued function.

In fact, the general quantum time scale $\{(-q)^n : q > 1, n \in \mathbb{Z}\}$ can be divided into two quantum-like time scales:

$$\mathbb{T}_1 = \{-q^{2n+1} : q > 1, n \in \mathbb{Z}\},\ \mathbb{T}_2 = \{q^{2n} : q > 1, n \in \mathbb{Z}\},\ \mathbb{T} = \mathbb{T}_1 \cup \mathbb{T}_2.$$

Therefore, we can introduce two derivatives on time scales $\mathbb{T}_1$ and $\mathbb{T}_2$, respectively, and divide (6.5) into the following two q-difference fuzzy dynamic equations:

$$V^{\Delta}(t) = D_q V(t) = \frac{f\left(\frac{t}{q^2}\right) \widetilde{-}_{gH} f(t)}{\left(\frac{1}{q^2} - 1\right)t} = A(t)V(t) + F(t),\ t \in \mathbb{T}_1, \tag{6.6}$$

$$V^{\Delta}(t) = D_q V(t) = \frac{f\left(q^2 t\right) \widetilde{-}_{gH} f(t)}{(q^2 - 1)t} = A(t)V(t) + F(t),\ t \in \mathbb{T}_2. \tag{6.7}$$

Hence, using time scales calculus, we can unify (6.6) and (6.7) by letting

$$\mu(t) = \begin{cases} \frac{1}{q^2}, & t \in \mathbb{T}_1, \\ q^2, & t \in \mathbb{T}_2, \end{cases} \qquad \sigma(t) = \begin{cases} \frac{t}{q^2}, & t \in \mathbb{T}_1, \\ q^2 t, & t \in \mathbb{T}_2, \end{cases}$$

and the following generalized fuzzy dynamic equations can be obtained:

$$V^{\Delta}(t) = D_{G_q} V(t) = \frac{f(\sigma(t)) \widetilde{-}_{gH} f(t)}{\mu(t)} = A(t)V(t) + F(t),$$

where D_{G_q} denote the generalized q-difference operator for fuzzy vector-valued functions.

Based on (6.5), we show the S-almost periodic fuzzy solutions for the following case. Suppose $F(t)$ is a two dimensional triangular fuzzy vector-valued function and $A(t) = \left(\chi_{\{c_1(t)\}}, \chi_{\{c_2(t)\}}\right)$ satisfies Corollary 6.1. Let a triangular fuzzy number $\widetilde{N} = (l, m, r)$ be

$$\widetilde{N}(t) = \begin{cases} 0, & t \leq l, \\ \frac{t-l}{m-l}, & l < t \leq m, \\ \frac{r-t}{r-n}, & m < t \leq r, \\ 0, & r < t, \end{cases} \quad \textit{where } l < m < r \textit{ and } l, m, r \in \mathbb{R}.$$

One can obtain that

$$[\widetilde{N}(t)]^1 = \{m\},\ [\widetilde{N}(t)]^0 = [l, r],\ [\widetilde{N}(t)]^\alpha = [m - (1-\alpha)(m-l), m + (1-\alpha)(r-m)].$$

Now, we construct the two dimensional S- almost periodic triangular fuzzy vector-valued function

$$F(t) = \big(|\sin\sqrt{5}t|\widetilde{N}(t), |\cos\sqrt{7}t|\widetilde{N}(t)\big)^T.$$

We obtain the α-level set of $F(t)$ as

$$\begin{aligned} [F(t)]^\alpha &= \big[|\sin\sqrt{5}t|\big(m-(1-\alpha)(m-l)\big), |\sin\sqrt{5}t|\big(m+(1-\alpha)(r-m)\big)\big] \\ &\quad \times\big[|\cos\sqrt{7}t|\big(m-(1-\alpha)(m-l)\big), |\cos\sqrt{7}t|\big(m+(1-\alpha)(r-m)\big)\big]. \end{aligned}$$

Hence, in this case, we obtain the α-level set of the S-almost periodic fuzzy solution for (6.5) *as follows:*

$$\begin{aligned} &\left[\int_{-\infty}^{t} e_A\big(t,\sigma(s)\big)F(s)\Delta s\right]^\alpha \\ &= \left[\int_{-\infty}^{t} e_{c_1}\big(t,\sigma(s)\big)|\sin\sqrt{5}s|\big(m-(1-\alpha)(m-l)\big)\Delta s,\right. \\ &\quad \left.\int_{-\infty}^{t} e_{c_1}\big(t,\sigma(s)\big)|\sin\sqrt{5}s|\big(m+(1-\alpha)(r-m)\big)\Delta s\right] \\ &\quad \times\left[\int_{-\infty}^{t} e_{c_2}\big(t,\sigma(s)\big)|\cos\sqrt{7}s|\big(m-(1-\alpha)(m-l)\big)\Delta s,\right. \\ &\quad \left.\int_{-\infty}^{t} e_{c_2}\big(t,\sigma(s)\big)|\cos\sqrt{7}s|\big(m+(1-\alpha)(r-m)\big)\Delta s\right] \\ &:= \big([\widetilde{N}(t)]^\alpha \times [\widetilde{N}(t)]^\alpha\big)\cdot\begin{pmatrix} \int_{-\infty}^{t} e_{c_1}\big(t,\sigma(s)\big)|\sin\sqrt{5}s|\Delta s \\ \int_{-\infty}^{t} e_{c_2}\big(t,\sigma(s)\big)|\cos\sqrt{7}s|\Delta s \end{pmatrix} \\ &= \begin{cases} \big([\widetilde{N}(t)]^\alpha \times [\widetilde{N}(t)]^\alpha\big)\cdot\begin{pmatrix} \int_{-\infty}^{t}\left(\frac{1}{1+c_1q^{-2}}\right)\prod_{s_0\in[s,t)}[1+(q^{-2}-1)c_1s_0]|\sin\sqrt{5}s|\Delta s \\ \int_{-\infty}^{t}\left(\frac{1}{1+c_2q^{-2}}\right)\prod_{s_0\in[s,t)}[1+(q^{-2}-1)c_2s_0]|\cos\sqrt{7}s|\Delta s \end{pmatrix}, \\ t\in\mathbb{T}_1, \\ \big([\widetilde{N}(t)]^\alpha \times [\widetilde{N}(t)]^\alpha\big)\cdot\begin{pmatrix} \int_{t}^{+\infty}\left(\frac{1}{1+c_1q^{2}}\right)\prod_{s_0\in[t,s)}[1+(q^{2}-1)c_1s_0]|\sin\sqrt{5}s|\Delta s \\ \int_{t}^{+\infty}\left(\frac{1}{1+c_2q^{2}}\right)\prod_{s_0\in[t,s)}[1+(q^{2}-1)c_2s_0]|\cos\sqrt{7}s|\Delta s \end{pmatrix}, \\ t\in\mathbb{T}_2. \end{cases} \end{aligned}$$

Application 6.2.4 *Consider the generalized two-dimensional difference fuzzy dynamic equations as follows:*

$$V^\Delta(t) = A(t)V(t) + F(t). \tag{6.8}$$

For $V(t) = [V_1(t), V_2(t)]^T$ *and*

$$A(t) = \left[\sin\frac{(t-1)\pi}{t} + 1\right]I, \quad F(t) = \left[\cos\frac{t(t-1)\pi}{2} I\right] V(t-1),$$

where I *is an identity matrix,* $t \in \mathbb{Z}$*, a Fibonacci sequence with the background of the generalized two-dimensional difference fuzzy dynamic equations will be established as follows:*

$$V(t+1) = \begin{bmatrix} \sin\frac{(t-1)\pi}{t} & 0 \\ 0 & \sin\frac{(t-1)\pi}{t} \end{bmatrix} V(t) + \begin{bmatrix} \cos\frac{t(t-1)\pi}{2} & 0 \\ 0 & \cos\frac{t(t-1)\pi}{2} \end{bmatrix} V(t-1),$$

with the initial condition $V(1) = V(2) = V_0$*, where* $t \in \mathbb{Z}$*. For* $\alpha \in [0, 1]$ *and* $\lambda \in \{1, 2\}$,

$$\left[V(t)\right]^\alpha = [V_1(t)]^\alpha \times [V_2(t)]^\alpha = \left[v_{1,\alpha}^-(t), v_{1,\alpha}^+(t)\right] \times \left[v_{2,\alpha}^-(t), v_{2,\alpha}^+(t)\right],$$

$$v_{\lambda,\alpha}^-(t) = w(t, \eta_\lambda^-, n_\lambda^-) + (1-\alpha)w(t-1, \eta_\lambda^-, n_\lambda^-),$$

$$v_{\lambda,\alpha}^+(t) = w(t, \eta_\lambda^+, n_\lambda^+) + (1-\alpha)w(t-1, \eta_\lambda^+, n_\lambda^+),$$

$$\begin{aligned} & w(t+1, \eta_\lambda^-, n_\lambda^-) \\ &= \sin\frac{(t-1)\pi}{t} w(t, \eta_\lambda^-, n_\lambda^-) + \cos\frac{t(t-1)\pi}{2} w(t-1, \eta_\lambda^-, n_\lambda^-), \\ & w(t+1, \eta_\lambda^+, n_\lambda^+) \\ &= \sin\frac{(t-1)\pi}{t} w(t, \eta_\lambda^+, n_\lambda^+) + \cos\frac{t(t-1)\pi}{2} w(t-1, \eta_\lambda^+, n_\lambda^+) \end{aligned} \tag{6.9}$$

for $w(2, \eta_\lambda^-, n_\lambda^-) = \eta_\lambda^-$, $w(1, \eta_\lambda^-, n_\lambda^-) = n_\lambda^-$, $w(2, \eta_\lambda^+, n_\lambda^+) = \eta_\lambda^+$, $w(1, \eta_\lambda^+, n_\lambda^+) = n_\lambda^+$, $\eta_\lambda^-, \eta_\lambda^+, n_\lambda^-, n_\lambda^+ \in \mathbb{N}$.

In the real practice, it is difficult to obtain the exact numbers of the rabbits in the wild. Assume that the numbers of the rabbits continuously increase at the time t, the interval $\left[v_{1,\alpha}^-(t), v_{1,\alpha}^+(t)\right]$ *is the range of the numbers of the male rabbits and* $\left[v_{2,\alpha}^-(t), v_{2,\alpha}^+(t)\right]$ *is the range of the numbers of the female rabbits.*

Since the limits of the living conditions in the wild, it cannot be guaranteed that all of rabbits are survival. Hence, assume also that the numbers of the rabbits at time $t+1$ *is the sum of the numbers of the rabbits at time t with the rate* $\sin\frac{(t-1)\pi}{t}$ *of survival and at time* $t-1$ *with the rate* $\cos\frac{t(t-1)\pi}{2}$ *of survival (maybe the rate of survival at time* $t-1$ *is negative), i.e., (6.9) holds.*

In particular, for $\eta_1^- = \eta_1^+ = 6$ *and* $n_1^- = n_1^+ = 3$*, the status of the numbers range of the male rabbits see Fig. 6.1, which is an unique S-almost periodic fuzzy vector-valued solution*

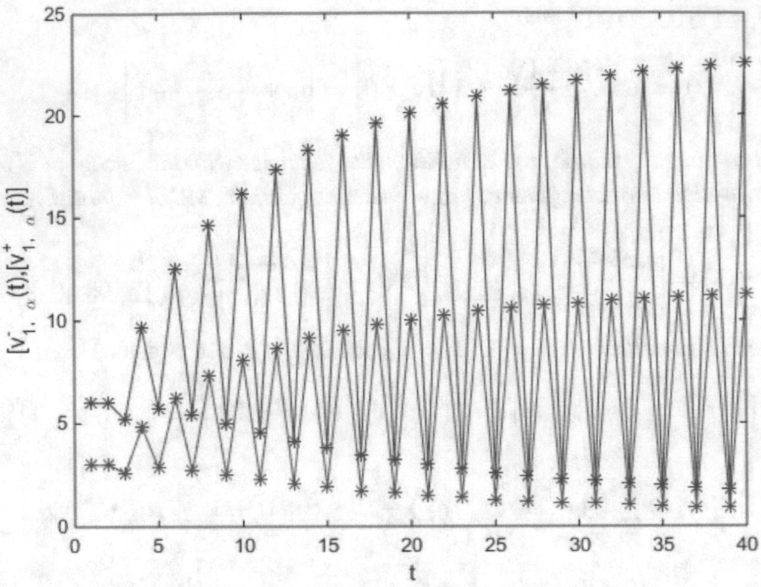

Fig. 6.1 The status of the numbers range of the male rabbits

for the component $V_1(t)$ *of* (6.8). *Similarly, for any other cases, we can obtain the status of the numbers range of the rabbits.*

Application 6.2.5 *Consider the generalized two-dimensional homogeneous fuzzy dynamic equations as follows:*

$$V^{\Delta}(t) = A(t)V(t). \tag{6.10}$$

For $\mathbb{T} = \mathbb{R}$, $V(t) = [V_1(t), V_2(t)]^T$, *where*

$$\big[V(t)\big]^{\alpha} = \big[V_1(t)\big]^{\alpha} \times \big[V_2(t)\big]^{\alpha} = \big[v_{1,\alpha}^-(t), v_{1,\alpha}^+(t)\big] \times \big[v_{2,\alpha}^-(t), v_{2,\alpha}^+(t)\big],$$

$$A(t) = \begin{bmatrix} k\big(\sin(5t\pi) + \cos t\big) & -a(\cos t + \sin 2t)V_2(t) \\ -l(\cos t + \sin 2t) & b(\cos t + \sin 2t)V_1(t) \end{bmatrix},$$

which is the fuzzy predator-prey dynamic system . Since $I_{V_1,V_2}^{\alpha,(II)} \geq 0$, *by Sect. 2.2, the fuzzy dynamic system* (6.10) *with the predator-prey application can be rewritten as*

$$\begin{bmatrix} V_1'(t) \\ V_2'(t) \end{bmatrix} = \begin{bmatrix} k\big(\sin(5t\pi) + \cos t\big)V_1(t)\widetilde{+}(-a)(\cos t + \sin 2t)V_2(t) \circledast V_1(t) \\ -l(\cos t + \sin 2t)V_2(t)\widetilde{+}b(\cos t + \sin 2t)V_1(t) \circledast V_2(t) \end{bmatrix}.$$

In another real practice, assume that there are some carps and pikes in the same pond, $[V_1(t)]^\alpha = [v_{1-}^{(t)}(\alpha), v_{1+}^{(t)}(\alpha)]$ *is the range of the numbers of the craps and* $[V_2(t)]^\alpha = [v_{2,\alpha}^-(t), v_{2,\alpha}^+(t)]$ *is the range of the numbers of the pikes with* $100(1-\alpha)$ *percent chance since the numbers of the carps and pikes cannot be measured exactly in general.*

In particular, for $k = 4$, $a = 2$, $l = 2$, $b = 3$, *we have*

$$[V_2(t) \circledast V_1(t)]^\alpha = [V_1(t)]^\alpha \circledcirc [V_2(t)]^\alpha = [v_{1,\alpha}^-(t)v_{2,\alpha}^-(t), v_{1,\alpha}^+(t)v_{2,\alpha}^+(t)].$$

Hence, for $v_{1,\alpha}^-(t)$ *and* $v_{2,\alpha}^-(t)$, *we have*

$$\begin{bmatrix} (v_{1,\alpha}^-(t))' \\ (v_{2,\alpha}^-(t))' \end{bmatrix} = \begin{bmatrix} 4(\sin(5t\pi) + \cos t)v_{1,\alpha}^-(t) - 2(\cos t + \sin 2t)v_{1,\alpha}^-(t)v_{2,\alpha}^-(t) \\ -2(\cos t + \sin 2t)v_{2,\alpha}^-(t) + 3(\cos t + \sin 2t)v_{1,\alpha}^-(t)v_{2,\alpha}^-(t) \end{bmatrix}. \quad (6.11)$$

Hence, the status of the S-almost periodic solution of the system (6.11) see Fig. 6.2. Similarly, for $v_{1,\alpha}^+(t)$ *and* $v_{2,\alpha}^+(t)$, *we can obtain its status.*

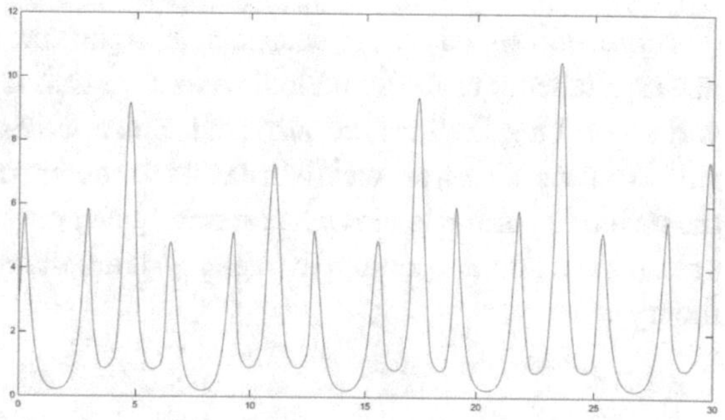

(a) the status of $v_{1,\alpha}^-(t)$

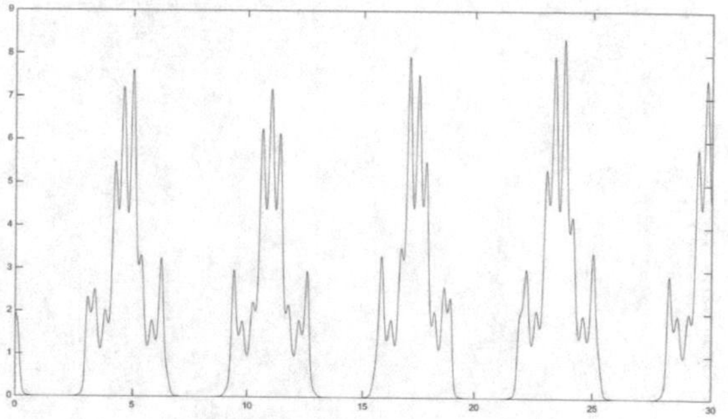

(b) the status of $v_{2,\alpha}^-(t)$

Fig. 6.2 The status of the S-almost periodic solution of the system (6.11)

Based on [104], we introduce a new division of fuzzy vectors induced by a determinant algorithm and obtain the relation between the division and the corresponding multiplication proposed in [104]. It is well known that interval analysis is a difficult task that plays a very important role in studying fuzzy calculus and dynamic equations (see [64]). In general, the classical fundamental formulas of calculus are not true and cannot be applied to the fuzzy situation directly. The proposed multiplication and division induced by a determinant algorithm is well defined and can be used with the gH-difference to form a traditional mixed operation, which lays a foundation for obtaining the basic formulas of calculus under a fuzzy background. Using the determinant algorithm, we suggest six types of division for fuzzy vectors in a fuzzy multidimensional space:

$$\left(\frac{u}{v}\right)_{\mathring{I}}, \left(\frac{u}{v}\right)_{\mathring{I}I}, \left(\frac{u}{v}\right)_{I\mathring{I}I}, \left(\frac{u}{v}\right)_{I\mathring{V}}, \left(\frac{u}{v}\right)_{\mathring{V}}, \left(\frac{u}{v}\right)_{\mathring{V}I},$$

and obtain the basic formulas of calculus such as

$$\left(\frac{f}{g}\right)_{\mathring{I}}^{\Delta_I}, \left(\frac{f}{g}\right)_{\mathring{I}}^{\Delta_{II}}, \left(\frac{f}{g}\right)_{\mathring{I}I}^{\Delta_I}, \left(\frac{f}{g}\right)_{\mathring{I}I}^{\Delta_{II}}, \ldots.$$

Moreover, based on the determinant algorithm and fuzzy calculus on time scales, we develop a theory of almost periodic fuzzy multidimensional dynamic systems on time scales and several applications are provided. Since we establish the theory on non-translational shifts of time scales, the results are valid for other fuzzy multidimensional dynamic systems on various hybrid domains and especially can cover the almost periodic fuzzy q-dynamic systems (i.e., fuzzy quantum dynamic systems which has important applications in quantum theory).

Almost Anti-periodic Discrete Oscillation

A

In 2017, M. Kostić introduced an interesting notion of almost anti-periodic functions in Banach space and studied the relationship between the types of anti-periodic functions and almost periodic functions in Banach spaces. After this, some related works were published (see [49–53]). It is natural to ask how to define the almost anti-periodic discrete process and explore what properties they will possess. The answer of this question will make it possible to study the almost anti-periodic discrete dynamic equation and contribute to establishing almost anti-periodic results on time scales.

In the literature [117], a theory of almost anti-periodic discrete oscillation was developed. The main aim of this appendix is to introduce the notion of the almost anti-periodic discrete process of the N-dimensional vector-valued and $N \times N$ matrix-valued functions and to establish the stability of the almost anti-periodic discrete solutions to the general N-dimensional mechanical system and underactuated Euler–Lagrange system. For some relating the finite-dimensional vector spaces, one may see [39] and the underactuated Euler–Lagrange system with N degrees of freedom and m independent controls (see [118]), as well as the general N-dimensional mechanical system (see [25]).

A.1 Almost Anti-periodic Discrete Functions

In this section, we will introduce the notions of the almost anti-periodic discrete functions for the $N \times N$ matrix-valued function and the N-dimensional vector-valued function and establish some of their basic properties.

Definition A.1 Let $M(\cdot) = [m_{ij}(\cdot)]_{N\times N} : \mathbb{Z} \to \mathbb{R}^{N\times N}$ be an $N \times N$ matrix-valued discrete function and $p(\cdot) = [p_1(\cdot), \ldots, p_N(\cdot)]^T : \mathbb{Z} \to \mathbb{R}^N$ be a N-dimensional vector-valued discrete function, we define

C. Wang and R. P. Agarwal, *Dynamic Equations and Almost Periodic Fuzzy Functions on Time Scales*, Synthesis Lectures on Mathematics & Statistics,
https://doi.org/10.1007/978-3-031-11236-2

$$\|M(n)\| = \frac{1}{N^2}\sum_{j=1}^{N}\sum_{i=1}^{N}|m_{ij}(n)|, \quad \|p(n)\| = \frac{1}{N}\sum_{j=1}^{N}|p_j(n)|.$$

If for $\forall\varepsilon > 0$, there exists a positive integer $l(\varepsilon)$ and $\tau \in [m, m+l]_{\mathbb{Z}}$, $\forall m \in \mathbb{Z}$, such that $p(n)$ satisfies the following condition

$$\|p(n+\tau) + p(n)\| < \varepsilon, \quad \forall n \in \mathbb{Z}.$$

Then, $p(n)$ is called the almost anti-periodic discrete function, τ is called the ε-almost anti-period of $p(n)$, and l is called an inclusion length. Similarly, $M(n)$ is an almost anti-periodic $N \times N$ matrix-valued discrete function if

$$\|M(n+\tau) + M(n)\| < \varepsilon, \quad \forall n \in \mathbb{Z}.$$

The set of all almost anti-periodic $N \times N$ matrix-valued discrete functions is denoted by $\mathfrak{G}$.

Definition A.2 ([53]) Let $\{q(n)\}$ be a discrete sequence, if for $\forall\varepsilon > 0$, there exists a positive integer $l(\varepsilon)$ such that $q(n)$ satisfies the following condition

$$|q(n+\tau) - q(n)| < \varepsilon, \quad \forall n \in \mathbb{Z}$$

for some $\tau \in [m, m+l]_{\mathbb{Z}}$, where $m \in \mathbb{Z}$. Then, $q(n)$ is called the almost periodic discrete function and τ is called the ε-almost period of $q(n)$.

Lemma A.1 *Let $M(n)$ be an almost anti-periodic discrete function; then, it is an almost periodic discrete function.*

Proof By Definition A.1, we have

$$\|M(n+2\tau) - M(n)\| \le |M(n+2\tau) + M(n+\tau)\| + \|M(n+\tau) + M(n)\| < 2\varepsilon.$$

The proof is completed. □

Through Lemma A.1, the following lemma is immediate.

Lemma A.2 ([105]) *Let $M(n)$ be an almost periodic discrete function (or an almost anti-periodic discrete function). Then, $\|M(\cdot)\| : \mathbb{Z} \to \mathbb{R}$ is bounded.*

In what follows, some basic properties of the almost anti-periodic discrete functions will be established.

Theorem A.1 *Let $M(\cdot) \in \mathfrak{G}$, then $cM(\cdot) \in \mathfrak{G}$ for all $c \in \mathbb{R}$.*

Proof By Definition A.1, one has

$$\|M(n+\tau) + M(n)\| < \frac{\varepsilon}{|c|}$$

for $c \neq 0$. Thus, we have

$$\|cM(n+\tau) + cM(n)\| \leq |c|\|M(n+\tau) + M(n)\| < |c|\frac{\varepsilon}{|c|} = \varepsilon,$$

which means $cM(\cdot) \in \mathfrak{G}$. For $c = 0$, one can obtain the desired result. The proof is completed. □

Theorem A.2 *Let $M(\cdot) \in \mathfrak{G}$ and $Q(n) = M(n+k)$ for some $k \in \mathbb{Z}$. Then, $Q(\cdot) \in \mathfrak{G}$.*

Proof Since $M(\cdot) \in \mathfrak{G}$, we have

$$\|M(n+\tau) + M(n)\| < \varepsilon \text{ for all } n \in \mathbb{Z}.$$

Hence,

$$\|M(n+k+\tau) + M(n+k)\| < \varepsilon, \quad n \in \mathbb{Z},$$

i.e.,

$$\|Q(n+\tau) + Q(n)\| < \varepsilon, \quad n \in \mathbb{Z},$$

which implies $Q(\cdot) \in \mathfrak{G}$. The proof is completed. □

Theorem A.3 *Let $B(\cdot) \in \mathfrak{G}$ and $S(n)$ be a summable $N \times N$ matrix sequence, i.e.,*

$$S_0 = \sum_{n\in\mathbb{Z}} \|S(n)\| < \infty.$$

If $M(n) = \sum_{m\in\mathbb{Z}} S(m)B(n-m)$ for $n \in \mathbb{Z}$. Then, $M(\cdot) \in \mathfrak{G}$.

Proof Since $S(n)$ is a summable matrix sequence, for $\forall\varepsilon > 0$, there exists $N_0 > 0$ such that

$$\sum_{|n|>N_0} \|S(n)\| < \varepsilon.$$

By Definition A.1 and Theorem A.2 , one has

$$\|B(n+\tau-m) + B(n-m)\| < \frac{\varepsilon}{S_0}.$$

By Lemma A.2, we have $\|B(n)\| < \lambda$ for some $\lambda > 0$ and any $n \in \mathbb{Z}$. Hence, we have

$$\begin{aligned}
\|M(n+\tau)+M(n)\| &= \left\| \sum_{m\in\mathbb{Z}} S(m)B(n+\tau-m) + \sum_{m\in\mathbb{Z}} S(m)B(n-m) \right\| \\
&\le \left\| \sum_{|m|\le N_0} S(m)B(n+\tau-m) + \sum_{|m|\le N_0} S(m)B(n-m) \right\| \\
&+ \left\| \sum_{|m|> N_0} S(m)B(n+\tau-m) + \sum_{|m|> N_0} S(m)B(n-m) \right\| \\
&\le \left\| \sum_{|m|\le N_0} S(m) \right\| \left\| B(n+\tau-m) + B(n-m) \right\| \\
&+ \left\| \sum_{|m|> N_0} S(m)B(n+\tau-m) \right\| + \left\| \sum_{|m|> N_0} S(m)B(n-m) \right\| \\
&\le S_0 \tfrac{\varepsilon}{S_0} + \left\| \sum_{|m|> N_0} S(m) \right\| \lambda + \left\| \sum_{|m|> N_0} S(m) \right\| \lambda \le \varepsilon + 2\lambda\epsilon,
\end{aligned}$$

which means $M(\cdot) \in \mathfrak{G}$. The proof is completed. □

Theorem A.4 *Let $M(\cdot) \in \mathfrak{G}$, then $\|M(n)\|$ is an almost periodic discrete function.*

Proof Since $M(\cdot) \in \mathfrak{G}$, we have

$$\|M(n+\tau)+M(n)\| < \varepsilon, \quad n \in \mathbb{Z}.$$

Hence,

$$\begin{aligned}
\big| \|M(n+\tau)\| - \|M(n)\| \big| &= \left| \frac{1}{N^2} \sum_{j=1}^{N} \sum_{i=1}^{N} |m_{ij}(n+\tau)| - \frac{1}{N^2} \sum_{j=1}^{N} \sum_{i=1}^{N} |m_{ij}(n)| \right| \\
&\le \frac{1}{N^2} \sum_{j=1}^{N} \sum_{i=1}^{N} \big| |m_{ij}(n+\tau)| - |m_{ij}(n)| \big| \\
&\le \frac{1}{N^2} \sum_{j=1}^{N} \sum_{i=1}^{N} \big| m_{ij}(n+\tau) + m_{ij}(n) \big| \\
&= \big\| M(n+\tau) + M(n) \big\| < \varepsilon,
\end{aligned}$$

which means that $\{\|M(n)\|\}$ is an almost periodic discrete function. The proof is completed. □

Based on the theorems above, the following result can be proved.

Proposition A.1 *Let $M(\cdot) \in \mathfrak{G}$, $c > 0$, if $\|M(n)\| \ge c > 0$. Then, the following limit exists:*

$$\lim_{n\to\infty} \left[\prod_{j=1}^{n} \|M(j)\| \right]^{\frac{1}{n}}. \tag{A.1}$$

Proof Let $h(n) = \left[\prod_{j=1}^{n} \|M(j)\|\right]^{\frac{1}{n}}$, we have

$$\ln[h(n)] = \frac{1}{n}\left[\sum_{j=1}^{n} \ln \|M(j)\|\right].$$

Step 1. We will prove that $\|M(n)\|$ is bounded. Since $M(n)$ is almost anti-periodic, by Theorem A.4, we obtain $\|M(n)\|$, which is almost periodic. Moreover, by Lemma A.2, one has $\|M(n)\|$, which is bounded; i.e., there exists $\lambda \in \mathbb{R}^+$ such that $\sup_{n\in\mathbb{Z}} |\ln \|M(n)\|| = \lambda$.

Step 2. We will prove that $\ln \|M(n)\|$ is an almost periodic discrete function. Since $f(x) = \ln x$ is uniformly continuous on $[c, e^{\lambda}]$, i.e., for $\forall \varepsilon > 0$, there exists $\delta(\varepsilon) > 0$ such that $|x_1 - x_2| < \delta$ implies

$$|f(x_1) - f(x_2)| < \varepsilon,$$

for $x_1, x_2 \in [c, e^{\lambda}]$. On the other hand, $M(\cdot) \in \mathfrak{G}$, i.e.,

$$\|M(\tau + n) + M(n)\| < \varepsilon.$$

By Theorem A.4, one has

$$\big|\|M(\tau + n)\| - \|M(n)\|\big| < \varepsilon.$$

By Step 1 and $\|M(n)\| \geq c > 0$, we have $\|M(n)\| \in [c, e^{\lambda}]$. Thus, one has

$$|\ln \|M(n + \tau)\| - \ln \|M(n)\|| < \varepsilon \quad \text{for all} \quad n \in \mathbb{Z},$$

which means that $\ln \|M(n)\|$ is almost periodic, i.e., for $\forall \varepsilon > 0$, there exists a positive integer $l_0(\varepsilon)$ such that $M(n)$ satisfies the following condition:

$$|\ln \|M(n + \tau)\| - \ln \|M(n)\|| < \varepsilon, \quad n \in \mathbb{Z},$$

for some $\tau \in [a, a + l_0]_{\mathbb{Z}}$ and $a \in \mathbb{Z}$.

Step 3. By Step 2, for $\tau \in [a, a + l_0]_{\mathbb{Z}}$ and $\forall \frac{\varepsilon}{4} > 0$, there exists an positive integer $l_0(\varepsilon)$ such that $M(n)$ satisfies the following condition:

$$|\ln \|M(n + \tau)\| - \ln \|M(n)\|| < \frac{\varepsilon}{4}, \quad n \in \mathbb{Z},$$

for some $\tau \in [a, a + l_0]_{\mathbb{Z}}$ and $a \in \mathbb{Z}$. Next, we will consider $\tau = a$ and $\tau \in (a, a + l_0]_{\mathbb{Z}}$ cases.

Cases I. $\tau = a$. Let $b \in \mathbb{Z}^+$, one has

$$\begin{aligned}\left|\sum_{j=a+1}^{a+b}\ln\|M(j)\|-\sum_{j=1}^{b}\ln\|M(j)\|\right| &= \left|\sum_{j=\tau+1}^{\tau+b}\ln\|M(j)\|-\sum_{j=1}^{b}\ln\|M(j)\|\right| \\ &\le \sum_{j=1}^{b}\left|\ln\|M(j+\tau)\|-\ln\|M(j)\|\right| \le \tfrac{\varepsilon b}{4}.\end{aligned}$$

Cases II. $\tau \in (a, a+l_0]_{\mathbb{Z}}$. By Step 2, for $b \in \mathbb{N}^+$, one has

$$\begin{aligned}&\left|\sum_{j=a+1}^{a+b}\ln\|M(j)\|-\sum_{j=1}^{b}\ln\|M(j)\|\right| \\ =&\left|\sum_{j=\tau+1}^{\tau+b}\ln\|M(j)\|-\sum_{j=1}^{b}\ln\|M(j)\|+\sum_{j=a+1}^{\tau}\ln\|M(j)\|-\sum_{j=a+b+1}^{\tau+b}\ln\|M(j)\|\right| \\ \le&\sum_{j=1}^{b}\left|\ln\|M(j+\tau)\|-\ln\|M(j)\|\right|+\sum_{j=a+1}^{\tau}\left|\ln\|M(j)\|\right|+\sum_{j=a+b+1}^{\tau+b}\left|\ln\|M(j)\|\right| \\ \le&\tfrac{\varepsilon b}{4}+2l_0\lambda.\end{aligned}$$

Moreover, for the case I, through taking $a = (k-1)n$, $b = n$, we have

$$\left|\sum_{j=(k-1)n+1}^{kn}\ln\|M(j)\|-\sum_{j=1}^{n}\ln\|M(j)\|\right| \le \frac{\varepsilon n}{4};$$

for the case II, through taking $a = (k-1)n$, $b = n$, one has

$$\left|\sum_{j=(k-1)n+1}^{kn}\ln\|M(j)\|-\sum_{j=1}^{n}\ln\|M(j)\|\right| \le \frac{\varepsilon n}{4}+2l_0\lambda.$$

Step 4. We will prove that the Limit (A.1) exists. For any $m, n \in \mathbb{N}^+$, one has

$$\begin{aligned}&\left|\tfrac{1}{n}\sum_{j=1}^{n}\ln\|M(j)\|-\tfrac{1}{m}\sum_{j=1}^{m}\ln\|M(j)\|\right| \\ \le&\tfrac{1}{mn}\left|m\sum_{j=1}^{n}\ln\|M(j)\|-\sum_{j=1}^{mn}\ln\|M(j)\|\right| \\ &+\tfrac{1}{mn}\left|\sum_{j=1}^{mn}\ln\|M(j)\|-n\sum_{j=1}^{m}\ln\|M(j)\|\right| \\ \le&\tfrac{1}{mn}\left[\sum_{k=1}^{m}\left|\sum_{j=(k-1)n+1}^{kn}\ln\|M(j)\|-\sum_{j=1}^{n}\ln\|M(j)\|\right|\right. \\ &\left.+\sum_{k=1}^{n}\left|\sum_{j=(k-1)m+1}^{km}\ln\|M(j)\|-\sum_{j=1}^{m}\ln\|M(j)\|\right|\right].\end{aligned}$$

On the other hand, by Step 3, if $\tau = a = (k-1)n$, then

$$\left|\frac{1}{n}\sum_{j=1}^{n}\ln\|M(j)\|-\frac{1}{m}\sum_{j=1}^{m}\ln\|M(j)\|\right| \le \frac{1}{mn}\left(m\frac{\varepsilon n}{4}+n\frac{\varepsilon m}{4}\right)=\frac{\varepsilon}{2}<\varepsilon;$$

if $\tau \in (a, a + l_0]_{\mathbb{Z}} = ((k-1)n, (k-1n) + l_0]_{\mathbb{Z}}$, then

$$\left| \frac{1}{n} \sum_{j=1}^{n} \ln \|M(j)\| - \frac{1}{m} \sum_{j=1}^{m} \ln \|M(j)\| \right| \leq \frac{1}{mn} \left[m(\tfrac{\varepsilon n}{4} + 2l_0\lambda) + n(\tfrac{m\varepsilon + 2l_0}{4}\lambda) \right]$$
$$= \tfrac{\varepsilon}{2} + 2l_0\lambda(\tfrac{1}{n} + \tfrac{1}{m}) < \varepsilon$$

for $m, n \geq [\frac{8l_0\lambda}{\varepsilon}] + 1$, which indicates that the Limit (A.1) exists. The proof is completed. □

Proposition A.2 *Let $M(\cdot) \in \mathfrak{G}$. Then, the following limit equalities hold:*

(i) $h = \lim_{n\to\infty} \left[\prod_{j=m+1}^{m+n} \|M(j)\| \right]^{\frac{1}{n}} = \lim_{n\to\infty} \left[\prod_{j=1}^{n} \|M(j)\| \right]^{\frac{1}{n}}.$

(ii) $h = \lim_{n\to-\infty} \left[\prod_{j=n}^{-1} \|M(j)\| \right]^{-\frac{1}{n}} = \lim_{n\to\infty} \left[\prod_{j=-n}^{-1} \|M(j)\| \right]^{\frac{1}{n}}.$

Proof (i) Through using the process of the proof of Proposition A.1, we can turn (i) into the following equality:

$$\lim_{n\to\infty} \frac{1}{n} \left[\sum_{j=m+1}^{m+n} \ln \|M(j)\| - \sum_{j=1}^{n} \ln \|M(j)\| \right] = 0. \tag{A.2}$$

By the proof of Proposition A.1 Step 2, for $\forall \frac{\varepsilon}{2} > 0$, there exists a positive integer $l_0(\varepsilon)$ such that $M(j)$ satisfies the following condition:

$$\left| \ln \|M(j+\tau)\| - \ln \|M(j)\| \right| \leq \frac{\varepsilon}{2} \quad \text{for all} \quad j \in \mathbb{Z},$$

where $\tau \in [m, m+l]_{\mathbb{Z}}$ and $m \in \mathbb{Z}$. Next, we will consider $\tau = m$ and $\tau \in (m, m+l]_{\mathbb{Z}}$ cases. By the proof of Proposition A.1 Step 3, taking $a = m$ and $b = n$, the desired results can be shown.

Cases I. $\tau = m$. For $\tau = m$, we have

$$\begin{aligned} &\frac{1}{n} \left| \sum_{j=m+1}^{m+n} \ln \|M(j)\| - \sum_{j=1}^{n} \ln \|M(j)\| \right| \\ &= \frac{1}{n} \left| \sum_{j=\tau+1}^{\tau+n} \ln \|M(j)\| - \sum_{j=1}^{n} \ln \|M(j)\| \right| \\ &\leq \frac{1}{n} \sum_{j=1}^{n} \left| \ln \|M(j+\tau)\| - \ln \|M(j)\| \right| \leq \frac{1}{n} \frac{\varepsilon n}{2} < \varepsilon. \end{aligned}$$

Therefore, Eq. (A.2) holds.

Cases II. $\tau \in (m, m+l]_{\mathbb{Z}}$. For $\tau \in (m, m+l]_{\mathbb{Z}}$, we have

$$
\begin{aligned}
&\frac{1}{n}\left|\sum_{j=m+1}^{m+n} \ln\|M(j)\| - \sum_{j=1}^{n} \ln\|M(j)\|\right| \\
&= \frac{1}{n}\left|\sum_{j=\tau+1}^{\tau+n} \ln\|M(j)\| - \sum_{j=1}^{n} \ln\|M(j)\| + \sum_{j=m+1}^{\tau} \ln\|M(j)\| - \sum_{j=m+n+1}^{\tau+n} \ln\|M(j)\|\right| \\
&\le \frac{1}{n}\left[\sum_{j=1}^{n} \big|\ln\|M(j+\tau)\| - \ln\|M(j)\|\big| + \sum_{j=m+1}^{\tau} \big|\ln\|M(j)\|\big| + \sum_{j=m+n+1}^{\tau+n} \big|\ln\|M(j)\|\big|\right] \\
&\le \frac{1}{n}\left(\frac{\varepsilon n}{2} + 2l\lambda\right) \le \varepsilon
\end{aligned}
$$

for $n \ge [\frac{4l\lambda}{\varepsilon}] + 1$. Hence, Eq. (A.2) holds.

(ii) Through using the proof process of Proposition A.1, we can turn (ii) into the following equality:

$$
\lim_{n\to-\infty} \frac{1}{-n} \sum_{j=-1}^{n} \ln\|M(j)\| = \lim_{n\to\infty} \frac{1}{n} \sum_{j=-1}^{-n} \ln\|M(j)\|.
$$

On the other hand, by the proof of Proposition A.1 Step 2, we obtain that $\ln\|M(n)\|$ is almost periodic discrete function. Hence,

$$
\lim_{n\to\infty} \frac{1}{n} \sum_{j=-1}^{-n} \ln\|M(j)\| = \lim_{n\to\infty} \frac{1}{n} \sum_{j=1}^{n} \ln\|M(j)\|,
$$

i.e.,

$$
\lim_{n\to\infty} \frac{1}{n}\left[\sum_{j=-n}^{-1} \ln\|M(j)\| - \sum_{j=1}^{n} \ln\|M(j)\|\right] = 0. \tag{A.3}
$$

Similarly, by the proof of Proposition A.1 Step 2, we have

$$
\big|\ln\|M(j+\tau)\| - \ln\|M(j)\|\big| \le \frac{\varepsilon}{2} \quad \text{for all} \quad j \in \mathbb{Z},
$$

where $\tau \in [-n-1, -n-1+l]_{\mathbb{Z}}$. Next, we will consider the case for $\tau = -n-1$ and $\tau \in (-n-1, -n-1+l]_{\mathbb{Z}}$. By the proof of Proposition A.1 Step 3, taking $a = -n-1$ and $b = n$, the desired results can be proved.

Cases I. $\tau = -n-1$. For $\tau = -n-1$, we have

$$
\begin{aligned}
\frac{1}{n}\left|\sum_{j=-n}^{-1} \ln\|M(j)\| - \sum_{j=1}^{n} \ln\|M(j)\|\right| &= \frac{1}{n}\left|\sum_{j=\tau+1}^{\tau+n} \ln\|M(j)\| - \sum_{j=1}^{n} \ln\|M(j)\|\right| \\
&\le \frac{1}{n}\left[\sum_{j=1}^{n} \big|\ln\|M(j+\tau)\| - \ln\|M(j)\|\big|\right] \\
&\le \frac{1}{n}\frac{\varepsilon}{2}n = \frac{\varepsilon}{2} < \varepsilon.
\end{aligned}
$$

Therefore, Eq. (A.3) holds.

Cases II. $\tau \in (-n-1, -n-1+l]_{\mathbb{Z}}$. For $\tau \in (-n-1, -n-1+l]_{\mathbb{Z}}$, it follows that

$$
\begin{aligned}
& \frac{1}{n}\left|\sum_{j=-n}^{-1} \ln \|M(j)| - \sum_{j=1}^{n} \ln \|M(j)\|\right| \\
&= \frac{1}{n}\left|\sum_{j=\tau+1}^{\tau+n} \ln \|M(j)\| - \sum_{j=1}^{n} \ln \|M(j)\| + \sum_{j=-n}^{\tau} \ln \|M(j)\| - \sum_{j=0}^{\tau+n} \ln \|M(j)\|\right| \\
&\le \frac{1}{n}\left[\sum_{j=1}^{n} \left|\ln \|M(j+\tau)\| - \ln \|M(j)\|\right| + \sum_{j=-n}^{\tau} \left|\ln \|M(j)\|\right| + \sum_{j=0}^{\tau+n} \left|\ln \|M(j)\|\right|\right] \\
&\le \frac{1}{n}\left(\frac{\varepsilon n}{2} + 2l\lambda\right) \le \varepsilon
\end{aligned}
$$

for $n \ge [\frac{4l\lambda}{\varepsilon}] + 1$. Hence, Eq. (A.3) holds. The proof is completed. □

A.2 Almost Anti-periodic Oscillation of the Mechanical System

In this section, we will consider the almost anti-periodic oscillation of the general mechanical systems.

A.2.1 Existence and Uniqueness of the Almost Anti-periodic Solution

Consider the general N-dimensional mechanical system as follows:

$$
B[(q(n+2) - 2q(n+1) + q(n)] + C[q(n+1) - q(n)] + S\big(q(n)\big) = f(n), \tag{A.4}
$$

where $q(\cdot) = [q_1(\cdot), \ldots, q_N(\cdot)] \in \mathbb{R}^N$, $B, C \in \mathbb{R}^{N\times N}$ is symmetric and B is positive definite, the vector $S(q)$ collects all position dependent forces, such a linear and nonlinear stiffness forces or non-potential forces, $f(n)$ is an almost anti-periodic discrete function. Moreover, Eq. (A.4) can be rewritten as

$$
Bq(n+2) + (C - 2B)q(n+1) + (B - C)q(n) + S\big(q(n)\big) = f(n).
$$

Assume that $x(n) = [q(n), q(n+1)]^T$, i.e.,

$$
x(n) = [q_1(n), \ldots, q_N(n), q_1(n+1), \ldots, q_N(n+1)]^T,
$$

Equation (A.4) can be rewritten as

$$
\begin{aligned}
x(n+1) &= \begin{bmatrix} q(n+1) \\ q(n+2) \end{bmatrix} = \begin{bmatrix} q(n+1) \\ \xi \end{bmatrix} + \begin{bmatrix} 0 \\ f(n) - S\big(q(n)\big) \end{bmatrix} \\
&= \begin{bmatrix} 0 & I \\ B^{-1}(C-B) & B^{-1}(2B-C) \end{bmatrix} \begin{bmatrix} q(n) \\ q(n+1) \end{bmatrix} \\
&\quad + \begin{bmatrix} 0 \\ B^{-1}f(n) - B^{-1}S\big(q(n)\big) \end{bmatrix} \\
&= Rx(n) + b(n),
\end{aligned} \tag{A.5}
$$

where $\xi = B^{-1}(C - B)q(n) + B^{-1}(2B - C)q(n + 1)$ and

$$R = \begin{bmatrix} 0 & I \\ B^{-1}(C - B) & B^{-1}(2B - C) \end{bmatrix}, \; b(n) = \begin{bmatrix} 0 \\ B^{-1}f(n) - B^{-1}S\big(q(n)\big) \end{bmatrix}.$$

In the sequel, we will establish some sufficient conditions of the existence and uniqueness of the almost anti-periodic solution of Eq. (A.5).

Theorem A.5 *Let $s < n$, the unique solution of Eq.* (A.5) *with the initial value condition $x(s) = x_0$ can be given as:*

$$x(n) = [q(n), q(n+1)]^T = R^{n-s}x(s) + \sum_{j=n-1-s}^{0} R^j b(n-1-j),$$

i.e.,

$$x(n) = [q(n), q(n+1)]^T = \begin{bmatrix} 0 & I \\ B^{-1}(C - B) & B^{-1}(2B - C) \end{bmatrix}^{n-s} x(s)$$
$$+ \sum_{j=n-1-s}^{0} \begin{bmatrix} 0 & I \\ B^{-1}(C - B) & B^{-1}(2B - C) \end{bmatrix}^{j} \begin{bmatrix} 0 \\ \eta \end{bmatrix},$$

where $R^0 = I$, $\eta = B^{-1}f(n-1-j) - B^{-1}S\big(q(n-1-j)\big)$.

Proof Step 1, we will prove the existence of the solution of Eq. (A.4). For $n > s$, we have

$$\begin{aligned} x(n+1) &= R^{n+1-s}x(s) + \sum_{j=n-s}^{0} R^j b(n-j) \\ &= R^{n+1-s}x(s) + \sum_{j=n-s}^{1} R^j b(n-j) + b(n) \\ &= R\left[R^{n-s}x(s) + \sum_{j=n-s}^{1} R^{j-1}b(n-j)\right] + b(n) = Rx(n) + b(n) \end{aligned}$$

and $x(s) = x_0$, which means that $x(n)$ is a solution of Eq. (A.4).

Step 2, we will prove the uniqueness of the solution of Eq. (A.5). Let $x(n)$ and $y(n)$ be two solutions of Eq. (A.5) with the initial value condition $x(s) = y(s) = x_0$, then we have

$$x(n) = Rx(n-1) + b(n-1) \quad \text{and} \quad y(n) = Ry(n-1) + b(n-1).$$

Assume that $V(n) = x(n) - y(n)$, then $V(n) = RV(n-1)$. Hence,

$$V(n) = R^{n-s}V(s).$$

On the other hand, $x(s) = y(s) = x_0$, i.e., $V(s) = 0$, which implies $V(n) = 0$, i.e., $x(n) = y(n)$. The proof is completed. □

Lemma A.3 *Let τ be a ε-almost anti-period of $f(n)$, if $\|R\| + \lambda\|B^{-1}\| < 1 - \lambda$, $\|R\| \le \lambda$, $\|S(q) + S(p)\| \le \frac{\lambda}{\|B^{-1}\|}\|p + q\|$ for $p, q \in \mathbb{R}^N$ and $0 < \lambda < 1$. Then,*

$$\|b(n+\tau) + b(n)\| \le \upsilon\|q(n+\tau) + q(n)\|$$
$$\|q(n+1+\tau) + q(n+1)\| < \zeta\|q(n+\tau) + q(n)\|$$

where $\upsilon \in (\lambda, 1)$ and $0 < \zeta = \frac{\|R\|+\lambda\|B^{-1}\|}{1-\|R\|} < 1$.

Proof Since $b(n) = \left[0, B^{-1}f(n) - B^{-1}S(q(n))\right]^T$, we have

$$\begin{aligned}
&\|b(n+\tau) + b(n)\| \\
&= \left\|B^{-1}f(n+\tau) - B^{-1}S(q(n+\tau)) + B^{-1}f(n) - B^{-1}S(q(n))\right\| \\
&\le \left\|B^{-1}f(n+\tau) + B^{-1}f(n)\right\| + \left\|B^{-1}S(q(n+\tau)) + B^{-1}S(q(n))\right\| \\
&\le \|B^{-1}\|\varepsilon + \|B^{-1}\|\frac{\lambda}{\|B^{-1}\|}\left\|q(n+\tau) + q(n)\right\| \\
&= \|B^{-1}\|\varepsilon + \lambda\left\|q(n+\tau) + q(n)\right\|.
\end{aligned}$$

Hence, there exists $\upsilon \in (\lambda, 1)$ such that

$$\|b(n+\tau) + b(n)\| \le \upsilon\|q(n+\tau) + q(n)\|.$$

On the other hand, τ is a ε-almost anti-period of $f(n)$ and $x(n) = [p(n), p(n+1)]^T$, we have

$$\begin{aligned}
&\|x(n+1+\tau) + x(n+1)\| \\
&= \|q(n+1+\tau) + q(n+1)\| + \|q(n+2+\tau) + q(n+2)\| \\
&= \|Rx(n+\tau) + b(n+\tau) + Rx(n) + b(n)\| \\
&\le \|R\|\|x(n+\tau) + x(n)\| + \|b(n+\tau) + b(n)\| \\
&\le \|R\|\big(\|q(n+\tau) + q(n)\| + \|q(n+1+\tau) + q(n+1)\|\big) \\
&+ \|B^{-1}\|\big(\|f(n+\tau) + f(n)\| + \|S(q(n+\tau)) + S(q(n))\|\big) \\
&\le (\|R\| + \lambda\|B^{-1}\|)\|q(n+\tau) + q(n)\| + \|R\|\|q(n+1+\tau) + q(n+1)\| + \|B^{-1}\|\varepsilon,
\end{aligned}$$

i.e.,

$$\begin{aligned}
(1 - \|R\|)\|q(n+1+\tau) + q(n+1)\| + \|q(n+2+\tau) + q(n+2)\| \\
\le (\|R\| + \lambda\|B^{-1}\|)\|q(n+\tau) + q(n)\| + \|B^{-1}\|\varepsilon.
\end{aligned}$$

Hence,

$$\|q(n+1+\tau) + q(n+1)\| < \frac{\|R\| + \lambda\|B^{-1}\|}{1 - \|R\|}\|q(n+\tau) + q(n)\|.$$

The proof is completed. □

Now, consider the underactuated Euler–Lagrange system with N degrees of freedom and m independent controls by a discrete dynamic equation as follows:

$$\begin{aligned}&B\big(q(n)\big)\big(q(n+2)-2q(n+1)+q(n)\big)\\&+C\big(q(n),q(n+1)-q(n)\big)[q(n+1)-q(n)]+G(q(n))=L\omega,\end{aligned} \tag{A.6}$$

where B is a positive definite matrix denoting the inertia matrix, C is the Coriolis matrix, G is the gravity vector, $L=[0,I_m]$, $q(\cdot)=[q_1(\cdot),\ldots,q_N(\cdot)]\in\mathbb{R}^N$ is the generalised coordinate vector, and $\omega\in\mathbb{R}^m$ is the control. Assume that $x(n)=[q(n),q(n+1)]^T$, Eq. (A.6) can be rewritten as

$$\begin{aligned}x(n+1)&=\begin{bmatrix}q(n+1)\\q(n+2)\end{bmatrix}\\&=\begin{bmatrix}q(n+1)\\\gamma_0q(n)+\gamma_1q(n+1)\end{bmatrix}+\begin{bmatrix}0\\B^{-1}\big(q(n)\big)\big[L\omega-G\big(q(n)\big)\big]\end{bmatrix}\\&=\begin{bmatrix}0&I\\\gamma_0&\gamma_1\end{bmatrix}x(n)+a(n)\\&=H(n)x(n)+a(n),\end{aligned} \tag{A.7}$$

where

$$\gamma_0=B^{-1}\big(q(n)\big)C(q(n),q(n+1)-q(n))-I,$$

$$\gamma_1=2I-B^{-1}\big(q(n)\big)C(q(n),q(n+1)-q(n)),$$

and

$$H(n)=\begin{bmatrix}0&I\\\gamma_0&\gamma_1\end{bmatrix},\quad a(n)=\begin{bmatrix}0\\B^{-1}\big(q(n)\big)\big[L\omega-G\big(q(n)\big)\big]\end{bmatrix}.$$

Theorem A.6 *Let $s<n$, the unique solution* (A.7) *with the initial value $x(s)=x_0$ can be given as*

$$x(n)=\left(\prod_{j=s}^{n-1}H(j)\right)x_0+\sum_{j=s}^{n-2}\left[\prod_{i=1}^{n-(j+1)}H(n-i)\right]a(j)+a(n-1).$$

Proof The proof process is similar to the proof process of Theorem A.5, we will not repeat it here. □

Lemma A.4 *Let $\tau\in\mathbb{Z}^+$,*

$$\begin{aligned}&\big\|B^{-1}(q)C(q,p)-I\big\|+\big\|2I-B^{-1}(q)C(q,p)\big\|<\tfrac{\eta-\lambda}{1+\eta}-\tfrac{1}{N}<1-\tfrac{1}{N},\\&\|B^{-1}(q)\big[L\omega-G(q)\big]+B^{-1}(p)\big[L\omega-G(p)\big]\|\le\lambda\|p+q\|,\\&\|H(n)a(n)+H(s)a(s)\|\le\max\{\|H(n)\|,\|H(s)\|\}\|a(n)+a(s)\|\end{aligned}$$

for any $p, q \in \mathbb{R}^N$, $n, s \in \mathbb{Z}$ *and* $0 < \lambda < \eta < 1$. *Then,*

$$\|a(n+\tau) + a(n)\| \leq \lambda \|q(n+\tau) + q(n)\|,$$
$$\|q(n+1+\tau) + q(n+1)\| < \eta \|q(n+\tau) + q(n)\|,$$
$$\|H(n+\tau)a(n+\tau) + H(n)a(n)\| \leq \max\{\|H(n+\tau)\|, \|H(n)\|\}\|a(n+\tau) + a(n)\|.$$

Proof Since

$$a(n) = \begin{bmatrix} 0 \\ B^{-1}\big(q(n)\big)\big[L\omega - G\big(q(n)\big)\big] \end{bmatrix},$$

we have

$$\begin{aligned} \|a(n+\tau) + a(n)\| &= \big\| B^{-1}\big(q(n+\tau)\big)\big[L\omega - G\big(q(n+\tau)\big)\big] \\ &\quad + B^{-1}\big(q(n)\big)\big[L\omega - G\big(q(n)\big)\big]\big\| \\ &\leq \lambda \big\|q(n+\tau) + q(n)\big\| \end{aligned}$$

and

$$\begin{aligned} &\|x(n+1+\tau) + x(n+1)\| \\ &= \|q(n+1+\tau) + q(n+1)\| + \|q(n+2+\tau) + q(n+2)\| \\ &= \|H(n)x(n+\tau) + a(n+\tau) + H(n)x(n) + a(n)\| \\ &\leq \|H(n)\|\|x(n+\tau) + x(n)\| + \|a(n+\tau) + a(n)\| \\ &\leq \|H(n)\|\big(\|q(n+\tau) + q(n)\| + \|q(n+1+\tau) + q(n+1)\|\big) + \lambda\|q(n+\tau) + q(n)\|\big), \end{aligned}$$

i.e.,

$$\begin{aligned} &(1 - \|H(n)\|)\|q(n+1+\tau) + q(n+1)\| + \|q(n+2+\tau) + q(n+2)\| \\ &\leq (\|H(n)\| + \lambda)\|q(n+\tau) + q(n)\|. \end{aligned}$$

Hence,

$$\|q(n+1+\tau) + q(n+1)\| < \frac{\|H(n)\| + \lambda}{1 - \|H(n)\|}\|q(n+\tau) + q(n)\|$$

and

$$\|q(n+2+\tau) + q(n+2)\| < (\|H(n)\| + \lambda\|B^{-1}\|)\|q(n+\tau) + q(n)\|.$$

On the other hand, since

$$H(n) = \begin{bmatrix} 0 & I \\ \gamma_0 & \gamma_1 \end{bmatrix},$$

where

$$\gamma_0 = B^{-1}\big(q(n)\big)C(q(n), q(n+1) - q(n)) - I,$$

$$\gamma_1 = 2I - B^{-1}\big(q(n)\big)C(q(n), q(n+1) - q(n)),$$

we have

$$\begin{aligned}\|H(n)\| &= \tfrac{1}{N^2}\big[N + \|B^{-1}(q(n))C(q(n), q(n+1)-q(n)) - I\| \\ &\quad + \|2I - B^{-1}(q(n))C(q(n), q(n+1)-q(n))\|\big] \\ &\leq \tfrac{1}{N} + \tfrac{\eta-\lambda}{1+\eta} - \tfrac{1}{N} = \tfrac{\eta-\lambda}{1+\eta} < 1,\end{aligned}$$

i.e., $\frac{\|H(n)\|+\lambda}{1-\|H(n)\|} < \eta$. Thus, $\|q(n+1+\tau) + q(n+1)\| < \eta\|q(n+\tau) + q(n)\|$. Since

$$\|H(n)a(n) + H(s)a(s)\| \leq \max\{\|H(n)\|, \|H(s)\|\}\|a(n) + a(s)\|,$$

we have

$$\|H(n+\tau)a(n+\tau) + H(n)a(n)\| \leq \max\{\|H(n+\tau)\|, \|H(n)\|\}\|a(n+\tau) + a(n)\|.$$

The proof is completed. □

A.2.2 Stability of the Almost Anti-periodic Solutions

In this section, we will establish the stability of the general N-dimensional mechanical system and the underactuated Euler–Lagrange system under the almost anti-periodic discrete process. Through Eqs. (A.5) and (A.7), the systems (A.4) and (A.6) can be turned into the following nonhomogeneous linear system

$$x(n+1) = M(n)x(n) + c(n), \quad n \in \mathbb{Z}, \tag{A.8}$$

which leads to the corresponding homogeneous linear system

$$x(n+1) = M(n)x(n), \quad n \in \mathbb{Z}, \tag{A.9}$$

where

$$M(n) = R = \begin{bmatrix} 0 & I \\ B^{-1}(C-B) & B^{-1}(2B-C) \end{bmatrix}$$

and

$$c(n) = b(n) = \begin{bmatrix} 0 \\ B^{-1}f(n) - B^{-1}S(q(n)) \end{bmatrix}$$

in Eq. (A.5),

$$M(n) = H(n) = \begin{bmatrix} 0 & I \\ \gamma_0 & \gamma_1 \end{bmatrix},$$

$$c(n) = a(n) = \begin{bmatrix} 0 \\ B^{-1}(q(n))[L\omega - G(q(n))] \end{bmatrix}$$

in Eq. (A.7), where

$$\gamma_0 = B^{-1}(q(n))C(q(n), q(n+1)-q(n)) - I,$$

$$\gamma_1 = 2I - B^{-1}\big(q(n)\big)C(q(n), q(n+1) - q(n)).$$

Next, we will establish the stability of Eq. (A.8) to obtain the stability of Eqs. (A.4) and (A.6). For convenience, denote $M_* := \inf\{\|M(n)\| : n \in \mathbb{Z}\}$, $M^* := \sup\{\|M(n)\| : n \in \mathbb{Z}\}$.

Theorem A.7 ([33]) *Let $n, s \in \mathbb{Z}$,*

$$\Psi(n,s) = \begin{cases} \prod_{j=1}^{n-s} M(n-j), & n > s, \\ I, & n = s, \\ 0, & n < s. \end{cases}$$

Then, the following results hold.

(i) The zero solution of Eq. (A.9) is stable if and only if

$$\big\|\Psi(n,s)\big\| \le K(s),$$

where $K(s)$ is a positive constant dependent on s.

(ii) The zero solution of Eq. (A.9) is uniformly stable if and only if

$$\big\|\Psi(n,s)\big\| \le K,$$

where K is a positive constant independent of s.

(iii) The zero solution of Eq. (A.9) is asymptotically stable if and only if

$$\lim_{n\to\infty} \big\|\Psi(n,s)\big\| = 0.$$

(iv) The zero solution of Eq. (A.9) is uniformly asymptotically stable if and only if

$$\big\|\Psi(n,s)\big\| \le K\eta^{n-s}$$

for some constants $K > 0$ and $\eta \in (0,1)$, $n \ge s$.

Theorem A.8 *If all the conditions of Lemma A.3 hold for $M(n) = R$ and $c(n) = b(n)$ (or all the conditions of Lemma A.4 hold for $M(n) = H(n)$ and $c(n) = a(n)$) and*

$$\lambda_0 = \lim_{n\to\infty} \left[\prod_{j=1}^{n} \|M(j)\|\right]^{\frac{1}{n}} < 1. \tag{A.10}$$

Then, Eq. (A.8) *has a unique uniformly asymptotically stable almost anti-periodic solution*

$$x_0(n) = \sum_{j=-\infty}^{n-2} \left[\prod_{i=1}^{n-j-1} M(n-i) \right] c(j) + c(n-1) \quad n \in \mathbb{Z}.$$

Moreover, Eq. (A.4) *(or Eq.* (A.6)*) has an unique uniformly asymptotically stable almost anti-periodic solution.*

Proof Step 1. We will prove that $x_0(n)$ is well-defined for all $n \in \mathbb{Z}$. By Proposition A.2 (ii) and Eq. (A.10), we can obtain

$$\lim_{n\to\infty} \frac{1}{n} \left[\sum_{j=-n}^{-1} \ln \|M(j)\| - \sum_{j=1}^{n} \ln \|M(j)\| \right] = 0,$$

i.e.,

$$\lim_{n\to\infty} \frac{1}{n} \left[\sum_{j=-n}^{-1} \ln \|M(j)\| \right] = \ln \lambda_0.$$

Hence,

$$\lim_{n\to\infty} \left[\prod_{j=-n}^{-1} \|M(j)\| \right]^{\frac{1}{n}} = \lambda_0,$$

which implies that for $\forall \varepsilon > 0$, there exists $N_2 > 0$ such that

$$\left| \left[\prod_{j=-n}^{-1} \|M(j)\| \right]^{\frac{1}{n}} - \lambda_0 \right| < \varepsilon$$

for $n > N_2$, i.e.,

$$\lambda_0 - \varepsilon < \left[\prod_{j=-n}^{-1} \|M(j)\| \right]^{\frac{1}{n}} < \lambda_0 + \varepsilon,$$

i.e.,

$$(\lambda_0 - \varepsilon)^n < \prod_{j=-n}^{-1} \|M(j)\| < (\lambda_0 + \varepsilon)^n,$$

Hence, there exist some $\lambda_1 \in (\lambda_0, 1)$ and $\lambda_2 \in (0, \lambda_0)$ such that

$$\lambda_2^n < \prod_{j=-n}^{-1} M_* \le \prod_{j=-n}^{-1} \|M(j)\| \le \prod_{j=-n}^{-1} M^* < \lambda_1^n \text{ for } n > N_2. \tag{A.11}$$

Hence,

$$\begin{aligned}
&\left\| \sum_{j<-N_2}^{n-2} \left[\prod_{i=1}^{n-j-1} M(n-i) \right] c(j) \right\| \\
&\leq \sum_{j<-N_2} \left\| \left[\prod_{i=1}^{n-j-1} M(n-i) \right] \right\| \|c(j)\| \leq \sum_{j<-N_2} \left[\prod_{i=1}^{n-j-1} \|M(n-i)\| \right] \|c(j)\| \\
&= \sum_{j<-N_2} \left[\prod_{i=1}^{n} \|M(n-i)\| \prod_{i=n+1}^{n-j-1} \|M(n-i)\| \right] \|c(j)\| \\
&= \sum_{j<-N_2} \left[\prod_{r=n-1}^{0} \|M(r)\| \prod_{i=-1}^{j+1} \|M(i)\| \right] \|c(j)\| \\
&= \prod_{r=n-1}^{0} \|M(r)\| \sum_{j<-N_2} \left[\prod_{i=j+1}^{-1} \|M(i)\| \right] \|c(j)\|.
\end{aligned}$$

By Eq. (A.11), one has

$$\begin{aligned}
\prod_{r=n-1}^{0} \|M(r)\| \sum_{j<-N_2} \lambda_2^{-j+1} \|c(j)\| &\leq \prod_{r=n-1}^{0} \|M(r)\| \sum_{j<-N_2} \left[\prod_{i=j+1}^{-1} \|M(i)\| \right] \|c(j)\| \\
&\leq \prod_{r=n-1}^{0} \|M(r)\| \sum_{j<-N_2} \lambda_1^{-j+1} \|c(j)\|.
\end{aligned}$$

Hence, the series $\prod_{r=n-1}^{0} \|M(r)\| \sum_{j<-N_2} \left[\prod_{i=j+1}^{-1} \|M(i)\| \right] \|c(j)\|$ converges, which implies that the following series converges:

$$\sum_{j=-\infty}^{n-2} \left[\prod_{i=1}^{n-j-1} M(n-i) \right] c(j).$$

Hence, for

$$x_0(n) = \sum_{j=-\infty}^{n-2} \left[\prod_{i=1}^{n-j-1} M(n-i) \right] c(j) + c(n-1),$$

this series converges for $n \in \mathbb{Z}$. Therefore, we prove that it is well-defined.

Step 2. We will prove that $x_0(n)$ is a solution of Eq. (A.8). Since

$$\begin{aligned}
x_0(n+1) &= \sum_{j=-\infty}^{n-1} \left[\prod_{i=1}^{n-j} M(n+1-i) \right] c(j) + c(n) \\
&= \sum_{j=-\infty}^{n-2} \left[\prod_{i=1}^{n-j} M(n+1-i) \right] c(j) + c(n) + \left[\prod_{i=1}^{n-(n-1)} M(n+1-i) \right] c(n-1) \\
&= \sum_{j=-\infty}^{n-2} \left[M(n) \prod_{i=2}^{n-j} M(n+1-i) \right] c(j) + c(n) + M(n)c(n-1)
\end{aligned}$$

$$= M(n) \sum_{j=-\infty}^{n-2} \left[\prod_{i=1}^{n-j-1} M(n-i) \right] c(j) + M(n)c(n-1) + c(n)$$

$$= M(n) \left(\sum_{j=-\infty}^{n-2} \left[\prod_{i=1}^{n-j-1} M(n-i) \right] c(j) + c(n-1) \right) + c(n)$$

$$= M(n)x_0(n) + c(n),$$

which means that $x_0(n)$ is a solution of Eq. (A.8).

Step 3. We will prove that the solution of Eq. (A.8) is an almost anti-periodic discrete process. By Step 1, we have

$$\lambda_2^n < \prod_{j=-n}^{-1} M_* \leq \prod_{j=-n}^{-1} \|M(j)\| \leq \prod_{j=-n}^{-1} M^* < \lambda_1^n$$

for $n > N_2$. Hence, there exists $N_3 > N_2$ such that

$$\left\| \sum_{j=-\infty}^{-N_3} \left[\prod_{i=1}^{n-j-1} M(n-i) \right] c(j) \right\| < \frac{\varepsilon}{4}.$$

Thus,

$$\|x_0(n+\tau) + x_0(n)\| = \|q(n+\tau+1) + q(n+1)\| + \|q(n+\tau) + q(n)\|$$

$$= \left\| \sum_{j=-\infty}^{n+\tau-2} \left[\prod_{i=1}^{n+\tau-j-1} M(n+\tau-i) \right] c(j) + c(n+\tau-1) \right.$$

$$\left. + \sum_{j=-\infty}^{n-2} \left[\prod_{i=1}^{n-j-1} M(n-i) \right] c(j) + c(n-1) \right\|$$

$$\leq \left\| \sum_{j=-N_3+\tau}^{n+\tau-2} \left[\prod_{i=1}^{n+\tau-j-1} M(n+\tau-i) \right] c(j) + c(n+\tau-1) \right.$$

$$\left. + \sum_{j=-N_3}^{n-2} \left[\prod_{i=1}^{n-j-1} M(n-i) \right] c(j) + c(n-1) \right\|$$

$$+ \left\| \sum_{j=-\infty}^{-N_3+\tau-1} \left[\prod_{i=1}^{n+\tau-j-1} M(n+\tau-i) \right] c(j) \right\| + \left\| \sum_{j=-\infty}^{-N_3-1} \left[\prod_{i=1}^{n+\tau-j-1} M(n-i) \right] c(j) \right\|$$

$$\leq \left\| \sum_{j=-N_3+\tau}^{n+\tau-2} \left[\prod_{i=1}^{n+\tau-j-1} M(n+\tau-i) \right] c(j) + \sum_{j=-N_3}^{n-2} \left[\prod_{i=1}^{n-j-1} M(n-i) \right] c(j) \right\|$$

$$+\|c(n-1)+c(n+\tau-1)\|+\frac{\varepsilon}{2}\leq\sum_{j=-N_3}^{n-2}\left\|\left[\prod_{i=1}^{n-j-1}M(n+\tau-i)\right]c(j+\tau)\right.$$

$$\left.+\left[\prod_{i=1}^{n-j-1}M(n-i)\right]c(j)\right\|+\|c(n-1)+c(n+\tau-1)\|+\frac{\varepsilon}{2}$$

$$\leq\sum_{j=-N_3}^{n-2}\lambda_1^{n-j-1}\|c(j+\tau)+c(j)\|+\|c(n-1)+c(n+\tau-1)\|+\frac{\varepsilon}{2}.$$

Hence, by Lemma A.3 (or Lemma A.4), we have

$$\|x_0(n+\tau)+x_0(n)\|=\|q(n+\tau+1)+q(n+1)\|+\|q(n+\tau)+q(n)\|$$

$$\leq\sum_{j=-N_3}^{n-2}\lambda_1^{n-j-1}\upsilon\|q(j+\tau)+q(j)\|+\upsilon\|q(n-1)+q(n+\tau-1)\|+\frac{\varepsilon}{2}$$

$$\leq\sum_{j=-N_3}^{n-2}\lambda_1^{n-j-1}\upsilon\eta^{N_3+j}\|q(-N_3+\tau)+q(-N_3)\|$$

$$+\upsilon\eta^{N_3-(n-1)}\|q(-N_3+\tau)+q(-N_3)\|+\frac{\varepsilon}{2}$$

$$=\|q(-N_3+\tau)+q(-N_3)\|\left(\sum_{j=-N_3}^{n-2}\lambda_1^{n-j-1}\upsilon\eta^{N_3+j}+\upsilon\eta^{N_3-(n-1)}\right)+\frac{\varepsilon}{2}$$

$$=\upsilon\|q(-N_3+\tau)+q(-N_3)\|\left(\sum_{j=-N_3}^{n-2}\lambda_1^{n-j-1}\eta^{N_3+j}+\eta^{N_3-(n-1)}\right)+\frac{\varepsilon}{2}$$

$$=\upsilon\|q(-N_3+\tau)+q(-N_3)\|\frac{\lambda_1^{N_3+n-2}-\eta^{N_3+n-2}}{1-\frac{\eta}{\lambda_1}}+\frac{\varepsilon}{2}$$

for $\upsilon, \eta, \lambda_1 \in (0, 1)$ and

$$\lim_{N_3\to\infty}\frac{\lambda_1^{N_3+n-2}-\eta^{N_3+n-2}}{1-\frac{\eta}{\lambda_1}}=0,$$

which means

$$\|q(n+\tau+1)+q(n+1)\|+\|q(n+\tau)+q(n)\|<\varepsilon.$$

Therefore, we obtain that the solution of Eq. (A.8) is an almost anti-periodic discrete process.

Step 4. We will prove that $x_0(n)$ is an unique uniformly asymptotically stable almost anti-periodic solution. By Step 1 and Eq. (A.10), we have

$$\lambda_2^n < \prod_{j=1}^{n} M_* \le \prod_{j=1}^{n} \|M(j) \le \prod_{j=1}^{n} M^* < \lambda_1^n$$

for some $\lambda_1 \in (\lambda_0, 1)$, $\lambda_2 \in (0, \lambda_0)$ and $n > N_0$. Hence, for $n > s$, we have

$$\|\Psi(n, s)\| = \left\| \prod_{j=1}^{n-s} M(n-j) \right\| < \lambda_1^n.$$

Thus, by Theorem A.7, $x_0(n)$ is an unique uniformly asymptotically stable almost anti-periodic solution. The proof is completed. □

Theorem A.9 *If*

$$\liminf_{n\to\infty} \left\| \left[\prod_{j=0}^{n-1} M(n-j) \right]^{-1} \right\|^{-\frac{1}{n}} > 1, \tag{A.12}$$

then the solution of Eq. (A.10) is unstable.

Proof Since

$$\left[\prod_{j=0}^{n-1} M(n-j) \right] \left[\prod_{j=0}^{n-1} M(n-j) \right]^{-1} = I,$$

we have

$$\left\| \prod_{j=0}^{n-1} M(n-j) \right\| \left\| \left[\prod_{j=0}^{n-1} M(n-j) \right]^{-1} \right\| \ge \|I\| = 1.$$

Hence,

$$\begin{aligned} \left\| \prod_{j=0}^{n-1} M(n-j) \right\| &\ge \frac{1}{\left\| \left[\prod_{j=0}^{n-1} M(n-j) \right]^{-1} \right\|} = \frac{1}{\left(\left\| \left[\prod_{j=0}^{n-1} M(n-j) \right]^{-1} \right\|^{\frac{1}{n}} \right)^n} \\ &= \left(\left\| \left[\prod_{j=0}^{n-1} M(n-j) \right]^{-1} \right\|^{\frac{1}{-n}} \right)^n. \end{aligned}$$

On the other hand, by Eq. (A.12), there exists some $\lambda_3 > 1$ such that

$$\left\| \left[\prod_{j=0}^{n-1} M(n-j) \right]^{-1} \right\|^{-\frac{1}{n}} > \lambda_3,$$

which means

$$\left\| \prod_{j=0}^{n-1} M(n-j) \right\| \ge \lambda_3^n.$$

Next, by contradiction, we will show the solution of Eq. (A.10) is unstable. Assume that the solution of Eq. (A.10) is stable, by Theorem A.7, there exists $K(s)$ such that

$$\|\Psi(n,s)\| = \left\|\prod_{j=1}^{n-1} M(n-j)\right\| \leq K(s),$$

it is a contradiction with

$$\lim_{n\to\infty}\left\|\prod_{j=0}^{n-1} M(n-j)\right\| \geq \lim_{n\to\infty} \lambda_3^n$$

for $\lambda_3 > 1$. Hence, the solution of Eq. (A.10) is unstable. The proof is completed. □

In what follows, we will provide two examples to support our obtained results of Theorem A.8.

Application A.2.1 In Eq. (A.4), let $s = 1$, $x(s) = [q(1), q(2)]^T$

$$B = C = \begin{bmatrix} 2 & 0 \\ 0 & 2 \end{bmatrix}, \quad S(q(n)) = 2q(n), \quad f(n) = \begin{bmatrix} \sin\frac{n\pi}{4} \\ \cos\frac{n\pi}{2} \end{bmatrix} \quad q(1) = \begin{bmatrix} 1 \\ \frac{1}{2} \end{bmatrix} \quad q(2) = \begin{bmatrix} -1 \\ 0 \end{bmatrix}.$$

In fact,

$$B^{-1} = \begin{bmatrix} \frac{1}{2} & 0 \\ 0 & \frac{1}{2} \end{bmatrix}, \quad B^{-1}f(n) - B^{-1}S(q(n)) = \begin{bmatrix} \frac{1}{2}\sin\frac{n\pi}{4} \\ \frac{1}{2}\cos\frac{n\pi}{2} \end{bmatrix} - q(n), \quad R = \begin{bmatrix} 0 & I \\ 0 & I \end{bmatrix}, \quad R^k = R.$$

Thus,

$$x(n) = [q(n), q(n+1)]^T = Rx(s) + R\sum_{j=n-1-s}^{0} b(n-1-j),$$

and

$$q(n) = \begin{bmatrix} q_1(n) \\ q_2(n) \end{bmatrix} = \begin{bmatrix} -1 + \sum\limits_{j=n-1-s}^{0} \frac{1}{2}\sin\frac{(n-1-j)\pi}{4} \\ \sum\limits_{j=n-1-s}^{0} \frac{1}{2}\cos\frac{(n-1-j)\pi}{2} \end{bmatrix}.$$

For the status of $q_1(n)$ and $q_2(n)$, see Fig. A.1. □

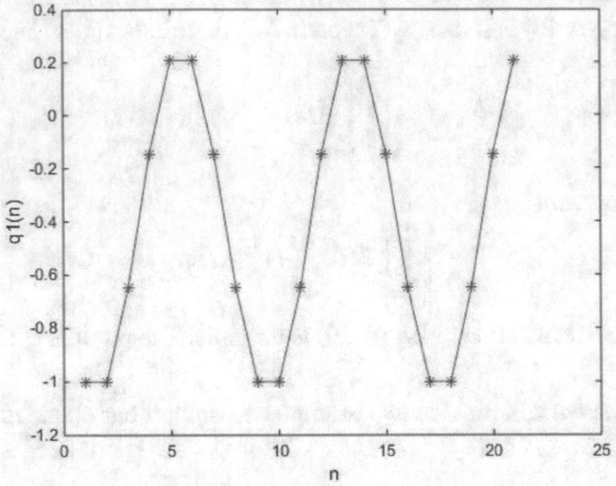

(a) The status of the almost anti−periodic solution of the system (A.4).

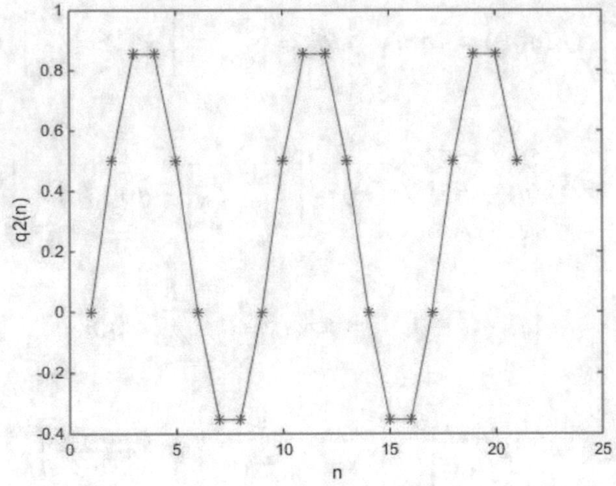

(b) The status of the almost anti−periodic solution of the system (A.4).

Fig. A.1 The status of the almost anti-periodic solution

Application A.2.2 In Eq. (A.6), let $s = 1$, $q(n) = [q_1(n), q_2(n)]^T$, $q(1) = [1, 1]^T$, $q(2) = [5, 2]^T$, $\omega = [0, 1]^T$

$$\begin{aligned} B(q(n)) =& C\big(q(n), q(n+1) - q(n)\big) \\ =& [2q(n), q(n)] + \begin{bmatrix} 1 & 0 \\ 0 & 0 \end{bmatrix} \quad G(q(n)) = \begin{bmatrix} -q_2(n)\cos\frac{n}{4\pi} \\ 1 - q_2 \sin\frac{n}{8\pi} \end{bmatrix}. \end{aligned}$$

In fact,

$$H(n) = H^k(n) = \begin{bmatrix} 0 & I \\ 0 & I \end{bmatrix}, \quad L\omega = \begin{bmatrix} 0 \\ \frac{1}{2} \end{bmatrix},$$

$$B^{-1}(q(n)) = \begin{bmatrix} 1 & \frac{-q_1(n)}{q_2(n)} \\ -\frac{1}{2} & \frac{\sin(q_1(n)) - q_1(n)\sin\frac{n}{8\pi}}{q_2(n)\sin\frac{n}{8\pi}} \end{bmatrix}, \quad L\omega - G(q(n)) = \begin{bmatrix} q_2(n)\cos\frac{n}{4\pi} \\ q_2\sin\frac{n}{8\pi} \end{bmatrix},$$

$$B^{-1}(q(n))\big(L\omega - G(q(n))\big) = \begin{bmatrix} q_2(n)\cos\frac{n}{4\pi} - q_1(n)\sin\frac{n}{8\pi} \\ -\frac{q_2(n)}{2}\cos\frac{n}{4\pi} + \sin(q_1(n)) - q_1(n)\sin\frac{n}{8\pi} \end{bmatrix}.$$

Hence,

$$\begin{aligned} x(n) &= \Big(\prod_{j=s}^{n-1} H(j)\Big)x_0 + \sum_{j=s}^{n-2}\Big[\prod_{i=1}^{n-(j+1)} H(n-i)\Big]a(j) + a(n-1) \\ &= H(n-1)x_0 + \sum_{j=s}^{n-2} H(n-i)a(j) + a(n-1) \end{aligned}$$

and

$$\begin{aligned} q(n+1) &= q(2) + \sum_{j=1}^{n-2}\begin{bmatrix} q_2(j)\cos\frac{j}{4\pi} - q_1(j)\sin\frac{j}{8\pi} \\ -\frac{q_2(j)}{2}\cos\frac{j}{4\pi} + \sin(q_1(j)) - q_1(j)\sin\frac{j}{8\pi} \end{bmatrix} \\ &+ \begin{bmatrix} q_2(n-1)\cos\frac{n-1}{4\pi} - q_1(n-1)\sin\frac{n-1}{8\pi} \\ -\frac{q_2(n-1)}{2}\cos\frac{n-1}{4\pi} + \sin(q_1(n)) - q_1(n-1)\sin\frac{n-1}{8\pi} \end{bmatrix} \\ &= q(n) + \begin{bmatrix} q_2(n-1)\cos\frac{n-1}{4\pi} - q_1(n-1)\sin\frac{n-1}{8\pi} \\ -\frac{q_2(n-1)}{2}\cos\frac{n-1}{4\pi} + \sin(q_1(n)) - q_1(n-1)\sin\frac{n-1}{8\pi} \end{bmatrix}. \end{aligned}$$

For the status of $q_1(n)$ and $q_2(n)$, see Fig. A.2. □

A.3 Almost Anti-periodic Functions on Time Scales

Anti-periodic phenomena always appear in various fields related to engineering science and technology. Almost anti-periodic phenomena is a natural extension of anti-periodic phenomena whose dynamical behavior can be reflected by almost anti-periodic process.

As an important tool of description of discrete process with almost anti-periodic circulation in the fields of engineering, biological chemistry, and neural computation, etc., a

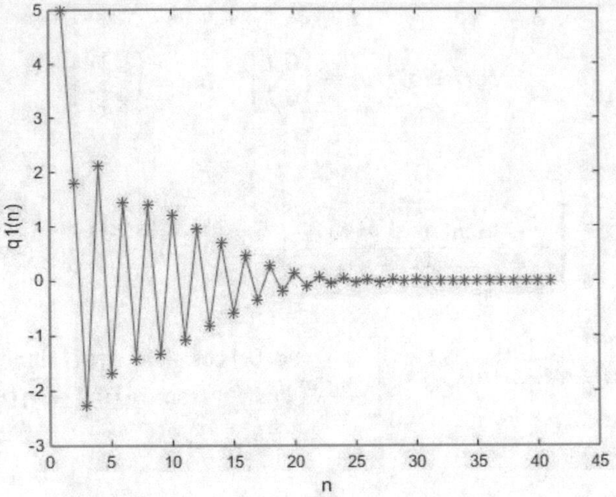

(a) The status of the almost anti−periodic solution of the system (A.6).

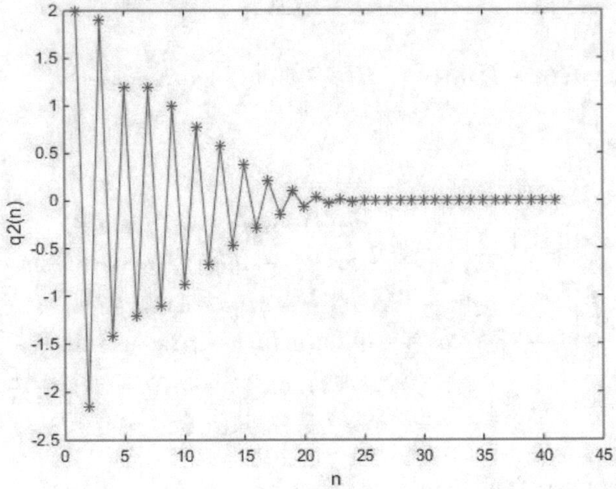

(b) The status of the almost anti−periodic solution of the system (A.6).

Fig. A.2 status of the almost anti−periodic solution

notion of discrete almost anti-periodic process has been proposed in this paper. Indeed, for an arbitrary almost anti-periodic sequence, the relatively dense set can also be applied to depict their accurate definitions. In the literature [105], Wang et al. addressed some basic notions to introduce a series of notions of almost periodic process on different types of time scales; the most convenient and precise statement is to adopt an intersection of intervals and subsets of real lines to portray the relative density distributed on time scales. The set with such an intersection property is called a relatively dense set. Naturally, it can be also applied to precisely describe an almost anti-periodic process on time scales.

The relatively dense set of the time scale version is presented as follows:

Assume that $\mathscr{T}$ is a time scale with periodicity, i.e., there is a nonempty set $\Omega \subset \mathbb{R}\backslash\{0\}$ such that $t + \gamma \in \mathscr{T}$ for all $t \in \mathscr{T}$ and $\gamma \in \Omega$. Now, a standard notion of relative density of a subset of $\mathbb{R}$ could be introduced.

A subset S of $\mathbb{R}$ is called a relatively dense set if and only if there is a positive number $\mathscr{L}$ such that $[b, b + \mathscr{L}]_{\mathscr{T}} \cap S \neq \emptyset$ for each $b \in \mathscr{T}$.

By using this basic notion, we can also address the notion of an almost anti-periodic process on time scales.

Let $\mathscr{M}(\cdot) = [\mu_{ij}(\cdot)]_{N\times N} : \mathscr{T} \to \mathbb{R}^{N\times N}$ be a $N \times N$ matrix-valued discrete function and $\omega(\cdot) = [\omega_1(\cdot), \ldots, \omega_N(\cdot)]^T : \mathscr{T} \to \mathbb{R}^N$ be a N-dimensional vector-valued function; then, we define

$$\|\mathscr{M}(t)\| = \frac{1}{N^2}\sum_{j=1}^{N}\sum_{i=1}^{N}|\mu_{ij}(t)|, \quad \|\omega(t)\| = \frac{1}{N}\sum_{j=1}^{N}|\omega_j(t)|.$$

If for $\forall\varepsilon > 0$, there exists a positive integer $l(\varepsilon)$ such that

$$\mathscr{E}(\varepsilon, \mu) := \big\{\tau \in \Omega : \|\omega(t + \tau) + \omega(t)\| < \varepsilon\big\}$$

is relatively dense in Ω. Then, $\omega(t)$ is called the almost anti-periodic function on time scales, and $\mathscr{E}(\varepsilon, \omega)$ is called the ε-almost anti-period of $\omega(t)$. Similarly, $\mathscr{M}(t)$ is an almost anti-periodic $N \times N$ matrix-valued function if

$$\mathscr{E}(\varepsilon, \mathscr{M}) = \big\{\tau \in \Omega : \|\mathscr{M}(t + \tau) + \mathscr{M}(t)\| < \varepsilon\big\}$$

is relatively dense in Ω.

The results of this appendix provide a new avenue to study almost the anti-periodic process on a discrete time case. Meanwhile, the definition mode of function adopted in the appendix could be extended to combine continuous cases and other complex time scales. Their comprehensive definitions and properties based on more complicated time situations can be applied to fuzzy dynamic equations and models in related research field.

References

1. Adívar, M., Raffoul, Y.N.: Stability and periodicity in dynamic delay equations. Comput. Math. Appl. **58**, 264–272 (2009)
2. Adívar, M.: A new periodic concept for time scales. Math. Slovaca **63**, 817–828 (2013)
3. Agarwal, R.P., O'Regan, D., Lakshmikantham, V.: Viability theory and fuzzy differential equations. Fuzzy Sets Syst. **151**, 563–580 (2005)
4. Agarwal, R.P., O'Regan, D., Saker, S.H.: Dynamic Inequalities on Time Scales. Springer International Publishing, Switzerland (2014)
5. Agarwal, R.P., O'Regan, D.: Some comments and notes on almost periodic functions and changing-periodic time scales. Electr. J. Math. Anal. Appl. **6**, 125–136 (2018)
6. Akhmet, M.U., Turan, M.: The differential equations on time scales through impulsive differential equations. Nonlinear Anal.: Theory Methods & Appl. **65**, 2043–2060 (2006)
7. Allahviranloo, T., Gholami, S.: Note on "Generalized Hukuhara differentiability of interval-valued functions and interval differential equations". J. Fuzzy Set Valued Anal. Article ID jfsva-00135, 2012, 1–4 (2012)
8. Atici, F.M., Biles, D.C.: First order dynamic inclusions on time scales. J. Math. Anal. Appl. **292**, 222–237 (2004)
9. Bede, B., Gal, S.G.: Almost periodic fuzzy-number-valued functions. Fuzzy Sets Syst. **147**, 385–403 (2004)
10. Bede, B., Gal, S.G.: Generalizations of the differentiability of fuzzy number valued functions with applications to fuzzy differential equations. Fuzzy Sets Syst. **151**, 581–599 (2005)
11. Bede, B.: A note on "two-point boundary value problems associated with non-linear fuzzy differential equations." Fuzzy Sets Syst. **157**, 986–989 (2006)
12. Bede, B., Stefanini, L.: Generalized differentiability of fuzzy-valued functions. Fuzzy Sets Syst. **230**, 119–141 (2013)
13. Bede, B.: Mathematics of Fuzzy Sets and Fuzzy Logic, Studies in Fuzziness and Soft Computing, vol. 295. Springer, Heidelberg (2013)
14. Benvenuti, P., Mesiar, R.: Pseudo arithmetical operations as a basis for the general measure and integration theory. Inf. Sci. **160**, 1–11 (2004)

C. Wang and R. P. Agarwal, *Dynamic Equations and Almost Periodic Fuzzy Functions on Time Scales*, Synthesis Lectures on Mathematics & Statistics,
https://doi.org/10.1007/978-3-031-11236-2

15. de Berg, M., Cheong, O., Van Kreveld, M., Overmars, M.: Computational Geometry: Algorithms and Applications, 3rd edn. Springer, Berlin (2008)
16. Bergstrom, H.: Weak Convergence of Measures. Academic Press, New York (1982)
17. Biacino, L., Lettieri, A.: Equations with fuzzy numbers. Inf. Sci. **47**, 63–76 (1989)
18. Bouchon-Meunier, B., Kosheleva, O., Kreinovich, V., Nguyen, H.T.: Fuzzy numbers are the only fuzzy sets that keep invertible operations invertible. Fuzzy Sets Syst. **91**, 155–163 (1997)
19. Buckley, J.J., Jowers, L.J.: Monte Carlo Methods in Fuzzy Optimization. Studies in Fuzziness and Soft Computing. Springer, Berlin (2008)
20. Bohner, M., Peterson, A.: Dynamic Equations on Time Scales. Birkh*ä*user Boston Inc., Boston (2001)
21. Bohner, M., Chieochan, R.: The Beverton-Holt q-difference equation. J. Biological Dyna. **7**, 86–95 (2013)
22. Bohr, H.: Zur Theorie der fastperiodischen Funktionen, I. Acta Math. **45**, 29–127 (1925)
23. Bohr, H.: Zur Theorie der fastperiodischen Funktionen, II. Acta Math. **46**, 101–214 (1925)
24. Boukezzoula, R., Galichet, S., Foulloy, L.: Inverse arithmetic operators for fuzzy intervals. In: Proceedings of EUSFLAT 2007 Conference, Ostrawa (2007)
25. Breunung, T.: Existence of quasi-periodic reponse in quasi-periodically forced nonlinear mechanical systems. Nonlinear Dyn. **105**, 1977–2004 (2021)
26. Corduneanu, C.: Almost Periodic Functions, 2nd edn. Chelsea, New York (1989)
27. Diagana, T.: Weighted pseudo almost periodic functions and applications. C.R. Math. **343**, 643–646 (2006)
28. Diamond, P., Kloeden, P.: Metric Spaces of Fuzzy Sets. World Scientific, Singapore (1994)
29. Diamond, P.: Stability and periodicity in fuzzy differential equations. IEEE Trans. Fuzzy Syst. **8**, 583–590 (2000)
30. Diamond, P.: Position paper on analysis submissions to fuzzy sets and systems. Fuzzy Sets Syst. **117**, 323–326 (2001)
31. Diamond, P.: Theory and applications of fuzzy Volterra integral equations. IEEE Trans. Fuzzy Syst. **10**, 97–102 (2002)
32. Diamond, P.: Brief note on the variation of constants formula for fuzzy differential equations. Fuzzy Sets Syst. **129**, 65–71 (2002)
33. Elaydi, S.N.: An Introduction to Difference Equations. Springer, New York (1996)
34. Fard, O.S., Bidgoli, T.A.: Calculus of fuzzy functions on time scales (I). Soft Compt. **19**, 293–305 (2015)
35. Fard, O.S., Torres, D.F.M., Zadeh, M.R.: Hukuhara approach to the study of hybrid fuzzy systems on time scales. Appl. Anal. Discr. Math. **10**, 152–167 (2016)
36. Fink, M.: Almost Periodic Differential Equations. Springer, Berlin (1974)
37. Guseinov, G.S., Kaymaklan, B.: Basics of Riemann delta and nabla integration on time scale. J. Differ. Equa. Appl. **8**, 1001–1017 (2002)
38. Gritzmann, P., Klee, V.: Computational complexity of inner and outer j-radii of polytopes in finite-dimensional normed spaces. Math. Program. **59**, 163–213 (1993)
39. Halmos, P.R.: Finite-Dimensional Vector Spaces. Undergraduate Texts in Mathematics. Springer, New York (1958)
40. Hausdorff, F.: Set Theory. Chelsea, New York (1957)
41. Hilger, S.: Ein Mafikettenkalk*ü*l mit Anwendung auf Zentrumsmannigfaltigkeiten. Ph.D. thesis, Universit*ä*t W*ü*rzburg (1988)
42. Hong, S.: Stability criteria for set dynamic equations on time scales. Comput. Math. Appl. **59**, 3444–3457 (2010)
43. Hong, S., Peng, Y.: Almost periodicity of set-valued functions and set dynamic equations on time scales. Inform. Sci. **330**, 157–174 (2016)

44. Hukuhara, M.: Integration des applications measurables dont la valeur est un compact convexe. Funkcialaj Ekvacioj **10**, 205–223 (1967)
45. Huellermeier, E.: An approach to modelling and simulation of uncertain dynamical systems. Int. J. Uncertain. Fuzziness Knowl.-Based Syst. **5**, 117–137 (1997)
46. Kac, V., Cheung, P.: Quantum Calculus, Universitext. Springer, New York (2002)
47. Kaufmann, E.R., Raffoul, Y.N.: Periodic solutions for a neutral nonlinear dynamical equation on a time scale. J. Math. Anal. Appl. **319**, 315–325 (2006)
48. Klir, G.J.: Fuzzy arithmetic with requisite constraints. Fuzzy Sets Syst. **91**, 165–175 (1997)
49. Kostić, M.: Existence of generalized almost periodic and asymptotic almost periodic solutions to abstract Volterra integro-differential equations. Electr. J. Diff. Equ. **239**, 1–30 (2017)
50. Kostić, M.: Almost Periodic and Almost Automorphic Type Solutions of Abstract Volterra Integro-Differential Equations. W. de Gruyter, Berlin (2019)
51. Kostić, M., Pilipović, S., Velinov, D.: Quasi-asymptotically almost periodic vector-valued generalized functions. Sarajevo J. Math. **15**, 181–199 (2019)
52. Kostić, M.: Asymptotically almost periodic and asymptotically almost automorphic vectorvalued generalized functions. Buletinul Academiei de Ştiinţe a Republicii Moldova. Matematica **3**, 34–53 (2019)
53. Kostić, M., Velinov, D.: A notion on almost anti-periodic function in basic spaces. Kargujevac Math. J. **44**, 287–297 (2020)
54. König, S.: Computational aspects of the Hausdorff distance in unbounded dimension. J. Comput. Geom. **5**, 250–274 (2014)
55. Laksmikantham, V., Mohapatra, R.N.: Theory of Fuzzy Differential Equations and Inclusions. Taylor & Francis, New York (2003)
56. Lawson, C.L., Hanson, R.J.: Solving Least Squares Problems. Prentice-Hall, Englewood Cliffs (1974)
57. Li, Z., Wang, C., Agarwal, R.P.: The non-eigenvalue form of Liouville's formula and α-matrix exponential solutions for combined matrix dynamic equations on time scales. Mathematics **7**, 962 (2019). https://doi.org/10.3390/math7100962
58. Li, Z., Wang, C.: Cauchy matrix and Liouville formula of quaternion impulsive dynamic equations on time scales. Open Math. **18**, 353–377 (2020)
59. Li, Z., Wang, C., Agarwal, R.P., O'Regan, D.: Commutativity of quaternion-matrix-valued functions and quaternion matrix dynamic equations on time scales. Stud. Appl. Math. **146**, 139–210 (2021)
60. Li, Z., Wang, C., Agarwal, R.P., Sakthivel, R.: Hyers-ulam-rassias stability of quaternion multidimensional fuzzy nonlinear difference equations with impulses. Iranian J. Fuzzy Syst. **18**, 143–160 (2021)
61. Liang, J., Maniar, L., N'Guérékata, G.M., Xiao, T.: Existence and uniqueness of $C^{(n)}$-almost periodic solutions to some ordinary differential equations. Nonlinear Anal.: TMA **66**, 1899–1910 (2007)
62. Ma, M., Friedman, M., Kandel, A.: A new fuzzy arithmetic. Fuzzy Sets Syst. **108**, 83–90 (1999)
63. Mares, M.: Weak arithmetic of fuzzy numbers. Fuzzy Sets Syst. **91**, 143–153 (1997)
64. Moore, R.E., Kearfott, R.B., Cloud, M.J.: Introduction to Interval Analysis. Society for Industrial and Applied Mathematics, Philadelphia (2009)
65. N'Guérékata, G.M.: Almost Automorphic and Almost Periodic Functions in Abstract Spaces. Kluwer Academic, New York (2001)
66. Nieto, J.J., Rodríguez-López, R.: Some results on boundary value problems for fuzzy differential equations with functional dependence. Fuzzy Sets Syst. **230**, 92–118 (2013)
67. Park, J.Y., Jung, I.H., Lee, M.J.: Almost periodic solutions of fuzzy systems. Fuzzy Sets Syst. **119**, 367–373 (2001)

68. Qin, G., Wang, C.: Lebesgue-Stieltjes combined $\Diamond_\alpha$-measure and integral on time scales. RACSAM **115**(2), 50 (2021)
69. Rodríguez-López, R.: On the existence of solutions to periodic boundary value problems for fuzzy linear differential equations. Fuzzy Sets Syst. **219**, 1–26 (2013)
70. Sambandham, M.: Hybrid fuzzy systems on time scales. Dynam. Syst. Appl. **12**, 217–227 (2003)
71. Stamov, G.T., Stamova, I.M.: Almost periodic solutions for impulsive neural networks with delay. Appl. Math. Modeling **31**, 1263–1270 (2007)
72. Stefanini, L., Guerra, M.L.: On fuzzy arithmetic operations: some properties and distributive approximations. Int. J. Appl. Math. **19**, 171–199 (2006). An extended version is available at the RePEc repository. http://ideas.repec.org/f/pst233.html
73. Stefanini, L.: On the generalied LU-fuzzy derivative and fuzzy differential equations. In: Proceedings of the FUZZIEEE 2007 Conference, London (2007). RePEc repository http://ideas.repec.org/f/pst233.html
74. Stefanini, L., Bede, B.: Generalized Hukuhara differentiability of interval valued functions and interval differential equations. Nonlinear Anal. **71**, 1311–1328 (2009)
75. Stefanini, L.: A generalization of Hukuhara difference and division for interval and fuzzy arithmetic. Fuzzy Sets Syst. **161**, 1564–1584 (2010)
76. Wang, C.: Almost periodic solutions of impulsive BAM neural networks with variable delays on time scales. Commun. Nonlinear Sci. Numer. Simul. **19**, 2828–2842 (2014)
77. Wang, C., Agarwal, R.P.: A further study of almost periodic time scales with some notes and applications. Abstr. Appl. Anal. **2014**, Article ID 267384, 1–11 (2014)
78. Wang, C.: Existence and exponential stability of piecewise mean-square almost periodic solutions for impulsive stochastic Nicholson's blowflies model on time scales. Appl. Math. Comput. **248**, 101–112 (2014)
79. Wang, C., Agarwal, R.P.: Weighted piecewise pseudo almost automorphic functions with applications to abstract impulsive ∇-dynamic equations on time scales. Adv. Differ. Equa. **2014**(153), 1–29 (2014)
80. Wang, C., Agarwal, R.P.: Exponential dichotomies of impulsive dynamic systems with applications on time scales. Math. Meth. Appl. Sci. **38**, 3879–3900 (2015)
81. Wang, C., Agarwal, R.P.: Changing-periodic time scales and decomposition theorems of time scales with applications to functions with local almost periodicity and automorphy. Adv. Differ. Equa. **2015**(296), 1–21 (2015)
82. Wang, C., Agarwal, R.P.: Uniformly rd-piecewise almost periodic functions with applications to the analysis of impulsive Δ-dynamic system on time scales. Appl. Math. Comput. **259**, 271–292 (2015)
83. Wang, C., Agarwal, R.P.: Relatively dense sets, corrected uniformly almost periodic functions on time scales, and generalizations. Adv. Differ. Equ. **312**, 1–9 (2015)
84. Wang, C., Agarwal, R.P.: Almost periodic dynamics for impulsive delay neural networks of a general type on almost periodic time scales. Commun. Nonlinear Sci. Numer. Simul. **36**, 238–251 (2016)
85. Wang, C., Agarwal, R.P., O'Regan, D.: Compactness criteria and new impulsive functional dynamic equations on time scales. Adv. Differ. Equa. **197**, 1–41 (2016)
86. Wang, C., Agarwal, R.P., O'Regan, D.: Matrix measure on time scales and almost periodic analysis of the impulsive Lasota-Wazewska model with patch structure and forced perturbations. Math. Meth. Appl. Sci. **39**, 5651–5669 (2016)
87. Wang, C., Agarwal, R.P., O'Regan, D.: Π-semigroup for invariant under translations time scales and abstract weighted pseudo almost periodic functions with applications. Dyna. Syst. Appl. **25**, 1–28 (2016)

88. Wang, C., Agarwal, R.P.: A classification of time scales and analysis of the general delays on time scales with applications. Math. Meth. Appl. Sci. **39**, 1568–1590 (2016)
89. Wang, C., Agarwal, R.P., O' Regan, D.: Periodicity, almost periodicity for time scales and related functions. Nonauton. Dyn. Syst. **3**, 24–41 (2016)
90. Wang, C.: Piecewise pseudo almost periodic solution for impulsive non-autonomous highorder Hopfield neural networks with variable delays. Neurocomputing **171**, 1291–1301 (2016)
91. Wang, C., Agarwal, R.P., O'Regan, D.: Weighted piecewise pseudo double-almost periodic solution for impulsive evolution equations. J. Nonlinear Sci. Appl. **10**, 3863–3886 (2017)
92. Wang, C., Agarwal, R.P., O'Regan, D.: A matched space for time scales and applications to the study on functions. Adv. Differ. Equ. **2017**(305), 1–28 (2017)
93. Wang, C., Agarwal, R.P., O'Regan, D.: The shift invariance of time scales and applications. International Workshop QUALITDE-2017. Tbilisi, Georgia (2017)
94. Wang, C., Agarwal, R.P.: Almost periodic solution for a new type of neutral impulsive stochastic Lasota-Wazewska timescale model. Appl. Math. Lett. **70**, 58–65 (2017)
95. Wang, C., Agarwal, R.P., Sakthivel, R.: Almost periodic oscillations for delay impulsive stochastic Nicholson's blowflies timescale model. Comp. Appl. Math. **37**, 3005–3026 (2018)
96. Wang, C., Agarwal, R.P., O'Regan, D.: Local-periodic solutions for functional dynamic equations with infinite delay on changing-periodic time scales. Math. Slovaca **68**, 1397–1420 (2018)
97. Wang, C., Sakthivel, R.: Double almost periodicity for high-order Hopfield neural networks with slight vibration in time variables. Neurocomputing **282**, 1–15 (2018)
98. Wang, C., Agarwal, R.P., O'Regan, D., N'Guérékata, G.M.: Complete-closed time scales under shifts and related functions. Adv. Differ. Equ. **2018**(429), 1–19 (2018)
99. Wang, C., Agarwal, R.P., O'Regan, D.: n_0-order Δ-almost periodic functions and dynamic equations. Applic. Anal. **97**, 2626–2654 (2018)
100. Wang, C., Agarwal, R.P., O'Regan, D.: δ-almost periodic functions and applications to dynamic equations. Mathematics **7**, 525 (2019). https://doi.org/10.3390/math7060525
101. Wang, C., Agarwal, R.P., O'Regan, D., N'Guérékata, G.M.: n_0-Order weighted pseudo Δ-almost automorphic functions and abstract dynamic equations. Mathematics **7**, 775 (2019). https://doi.org/10.3390/math7090775
102. Wang, C., Agarwal, R.P., O'Regan, D., Sakthivel, R.: Local pseudo almost automorphic functions with applications to semilinear dynamic equations on changing-periodic time scales. Bound Value Probl. **133** (2019). https://doi.org/10.1186/s13661-019-1247-4
103. Wang, C., Agarwal, R.P., O'Regan, D., Sakthivel, R.: Discontinuous generalized doublealmost-periodic functions on almost-complete-closed time scales. Bound Value Probl. **165** (2019). https://doi.org/10.1186/s13661-019-1283-0
104. Wang, C., Agarwal, R.P., O'Regan, D.: Calculus of fuzzy vector-valued functions and almost periodic fuzzy vector-valued functions on time scales. Fuzzy Sets Syst. **375**, 1–52 (2019)
105. Wang, C., Agarwal, R.P., O'Regan, D., Sakthivel, R.: Theory of Translation Closedness for Time Scales. Developments in Mathematics, vol. 62. Springer, Switzerland (2020)
106. Wang, C., Agarwal, R.P., O'Regan, D., Sakthivel, R.: A computation method of Hausdorff distance for translation time scales. Applic. Anal. **99**, 1218–1247 (2020)
107. Wang, C., Agarwal, R.P.: Almost automorphic functions on semigroups induced by complete-closed time scales and application to dynamic equations. Discrt. Contin. Dynam. Syst. B **25**, 781–798 (2020)
108. Wang, C., Qin, G., Agarwal, R.P., O'Regan, D.: $\Diamond_\alpha$-Measurability and combined measure theory on time scales. Appl. Anal. (2020). https://doi.org/10.1080/00036811.2020.1820997
109. Wang, C., Agarwal, R.P., O'Regan, D.: Weighted pseudo d-almost automorphic functions and abstract dynamic equations. Georgian Math. J. (2021). https://doi.org/10.1515/gmj-2019-2066

110. Wang, C., N'Guérékata, G.: C_0-semigroup and Stepanov-like almost automorphic functions in matched spaces of time scales. In: N'Guérékata, G., Toni, B. (eds.) Studies in Evolution Equations and Related Topics. Springer, Berlin (2021)
111. Wang, C., Agarwal, R.P.: A survey of function analysis and applied dynamic equations on hybrid time scales. Entropy **23**, 450 (2021). https://doi.org/10.3390/e23040450
112. Wang, C., Chen, D., Li, Z.: General theory of the higher-order quaternion linear difference equations via the complex adjoint matrix and the quaternion characteristic polynomial. J. Diff. Equa. Appl. **27**, 787–857 (2021)
113. Wang, C., Li, Z., Agarwal, R.P.: Fundamental solution matrix and Cauchy properties of quaternion combined impulsive matrix dynamic equation on time scales. Analele Stiintifice ale Universitatii Ovidius Constanta, Seria Matematica **29**, 107–130 (2021)
114. Wang, C., Li, Z., Agarwal, R.P.: Hyers-Ulam-Rassias stability of high-dimensional quaternion impulsive fuzzy dynamic equations on time scales. Discrt. Contin. Dynam. Syst. S **15**, 359–386 (2022)
115. Wang, C., Li, Z., Agarwal, R.P.: A new quaternion hyper-complex space with hyper argument and basic functions via quaternion dynamic equations. J. Geom. Anal. 32, Article number: 67 (2022)
116. Wang, C., Agarwal, R.P., O'Regan, D.: Almost periodic fuzzy multidimensional dynamic systems and applications on time scales. Chaos, Solitons & Fract. **156**, 111781 (2022)
117. Wang, C., Wang, J., Agarwal, R.P., Li, Z.:Almost anti-periodic discrete oscillation of general n-dimensional mechanical system and underactuated Euler-Lagrange system. Appl. Sci. **2022**, 12 (1991). https://doi.org/10.3390/app12041991 (2022)
118. Wang, Z., Freidovich, L.B., Zhang, H.: Periodic motion planning and control for underactuated mechanical systems. Int. J. Control **91**, 1350–1362 (2018)
119. Wu, C., Ma, M.: Embedding problem on fuzzy number space: Part I*. Fuzzy Sets Syst. **44**, 33–38 (1991)
120. Wu, C., Ma, M.: On embedding problem on fuzzy number space: part 2. Fuzzy Sets Syst. **45**, 189–202 (1992)
121. Zadeh, L.A.: Fuzzy sets. Information Contr. **8**, 338–353 (1965)
122. Zadeh, L.A.: Concept of a linguistic variable and its application to approximate reasoning, I. Information Sci. **8**, 199–249 (1975)
123. Zadeh, L.A.: Fuzzy sets as a basis for a theory of possibility. Fuzzy Sets Syst. **1**, 3–28 (1978)

Index

C. Wang and R. P. Agarwal, *Dynamic Equations and Almost Periodic Fuzzy Functions on Time Scales*, Synthesis Lectures on Mathematics & Statistics,
https://doi.org/10.1007/978-3-031-11236-2

Printed in the United States
by Baker & Taylor Publisher Services